DS IN ENZYMOLOGY

E PYLE
Molecular, Cellular and Developmental
partment of Chemistry
ward Hughes Medical Institute

IRISTIANSON
Vagelos Laboratories
Chemistry
nsylvania

LOWICK and NATHAN O. KAPLAN

METHO

Editors-in-Chief

VOLUME SIX HUNDRED AND SEVENTY TWO

METHODS IN ENZYMOLOGY

Helicase Enzymes Part A

Edited by

MICHAEL A. TRAKSELIS
Department of Chemistry and Biochemistry, Baylor University, Waco, TX, United States

ELSEVIER

ACADEMIC PRESS
An imprint of Elsevier

Academic Press is an imprint of Elsevier
50 Hampshire Street, 5th Floor, Cambridge, MA 02139, United States
525 B Street, Suite 1650, San Diego, CA 92101, United States
The Boulevard, Langford Lane, Kidlington, Oxford OX5 1GB, United Kingdom
125 London Wall, London, EC2Y 5AS, United Kingdom

First edition 2022

Notices
Knowledge and best practice in this field are constantly changing. As new research and experience broaden our understanding, changes in research methods, professional practices, or medical treatment may become necessary.

Practitioners and researchers must always rely on their own experience and knowledge in evaluating and using any information, methods, compounds, or experiments described herein. In using such information or methods they should be mindful of their own safety and the safety of others, including parties for whom they have a professional responsibility.

To the fullest extent of the law, neither the Publisher nor the authors, contributors, or editors, assume any liability for any injury and/or damage to persons or property as a matter of products liability, negligence or otherwise, or from any use or operation of any methods, products, instructions, or ideas contained in the material herein.

ISBN: 978-0-323-91476-5
ISSN: 0076-6879

For information on all Academic Press publications
visit our website at https://www.elsevier.com/books-and-journals

Publisher: Zoe Kruze
Developmental Editor: Federico Paulo S. Mendoza
Production Project Manager: James Selvam
Cover Designer: Greg Harris

Typeset by STRAIVE, India

Contents

Contributors

Sahil Batra
Molecular Biology Program, Memorial Sloan Kettering Cancer Center, New York, NY, United States

Megan S. Behrmann
Department of Chemistry and Biochemistry, Baylor University, Waco, TX, United States

Binyam Belachew
Department of Biochemistry and Molecular Biology, University of Arkansas for Medical Sciences, Little Rock, AR, United States

James M. Berger
Department of Biophysics and Biophysical Chemistry, Johns Hopkins School of Medicine, Baltimore, MD, United States

Robert M. Brosh Jr.
Helicases and Genomic Integrity Section, Translational Gerontology Branch, National Institute on Aging, NIH, Baltimore, MD, United States

Alicia K. Byrd
Department of Biochemistry and Molecular Biology; Winthrop P. Rockefeller Cancer Institute, University of Arkansas for Medical Sciences, Little Rock, AR, United States

Sara C. Conwell
Department of Biochemistry, Vanderbilt University School of Medicine, Nashville, TN, United States

Alessandro Costa
Macromolecular Machines Laboratory, The Francis Crick Institute, London, United Kingdom

Matthew T. Cranford
Department of Biochemistry, Vanderbilt University School of Medicine, Nashville, TN, United States

Arindam Datta
Helicases and Genomic Integrity Section, Translational Gerontology Branch, National Institute on Aging, NIH, Baltimore, MD, United States

Sujan Devbhandari
Molecular Biology Program, Memorial Sloan Kettering Cancer Center, New York, NY, United States

James M. Dewar
Department of Biochemistry, Vanderbilt University School of Medicine, Nashville, TN, United States

Nicholas E. Dixon
Molecular Horizons and School of Chemistry and Molecular Bioscience, University of Wollongong; Illawarra Health and Medical Research Institute, Wollongong, NSW, Australia

Alexander T. Duckworth
Department of Biomolecular Chemistry, University of Wisconsin-Madison School of Medicine and Public Health, Madison, WI, United States

Amy J. Fernandez
Department of Biophysics and Biophysical Chemistry, Johns Hopkins School of Medicine, Baltimore, MD, United States

Yang Gao
Department of BioSciences, Rice University, Houston, Texas, United States

Julia F. Greiwe
Macromolecular Machines Laboratory, The Francis Crick Institute, London, United Kingdom

Grzegorz Ira
Department of Molecular and Human Genetics, Baylor College of Medicine, Houston, TX, United States

Slobodan Jergic
Molecular Horizons and School of Chemistry and Molecular Bioscience, University of Wollongong; Illawarra Health and Medical Research Institute, Wollongong, NSW, Australia

Stefan Juranek
Department of Oncology, Hematology and Rheumatology, University Hospital Bonn, Bonn, Germany

Gurleen Kaur
Molecular Horizons and School of Chemistry and Molecular Bioscience, University of Wollongong; Illawarra Health and Medical Research Institute, Wollongong, NSW, Australia

Tamar Kavlashvili
Department of Biochemistry, Vanderbilt University School of Medicine, Nashville, TN, United States

James L. Keck
Department of Biomolecular Chemistry, University of Wisconsin-Madison School of Medicine and Public Health, Madison, WI, United States

Jacob S. Lewis
Molecular Horizons and School of Chemistry and Molecular Bioscience, University of Wollongong; Illawarra Health and Medical Research Institute, Wollongong, NSW, Australia

Liping Liu
Department of Biology, University of Iowa, Iowa City, IA, United States

Chen-Yu Lo
Department of BioSciences, Rice University, Houston, Texas, United States

Anna Malkova
Department of Biology, University of Iowa, Iowa City, IA, United States

Emory G. Malone
Department of Biochemistry and Molecular Biology, University of Arkansas for Medical Sciences, Little Rock, AR, United States

Alice Meroni
Division of Oncology, Department of Medicine, Washington University in St. Louis, St. Louis, MO, United States

Thomas C.R. Miller
Center for Chromosome Stability, Department of Cellular and Molecular Medicine, University of Copenhagen, Copenhagen, Denmark

Sua Myong
Department of Biophysics, Johns Hopkins University, Baltimore, MD; Physics Frontier Center (Center for Physics of Living Cells), University of Illinois, Urbana, IL, United States

Mike E. O'Donnell
Laboratory of DNA Replication, The Rockefeller University; Howard Hughes Medical Institute at Rockefeller University, New York, NY, United States

Deniz Ozaslan
Department of Biochemistry and Molecular Biology, University of Arkansas for Medical Sciences, Little Rock, AR, United States

Katrin Paeschke
Department of Oncology, Hematology and Rheumatology, University Hospital Bonn, Bonn, Germany

Smita S. Patel
Department of Biochemistry and Molecular Biology, Rutgers University, Robert Wood Johnson Medical School, Piscataway, NJ, United States

Tapas Paul
Department of Biophysics, Johns Hopkins University, Baltimore, MD, United States

Nhung Pham
Department of Molecular and Human Genetics, Baylor College of Medicine, Houston, TX, United States

Thomas Pühringer
Macromolecular Machines Laboratory, The Francis Crick Institute, London, United Kingdom

Kevin D. Raney
Department of Biochemistry and Molecular Biology, University of Arkansas for Medical Sciences, Little Rock, AR, United States

Dirk Remus
Molecular Biology Program, Memorial Sloan Kettering Cancer Center, New York, NY, United States

Anupam Singh
Department of Biochemistry and Molecular Biology, Rutgers University, Robert Wood Johnson Medical School, Piscataway, NJ, United States

Lisanne M. Spenkelink
Molecular Horizons and School of Chemistry and Molecular Bioscience, University of Wollongong; Illawarra Health and Medical Research Institute, Wollongong, NSW, Australia

Pilendra K. Thakre
Department of Molecular and Human Genetics, Baylor College of Medicine, Houston, TX, United States

Matthew D. Thompson
Department of Biochemistry and Molecular Biology, University of Arkansas for Medical Sciences, Little Rock, AR, United States

Michael A. Trakselis
Department of Chemistry and Biochemistry, Baylor University, Waco, TX, United States

Antoine M. van Oijen
Molecular Horizons and School of Chemistry and Molecular Bioscience, University of Wollongong; Illawarra Health and Medical Research Institute, Wollongong, NSW, Australia

Alessandro Vindigni
Division of Oncology, Department of Medicine, Washington University in St. Louis, St. Louis, MO, United States

Zhenxin Yan
Department of Molecular and Human Genetics, Baylor College of Medicine, Houston, TX, United States

Nina Y. Yao
Laboratory of DNA Replication, The Rockefeller University, New York, NY, United States

Preface

Nucleic acid polymers comprised of DNA or RNA have a spectrum of secondary structure stabilities including extremely dynamic and transient species, helices, loops and stems, and various more stable G-quadruplexes. The range of properties may have been critical to the evolution and selection of nucleic acids as early biological materials, as duplex stability is important for the protection of the genome, while other highly variant structures are utilized for specific purposes within the central dogma. A consequence of these secondary and tertiary nucleic acid structures is that transient separation of the base pairing interactions is required for various informational transactions that occur within the cell. This is accomplished by active and passive helicases, translocases, and ATP-dependent motor proteins and enzymes that act to partially or fully separate or remodel DNA or RNA strands. Nucleic acid remodeling processes are required for DNA replication, DNA repair, DNA recombination, transcription, ribosome biogenesis, translation, RNA splicing, RNA editing, RNA transport, RNA degradation, bacterial conjugation, chromosome segregation, and viral packaging and infection. Therefore, it is not surprising that many inherited human diseases including Bloom syndrome, Werner's syndrome, Rothmund-Thomson syndrome, Warsaw breakage syndrome, Fanconi anemia, Cockayne syndrome, xeroderma pigmentosum, trichothiodystrophy, Crohn's disease, dyskeratosis congenita, and ophthalmoplegia have been linked to defects in specific helicases. Moreover, defects or dysregulation in any 1 of 100 helicases in humans can lead to genomic instability, resulting in increased incidences of cancer, ovarian deficiency, infertility, aging, and neurodegeneration. Furthermore, many viruses encode helicases that play critical roles in viral life cycles, which have become targets for antiviral therapies.

While gel-based unwinding methods have provided for the bulk of early studies on the mechanisms of helicase action, the development and expansion of more recent technologies including single-molecule analyses, chemical biology tools, multiplex fluorescence approaches, advances in structural biology, in vivo DNA labeling and tracking, more advanced genetic methods, various next-generation sequencing isolation methods, and integrated mass spectrometry have greatly contributed to a broader spectrum of interdisciplinary approaches. The outcomes of these approaches have greatly expanded our mechanistic understanding of a variety of helicases, translocases, and remodeling/motor enzymes. Given the diversity of enzymes, their substrates, and the many information processing pathways within a cell, the variety of tools and techniques developed for the examination of helicase/translocase function has become extensive. The breadth of these techniques, as they have been applied to both DNA (Volume 672) and RNA (Volume 673) helicase/translocase substrates, is presented in this two-volume collection of *Methods in Enzymology*.

Volume 672 includes a collection of recent methods and developments that have been primarily utilized to examine DNA helicases and translocases. Chapter 1, by Bryd and colleagues, kicks it off by multiplexing DNA unwinding through multiple fluorophores simultaneously detecting intermediates and products in a single reaction through resolving and imaging of a gel. In Chapter 2, Raney and coworkers utilize methods to investigate the assembly of interacting multiple helicase subunits on a template to enhance unwinding. Chapter 3, by Keck and Duckworth, combines chemical biology tools for incorporating unnatural amino acids with mapping of DNA-binding sites by photoactivated crosslinking. Chapters 4 and 5 examine the assemblies and actions of other replisome proteins in context with hexameric replication helicases by rapid kinetic methods from the Patel laboratory and structural biology methods from the Gao laboratory. Chapter 6, from my own laboratory, takes advantage of TUNEL methodologies to examine DNA gaps and repair processes in bacteria enhanced by targeted genetic mutations in the replication helicase, *dnaB*. The loading and activation of hexameric helicases is a complex process, and Chapter 7 from the Berger laboratory utilizes biochemical tools to quantify these events. The Brosh laboratory, in Chapter 8, has detailed methods that have become popular recently to quantify DNA fiber lengths in cells to examine helicase-associated DNA replication or repair processes. The design of DNA substrates and utilization of various nucleotide substrates have become more important to trap and solve higher-order helicase structures, and the O'Donnell laboratory has outlined their strategies that have provided for great successes in Chapter 9. Speaking of high-resolution snapshots of hexameric helicase loading, the Costa laboratory has been leading this effort and highlights in silico methods to extract structural intermediates from CryoEM grids in Chapter 10. Chapters 11–13 by Remus, Paeschke, and Myong examine the unique and difficult process of melting very stable G-quadruplex structures, which are only just being recognized for their importance in controlling a variety of genetic information processes. Chapter 14, by the van Oijen laboratory, details their method of creating extremely long DNA substrates with site-specific lesions for single-molecule analysis of coupled replication and repair events directed by helicases. Replication fork collapse from helicase blocks or translocation of single-stranded nicks can be studied using novel plasmids and by examining the consequences in the robust *Xenopus* egg extract system as outlined by the Dewar laboratory in Chapter 15. Genetic approaches have been useful in teasing out subpathways of DNA recombination, and Ira, Malkova, and colleagues highlight yeast reporter systems to quantify BIR outcomes in Chapter 16. Finally, Volume 1 concludes with an assay to detect topoisomerase trapped intermediates within the whole genome that may be the result of specific inhibitors of defects in repair processing in Chapter 17 described by the Vindigni laboratory.

Volume 673 focuses primarily on the methods and techniques developed and adapted to examine the mechanisms and actions of RNA helicases, translocases, and remodelers. In Chapter 1, the Peterson laboratory describes quantitative fluorescence approaches to monitor enzymes that can remodel nucleosomes within a chromatin architecture. Katrin Karbstein outlines strategies and approaches for tackling elusive DEAD-box ATPases that act on many different RNA substrates including ribosome assembly in Chapter 2. Chapter 3, from the Pyle laboratory, takes a traditional ATPase assay and adapts it for lower volumes and higher throughput using the Rig-I helicase involved in adaptive immunity as an example. Assays to examine various DEAD-box RNA helicases continue in Chapters 4–7, with Russell, Johnson, Correll, Bolger, and Floor further detailing ATPase, RNA-folding intermediates, protein purifications, assessing cellular phenotypes in yeast, and utilizing human cells and extracts for translational luciferase read outs. RecQ and Pif1 family helicases are utilized in many aspects of DNA replication progression to resolve transcription conflicts or G-quadruplex structures, and the methods showing their activities on properly designed templates in bulk and single-molecule methods are presented in Chapters 8 and 9 by Bochman and Schauer, and genomic binding and transcriptional profiling related to disease in Chapter 10 by Sharma. Single-molecule FRET assays have been a hallmark of helicase studies for several decades now, and Klostermeier details the approaches for proper protein labeling and RNA substrate design in Chapter 11. Longer custom-designed DNA substrates are required for both optical and magnetic tweezer experiments with helicases, and the creation and validation of these are expertly laid out in Chapters 12 and 13 by Dillingham, Moreno-Herrero, Croquette, and Le Hir. CRISPR-Cas systems also rely on helicase-nuclease enzymes to achieve adaptive immunity, and Ke outlines the methods to reconstitute the biochemistry of a type 1A system and utilize it for gene editing in human cells in Chapter 14. The final three chapters (Chapters 15–17) of Volume 673, by Johnson, Lima, and D'Arcy, detail the purification, reconstitution, substrate design, and HDX-MS structural characterization of RNA helicases and their substrates associated with exosomes.

Let me take this opportunity to extend my sincere thanks to all the authors in these two volumes for providing insights and explicit instructions on how to utilize their methods to examine diverse helicase mechanisms. I commend Paulo Mendoza and the Elsevier staff for their administrative help and guidance. I am grateful to Anna Pyle (editor-in-chief of *Methods in Enzymology*) for the invitation and for being a strong representative within the helicase field. My hope is that other laboratories will find these methods helpful as a starting point to adapt for their own enzymes of interest or extend these methods further for more in-depth understanding of helicase function. As I reflect on these chapters, it is a real honor to be the editor of both volumes, where long-standing leaders in the field, whom I have looked up to, and eager younger investigators bound to take on the mantel someday continue to provide inspiration to me and others working on this broad family of enzymes.

If you enjoyed these two volumes or found them helpful for your own research, I encourage you to consider joining us at the FASEB conference "Helicases and nucleic acid-based machines: Structure, mechanism, regulation, and roles in human diseases" in July 2023, where many of these authors will be attending and presenting and further refining their methods.

Michael A. Trakselis
Editor

CHAPTER ONE

Monitoring helicase-catalyzed unwinding of multiple duplexes simultaneously

Matthew D. Thompson[a], Emory G. Malone[a], and Alicia K. Byrd[a,b,*]
[a]Department of Biochemistry and Molecular Biology, University of Arkansas for Medical Sciences, Little Rock, AR, United States
[b]Winthrop P. Rockefeller Cancer Institute, University of Arkansas for Medical Sciences, Little Rock, AR, United States
*Corresponding author: e-mail address: akbyrd@uams.edu

Contents

Methods in Enzymology, Volume 672
ISSN 0076-6879
https://doi.org/10.1016/bs.mie.2022.02.018

Abstract

Helicases catalyze the unwinding of duplex nucleic acids to aid a variety of cellular processes. Although helicases unwind duplex DNA in the same direction that they translocate on single-stranded DNA, forked duplexes provide opportunities to monitor unwinding by helicase monomers bound to each arm of the fork. The activity of the helicase bound to the displaced strand can be discerned alongside the helicase bound to the translocase strand using a forked substrate with accessible duplexes on both strands labeled with different fluorophores. In order to quantify the effect of protein-protein interactions on the activity of multiple monomers of the *Bacteroides fragilis* Pif1 helicase bound to separate strands of a forked DNA junction, an ensemble gel-based assay for monitoring simultaneous duplex unwinding was developed (Su et al., 2019). Here, the use of that assay is described for measuring the total product formation and rate constants of product formation of multiple duplexes on a single nucleic acid substrate. Use of this assay may aid characterization of protein-protein interactions between multiple helicase monomers at forked nucleic acid junctions and can assist with the characterization of helicase action on the displaced strand of forked duplexes.

1. Introduction

Helicases catalyze the unwinding of duplex nucleic acids to form single-stranded products for a diverse array of cellular processes (Brosh & Matson, 2020). Duplex DNA unwinding catalyzed by ATP hydrolysis proceeds in the direction of single-stranded DNA (ssDNA) translocation, either 5′-to-3′ or 3′-to-5′, giving helicases a specified direction of unwinding (Dillingham, Wigley, & Webb, 1999; Morris & Raney, 1999; Raney & Benkovic, 1995; Saikrishnan, Powell, Cook, Webb, & Wigley, 2009). Ensemble unwinding assays are frequently performed using gel electrophoresis to resolve nucleic acid products of helicase-catalyzed unwinding of a single duplex with a radio- or fluorophore-labeled strand (Matson & Kaiser-Rogers, 1990; Matson, Tabor, & Richardson, 1983; Venkatesan, Silver, & Nossal, 1982). Ensemble duplex DNA unwinding can also be monitored over time using real-time fluorescence measurement (e.g. stopped-flow fluorescence spectroscopy) with fluorophore-labeled DNA duplexes (Belon & Frick, 2008; Bjomson, Amaratunga, Moore, & Lohman, 1994; Donmez & Patel, 2008; Jeong, Levin, & Patel, 2004; Özeş, Feoktistova, Avanzino, Baldwin, & Fraser, 2014; Raney et al., 1994). Additionally, single-molecule Förster resonance energy transfer (smFRET) and magnetic or optical tweezers can be used to monitor unwinding of individual DNA molecules (Amit, Gileadi, & Stavans, 2004; Bianco et al., 2001; Dawid, Croquette, Grigoriev, & Heslot, 2004; Myong, Bruno, Pyle, & Ha, 2007; Rasnik et al., 2008; Yodh, Stevens, Kanagaraj, Janscak, & Ha, 2009). These assays have uncovered novel unwinding events such as strand switching, where a helicase switches strands while unwinding a duplex and consequently reanneals the duplex at a similar rate by translocating down the opposite strand (Byrd et al., 2012; Comstock et al., 2015; Dessinges, Lionnet, Xi, Bensimon, & Croquette, 2004), and reeling (also called patrolling, shuttling, looping, and repetitive activity), where a helicase remains fixed in place and can pull in and unwind DNA without translocating relative to a fixed point on the substrate (Carney et al., 2021; Lu et al., 2019; Park et al., 2010; Zhou, Zhang, Bochman, Zakian, & Ha, 2014).

Robust experimental design in ensemble and single-molecule duplex unwinding assays has enabled detailed mechanistic descriptions of nucleic acid unwinding by helicases such as the bacteriophage T4 superfamily 1B helicase Dda (Aarattuthodiyil, Byrd, & Raney, 2014; Byrd et al., 2012; Byrd & Raney, 2005; Eoff & Raney, 2006; He et al., 2012; Nanduri, Byrd, Eoff, Tackett, & Raney, 2002), *E. coli* superfamily 1A helicase UvrD (Ali & Lohman, 1997; Ali, Maluf, & Lohman, 1999; Carney et al., 2021; Comstock et al., 2015; Maluf, Fischer, & Lohman, 2003) and the Hepatitis C Virus superfamily 2 helicase NS3 (Jennings et al., 2009; Levin, Wang, & Patel, 2004; Myong et al., 2007; Rajagopal & Patel, 2008; Sikora et al., 2008; Zhou, Ren, Adams, & Pyle, 2018). DNA footprinting, protein-DNA crosslinking, and X-ray crystallography have established that helicases such as Dda (Aarattuthodiyil et al., 2014), *E. coli* PriA (Windgassen & Keck, 2016; Windgassen, Leroux, Satyshur, Sandler, & Keck, 2018), and *Bacteroides fragilis* Pif1 (BaPif1) can bind to the displaced strand of a forked duplex opposite an active helicase on the translocase strand. However, use of forked duplex unwinding substrates with only a single labeled duplex prevented the investigation of simultaneous unwinding of duplexes by helicases monomers translocating on opposite strands of the same forked duplex. The need to assess simultaneous unwinding of multiple duplexes on a forked substrate was established by the solution of a crystal structure with two BaPif1 monomers bound to a forked duplex (Su et al., 2019). Unexpectedly, the structure illustrated that the BaPif1 bound to the 3′-ssDNA arm of the displaced strand was situated closer to the forked junction than the BaPif1 bound to the

5′-ssDNA arm of the translocase strand, so the activity of both helicases on the duplex was assessed to confirm the relevance of the structure for forked duplex unwinding by BaPif1 in vitro (Su et al., 2019). To accomplish this, a traditional duplex unwinding assay was modified to include multiple fluorophore-labeled duplexes on a forked duplex substrate (Su et al., 2019). This enabled simultaneous monitoring of unwinding of the duplex on each strand of the forked duplex by each BaPif1 monomer and demonstrated the existence of forked duplexes with active BaPif1 bound to each strand during in vitro unwinding experiments.

The multiple duplex unwinding assay can further be used to assess features of unwinding mechanisms of helicases on forked duplexes, as mechanisms for the increased unwinding of forked duplexes by helicases such as *Saccharomyces* Pif1 and human FANCJ remain unclear (Gupta et al., 2005; Lahaye, Leterme, & Foury, 1993; Ramanagoudr-Bhojappa et al., 2013). For example, an increased rate of unwinding of a forked duplex could be explained by either interactions between a helicase bound to the translocase strand with structural features of a forked duplex such as the 3′-ssDNA arm of the displaced strand or the ssDNA-dsDNA fork junction, as demonstrated for Dda (Aarattuthodiyil et al., 2014), or could be explained by protein-protein interactions between active or inactive helicases bound on both the translocase and displaced strand. To distinguish between these mechanisms, the multiple duplex unwinding assay could be used alongside DNA footprinting to assess whether simultaneous binding and/or unwinding of a displaced strand duplex occurs alongside unwinding of the translocase strand and whether protein-protein interactions between the helicases on each strand influence unwinding, as demonstrated for BaPif1 (Su et al., 2019).

Here, we describe and generalize the ensemble multiple duplex unwinding assay originally developed for use with BaPif1 (Su et al., 2019) for use with other helicases, and we demonstrate its utility for determining both total product formation and rate constants of product formation for multiple duplexes on a single substrate.

2. Key resource tables

See Table 1.

Table 1 Key oligonucleotide substrates and sequences.

Oligonucleotide	Source	Sequence (5′-3′)
T_{50}	Integrated DNA Technologies	TTT TTT TTT TTT TTT TTT TTT TTT TTT TTT TTT TTT TTT TTT TTT TTT TT
Parent16	Integrated DNA Technologies	CCA GGC GAC ATC AGC G
Lead16	Integrated DNA Technologies	GCA GTG ACC AGA CAG G
Parent16_10T_Lead16	Integrated DNA Technologies	CCA GGC GAC ATC AGC GTT TTT TTT TTG CAG TGA CCA GAC AGG
5′F-10TParent16comp	Integrated DNA Technologies	6-FAM-TTT TTT TTT TCG CTG ATG TCG CCT GG
5′Cy5-Lead16comp	Integrated DNA Technologies	Cy5-CCT GTC TGG TCA CTG C

3. Safety considerations and standards

Care should be taken when performing this protocol. Gloves, a lab coat, and eye protection should be worn as the transport and mixing of large volumes of liquid is required. A chemical fume hood is also required for the handling of volatile substances.

The relevant safety data sheet (SDS) for each reagent should be reviewed before performing this protocol. Dichlorodimethylsilane is a respiratory hazard, causes eye damage, and is flammable, so it should only be used in a fume hood. Acrylamide is a neurotoxin, carcinogen, and may cause reproductive toxicity, so precautions should be taken to avoid contact with acrylamide. Wash hands well after using either of these reagents.

Relevant synthetic DNA safety standards and regulations should be followed at all times when performing this protocol. Synthetic oligonucleotides as used in this protocol are exempt from the *NIH Guidelines for Research Involving Recombinant or Synthetic Nucleic Acid Molecules* (see Sections III-F-1, III-F-2, and III-F-8).

4. Materials and equipment

4.1 Equipment and materials

Equipment

- Heat block set at 95 °C
- Heat block set at 25 °C (or the appropriate temperature for your enzyme)
- Fluorescent gel imager such as Amersham™ Typhoon™ RGB Imager
- High voltage power supply (capable of operating at ~1000 V)
- Sequencing gel system such as Life Technologies/BRL Sequencing System, Model S2 gel system including glass plates, 0.4 mm spacers, 32-well comb, casting clamp, and gel stand
- Fume hood
- Magnetic stir plate
- Benchtop centrifuge
- Vortex

Materials

- Six large binder clips
- Plastic wedge
- Low autofluorescence plastic wrap such as Anchor wrap All-Purpose Foodservice Film
- KimWipes
- 1.5 mL tubes; clear and amber
- Absorbent bench paper
- Software for analysis of band intensities such as ImageQuant TL or ImageJ

4.2 Reagents

- Fluorophore-labeled and unlabeled oligonucleotide substrates (see Table 1)
- 10 mM HEPES, pH 7.5, 1 mM EDTA solution
- ddH_2O
- 10% (v/v) Dichlorodimethylsilane in hexanes—Dichlorodimethylsilane is corrosive. Store at 4 °C with a tightly closed cap wrapped with parafilm
- 40% (w/v) 19:1 acrylamide/bis-acrylamide—Store at 4 °C
- 10% (w/v) Ammonium persulfate (APS) in ddH_2O—Prepare fresh immediately before use
- Tetramethylethylenediamine (TEMED)
- 10 × Tris-borate-EDTA (TBE), pH 8.0

- 2 × Quenching solution—200 mM EDTA, 0.2% sodium dodecyl sulfate (SDS), 5% glycerol, 0.1% Orange G in ddH_2O
- 5 × helicase assay buffer—We used 50 mM Tris, pH 7.5, 250 mM NaCl, 0.5 mM EDTA, 10 mM DTT, 0.5 mg/mL bovine serum albumin (BSA), 25% glycerol for BaPif1 and human PIF1 (hPIF1), and we used 125 mM HEPES, pH 7.5, 50 mM potassium acetate (KOAc), 0.5 mM EDTA, 10 mM BME, and 0.5 mg/mL bovine serum albumin (BSA) for Dda
- 100 mM MgCl2 or magnesium acetate ($Mg(OAc)_2$)
- 100 mM ATP—Prepare in ddH_2O and bring pH to 7 by adding 1 M NaOH dropwise to reduce spontaneous hydrolysis. Store at −20 °C
- Purified helicase—We used Dda, BaPif1, and human PIF1. Purifications are described elsewhere (Byrd et al., 2021; George et al., 2009; Zhou et al., 2016)

Alternatives:

- Depending on the lasers and filters available on the fluorescent gel imager, it may be necessary to use different fluorophores whose excitation and emission spectra overlap the available lasers and filters. Any combination of fluorophores with well separated spectra can be used
- Any vertical electrophoresis system can be utilized instead of a sequencing gel system. However, the gel will have to be carefully monitored to ensure separation of substrate, intermediates, and products without running the smallest products off the gel
- The preferred reaction buffer will vary based on the specific helicase being studied. 5 × helicase assay buffer should be modified to match the appropriate reaction conditions for the helicase of interest

5. Preparation of dual duplex DNA substrates

Substrates should be designed to match the characteristics of the helicase of interest. For example, as illustrated in Fig. 1A, the substrates used for Pif1 and Dda contained 10 thymidines 5′ to each duplex to allow for binding and unwinding by these 5′-to-3′ helicases. Additionally a short ssDNA tail can be added to the 3′-end of the leading duplex displaced strand to create a forked leading duplex for enzymes that prefer to unwind a forked duplex such as *Saccharomyces cerevisiae* Pif1 (Su et al., 2019).

Each duplex is labeled with a fluorophore (FAM and Cy5, in this example) so that unwinding of each duplex can be monitored simultaneously.

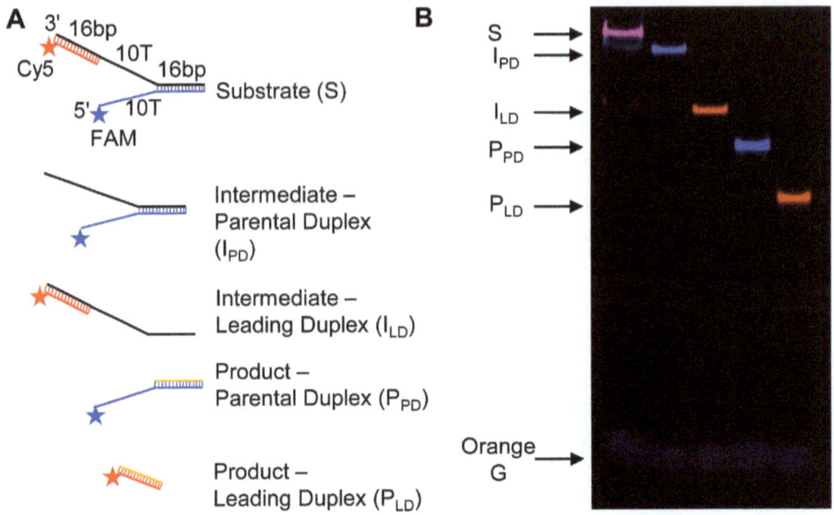

Fig. 1 Expected separation of reaction products following PAGE. (A) Schematic of the FAM- (blue star) and Cy5-labeled (red star) dual duplex unwinding substrate (S), the unwinding intermediates containing the FAM-labeled parental duplex (I_{PD}) and the Cy5-labeled leading strand duplex (I_{LD}), and the trapped unwinding products of FAM-labeled parental duplex (P_{PD}) and Cy5-labeled leading duplex (P_{LD}). (B) The fluorophore-labeled duplexes were separated on a 20% acrylamide gel by electrophoresis. The FAM-labeled duplexes appear in blue, Cy5-labeled duplexes in red, and the dual-labeled substrate in magenta.

Fluorophore pairs other than those utilized here can be used instead, provided that imaging equipment is available, and that spectral overlap and background fluorescence are not encountered.

Before an unwinding reaction can be carried out with dual duplex DNA substrates, the appropriate oligonucleotides must be annealed, and successful annealing must be confirmed with polyacrylamide gel electrophoresis (PAGE). The expected outcome of this step is the verification that the dual-duplex substrate, the intermediates, and the reaction products can be properly separated and identified with PAGE following the completion of an unwinding reaction.

5.1 Procedure

5.1.1 Annealing oligonucleotides to form duplexes

Timing: 2h-overnight

1. Lyophilized oligonucleotides were resuspended in 10mM HEPES, 1mM EDTA. Absorbance was measured at 260nm, and the concentration of

each oligonucleotide was calculated using the extinction coefficient. Oligonucleotides were diluted to 50 μM in 10 mM HEPES, 1 mM EDTA and stored at −20 °C until further use (see note 5.1).

2. To prepare substrate, intermediates, and products, dilute appropriate oligonucleotides in 10 mM HEPES, 1 mM EDTA in an amber tube
   (a) Dual duplex substrate (S)
   - Mix equimolar amounts of Parent16_10T_Lead16, 5′F-10TParent16comp, and 5′Cy5-Lead16comp. See note 5.2 for considerations on substrate design. See critical note 5.1 for important considerations when annealing the dual duplex substrate.

   (b) Intermediate-Parental Duplex (I_{PD})
   - Mix 1 equiv. of 5′F-10TParent16comp with a 1.2-fold excess of Parent16_10T_Lead16. Addition of excess unlabeled strand will ensure complete duplex formation of the FAM-labeled duplex

   (c) Intermediate-Leading Duplex (I_{LD})
   - Mix 1 equiv. of 5′Cy5-Lead16comp with a 1.2-fold excess of Parent16_10T_Lead16. Addition of excess unlabeled strand will ensure complete duplex formation of the Cy5-labeled duplex

   (d) Product-Parental Duplex (P_{PD})
   - Mix 1 equiv. of 5′F-10TParent16comp with a 1.2-fold excess of Parent16. Addition of excess unlabeled strand will ensure complete duplex formation of the FAM-labeled duplex

   (e) Product-Leading Duplex (P_{LD})
   - Mix 1 equiv. of 5′Cy5-Lead16comp with a 1.2-fold excess of Lead16. Addition of excess unlabeled strand will ensure complete duplex formation of the Cy5-labeled duplex

3. Heat oligonucleotides to 95 °C for 10 min followed by slow cooling to room temperature to anneal duplexes
4. Vortex and briefly centrifuge duplexes in a benchtop centrifuge before using

   Pause Point: Annealed oligonucleotides can be stored at −20 °C protected from light until further use.

5.1.2 Preparing a non-denaturing 20% polyacrylamide gel

Timing: 1.5–2 h

1. Move glass plates into a chemical fume hood and apply a thin layer 10% dichlorodimethylsilane in hexanes with a KimWipe. Wash excess silane off using ddH_2O and a KimWipe. It is critical that this step be

performed in a fume hood as dichlorodimethylsilane is a respiratory hazard. See critical note 5.2 for storage and safety considerations when working with dichlorodimethylsilane

2. Remove the plates from the chemical fume hood and place the large glass plate on a flat surface covered with bench paper. Place the 0.4 mm spacers along each vertical (long) edge of the large glass plate
3. Invert the short glass plate onto the spacers and the long glass plate, so that the silanized surfaces face each other. Adjust the spacers so that they run parallel to the vertical edges of both plates
4. Hold the two plates in position and wrap a rubber casting clamp around the corners and sides of the plates to create a sealed chamber with the closed drainage holes in the casting clamp on the top (closest to the shorter plate)
5. Gently adjust the spacers to ensure that the foam seal of each spacer presses down onto the top edge of the shorter plate. This seals the acrylamide chamber against leakage of electrophoresis buffer while running the gel
6. For a 20% gel, mix 37.5 mL 20% 19:1 acrylamide:bis-acrylamide, 7.5 mL 10× TBE, 30 mL ddH_2O, 1 mL 10% ammonium persulfate, and 75 μL TEMED in a plastic beaker with stirring. Add ammonium persulfate and TEMED immediately before pouring gel as this will initiate polymerization. Acrylamide is toxic; see critical note 5.3 for safety information
7. Hold the sealed glass plates at a gentle angle and pour the acrylamide solution onto a corner where the large plate, short plate, and spacer meet. Allow the acrylamide solution to flow between the plates to the bottom of the gel and maintain a pool of acrylamide solution that has not yet flowed between the plates to prevent the introduction of air bubbles into the gel. See critical note 5.3 for safety and disposal considerations when working with acrylamide
8. Prop the gel at a gentle angle by placing an empty rack for 15 mL tubes or a 200 μL pipette tip box under the top of the gel. Insert a 32 well comb and clamp the glass plates with 6 large binder clips along the top of the gel
9. Wait until the acrylamide has just polymerized (~20–30 min).
10. Remove the binder clips and use a clean razor blade to carefully remove the polymerized acrylamide between the top of the comb and the top of the short plate

11. Apply ddH_2O to the newly exposed comb surface and use a razor blade to gently lift the comb
12. Apply ddH_2O under the comb to loosen it from the glass plate
13. Gently remove the comb from between the plates and ensure that the wells are intact
14. Remove the casting clamp and place the gel and plates into the gel stand with the longer plate facing out
15. Secure the gel into the gel stand and ensure that a tight seal is formed in the top buffer reservoir
16. Add 1 l of 1 × TBE split between the top and bottom buffer reservoirs. Flush any air bubbles from the wells using a 30 ml syringe filled with 1 × TBE from the top buffer reservoir
17. Pre-run the gel at 22 mA for 60 min
18. Flush any air bubbles from the wells using a 30 mL syringe filled with 1 × TBE from the top buffer reservoir before loading samples

5.1.3 Separating samples by electrophoresis

Timing: 3 h

1. Dilute duplex oligonucleotides to a final concentration of 50 nM with ddH_2O and 1 × freshly made quenching solution. Vortex to mix, and spin down oligonucleotides before using. See note 5.3 for considerations about loading dye selection for this procedure
2. Load 5 μL of each substrate into an individual well in the pre-run gel
3. Run the gel at 22 mA for 2.5 h. See critical note 5.4 for safety considerations while running the gel, and see note 5.4 for modifications for duplexes shorter than 14 base pairs
4. Remove the gel from the gel stand, and gently slide the spacers out from between the glass plates
5. Use a plastic wedge to gently pry one glass plate off of the other, and ensure that the gel remains attached only to a single plate
6. Carefully place a sheet of plastic wrap onto the exposed gel surface. Wrinkles and bubbles in the plastic will be visible in the scan so take care not to trap air bubbles or wrinkle the plastic wrap in the area of interest. See note 5.5 for considerations on plastic wrap selection
7. Wipe the back of the glass plate with 100% ethanol before scanning
8. Scan the gel, plastic-side down, with the FAM and Cy5 channels of a fluorescent gel imager such as the Amersham™ Typhoon™ RGB

Imager (Cytiva) using a 488 nm laser and 525BP20 emission filter for FAM and a 635 nm laser and a 670BP30 filter for Cy5

9. Open the image file in the software provided with your imager or in ImageJ

5.2 Expected outcomes

As illustrated in Fig. 1B, the success of oligonucleotide annealing to form the fluorescently-labeled substrate can be assessed after electrophoresis. The dual-duplex substrate for use in unwinding reactions appears as the highest molecular weight band in both FAM and Cy5 detection channels. The expected intermediates and products of a helicase-catalyzed unwinding reaction also separate by molecular weight. The presence of a single band with the appropriate fluorescent label(s) in each lane is expected and confirms that the duplexes were annealed with equivalent quantities of all three strands and were not denatured before or during electrophoresis.

5.3 Optimization and troubleshooting

5.3.1 Presence of single-fluorophore-labeled species after substrate annealing along with annealed substrate band

An imbalance of concentrations of fluorophore-labeled oligonucleotides and unlabeled strand can lead to the presence of singly-labeled duplexes that are structurally identical to intermediate (i.e. I_{PD} or I_{LD}) bands or the presence of bands at low molecular weight corresponding to unannealed fluorophore-labeled oligonucleotides.

If singly-labeled duplexes labeled with both FAM and Cy5 are present after substrate annealing but no free fluorophore-labeled oligonucleotides are present, too much unlabeled strand was added before annealing.

If singly-labeled duplexes labeled with both FAM and Cy5 are present after substrate annealing and free fluorophore-labeled oligonucleotides are also present, too little unlabeled strand was added before annealing.

If a singly-labeled duplex and unannealed fluorophore-labeled oligonucleotides with the same fluorophore are present, an excess of that oligonucleotide was added before annealing.

5.3.2 Potential solution

Check the concentration of each oligonucleotide by measuring the absorbance at 260 nm and calculating the concentration using the extinction coefficient. Ensure careful pipetting and proper mixing when creating dilutions and solutions.

5.3.3 Presence of smeared bands starting from the expected size of the appropriate duplex and ending at the expected size of the fluorophore-labeled ssDNA

Melting of the duplexes while the gel is running will lead to smearing of each band.

5.3.4 Potential solution

Ensure that the polyacrylamide gel components and run conditions (e.g. gel temperature exceeding T_m of the duplex) do not promote denaturation of the substrate. See note 5.4 for substrate design considerations.

5.3.5 Presence of multiple species of a lower-than-expected molecular weight

Degradation of any of the oligonucleotides or the presence of partial synthesis products can result in the appearance of a ladder of products on the gel. The presence of extra bands can interfere with proper quantitation of unwinding reactions.

5.3.6 Potential solution

Gel purify the oligonucleotides, and select the slowest migrating species for each oligonucleotide because the full length synthesis product that has not been degraded will be the largest species. To prevent this problem from occurring in the first place, gel purify oligonucleotides before beginning experimentation and prepare all solutions using reagents that are free from DNase contamination.

5.4 Notes

Note 5.1: Protect FAM- and Cy5-labeled oligonucleotides from light to prevent photobleaching by storing stocks and dilutions in amber tubes.

Note 5.2: The Pif1 and Dda helicases that we used translocate 5′-to-3′ on ssDNA. If the helicase being studied translocates 3′-to-5′ on the ssDNA, the substrate will need to be inverted so that both helicases would translocate 3′-to-5′ towards a duplex.

Note 5.3: Orange G dye is used because, although it is visible in the FAM fluorescence scan at 525 nm, it migrates faster than a 12mer oligonucleotide so it does not interfere with quantitation. Xylene cyanol is also visible in a scan for FAM fluorescence at 525 nm, and bromophenol blue is visible in a scan for Cy5 fluorescence at 670 nm. Both xylene cyanol and bromophenol blue migrate slower than Orange G and have the potential to interfere with imaging of FAM or Cy5 labeled oligonucleotides.

Note 5.4: If either duplex is shorter than 14 base pairs, run the gel at 13 mA for 5 h to prevent the gel from heating enough that the duplex(es) melt(s) while the gel runs.

Note 5.5: We use Anchor Wrap All-Purpose Foodservice Film because it does not display background fluorescence with FAM or Cy5. Other plastic film options would need to be evaluated for background fluorescence during imaging for optimal quantitation. We have observed high background fluorescence from AEP Industries Inc. Sealwrap brand plastic wrap and do not recommend this brand. The background will be affected by the sensitivity of the imager used and the concentration of fluorescently labeled DNA.

5.5 Critical notes

Critical Note 5.1: It is essential that equimolar quantities of each of the 3 oligonucleotides are mixed when preparing the substrate. If an excess of any of the oligonucleotides is added, the substrate will contain intermediates or free fluorescently labeled oligonucleotides.

Critical Note 5.2: Dichlorodimethylsilane is corrosive and toxic. Apply silane by briefly inverting bottle over a KimWipe, not by pipetting the dichlorodimethylsilane solution, because it will corrode the metal components of the pipette. Always use dichlorodimethylsilane in a chemical fume hood.

Critical Note 5.3: Acrylamide is a neurotoxin. Disposable gloves, lab coat, and eye protection should be worn. When inserting the comb, the acrylamide solution can spray upwards. Let extra acrylamide solution polymerize in a beaker before disposal as polyacrylamide is non-toxic.

Critical Note 5.4: Ensure that the buffer does not leak from the upper buffer chamber and that the gel does not float up out of the silanized plates. If loss of contact between the gel and the buffer in either chamber occurs, the high voltage across the buffer reservoirs creates an electrical fire hazard.

6. Measuring helicase-catalyzed unwinding of dual-duplex DNA substrates

Once the dual-labeled substrate has been prepared, it can be used to measure product formation of both duplexes simultaneously. As illustrated in Figs. 3 and 4, helicase-catalyzed unwinding of each duplex creates single-fluorophore-labeled products that can be separated with gel electrophoresis and quantified. This method allows for the measurement of duplex unwinding preference for the helicase of interest and can be used to examine how factors such as protein-protein interactions regulate unwinding of each duplex (Su et al., 2019). The same basic experimental design can be used to determine the quantity of product formed (amplitude) at the completion of a single cycle reaction (Fig. 3) or to measure product formation over time (Fig. 4). Optional addition of a T_{50} oligonucleotide to the reaction in 30-fold excess relative to the unwinding substrate will create a competing binding site for helicases that have dissociated from the unwinding substrate, so product formation will assess the result of a single unwinding attempt (i.e. single-turnover conditions) per substrate. Omission of the T_{50} oligonucleotide from the reaction will allow for rebinding of helicases to the substrate after dissociation, so product formation will assess the result of multiple unwinding attempts (i.e. multi-turnover conditions) per substrate.

6.1 Procedure

6.1.1 Unwinding reaction

Timing: 4–5 h

1. Prepare a native 20% polyacrylamide gel and pre-run it for 60 min at 22 mA (see *Preparing a native 20% polyacrylamide gel* in Section 5.1.2).
2. Prepare and label a 0.6 mL microcentrifuge tube for each condition and replicate of the experiment. These tubes will be referred to as *reaction tubes* (blue tube in Fig. 2). Reactions should be performed in triplicate.

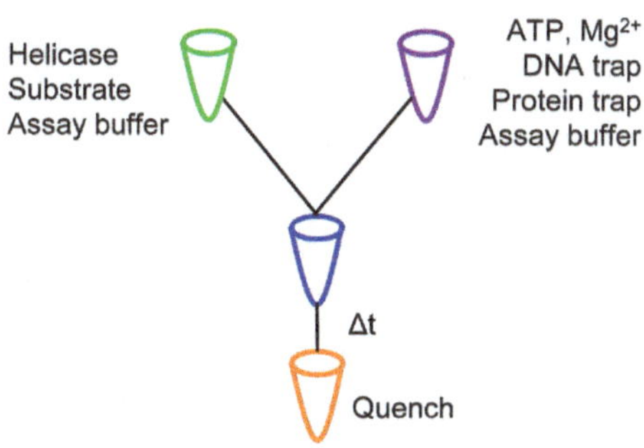

Fig. 2 Illustration of reaction setup. The reaction is prepared in two halves, one containing 2× helicase and substrate, and the other containing 2× ATP, Mg^{2+}, and trapping strands. The reaction is initiated by mixing equivalent quantities of each 2× reaction half. At various times, 10 μL are removed from the reaction tube and mixed with an equal volume of 2× quench solution to stop the reaction.

3. For each reaction, label a microcentrifuge tube for each time point, including a zero-second time point and a heated control. Add 10 μL of 2 × quenching solution to each tube. These tubes will be referred to as *time point tubes* (orange tube in Fig. 2). See note 6.1 for considerations of the number of time points to collect and for considerations on the duration of each time point depending on the desired experimental outcome.
4. Prepare an adequate amount of 2 × ATP/trap solution (1 × helicase assay buffer, 8 mM ATP, 10 mM $Mg(OAc)_2$, 0.8 μM Parent16, 0.8 μM Lead16, and 24 μM T_{50} in ddH_2O) (purple tube in Fig. 2) to have a total volume five times the number of *time point tubes* plus an additional 10 μL. This will provide sufficient volume to use 5 μL for each time point and accounts for sample loss due to sticking to tubes. See note 6.2 for considerations on inclusion of the T_{50} protein trapping strand. See note 6.3 for considerations on the Parent16 and Lead16 annealing traps.
5. Prepare an equivalent volume of 2 × helicase/substrate solution (1 × helicase assay buffer, 40 nM dual-fluorophore-labeled substrate, and 2 × helicase in ddH_2O (green tube in Fig. 2). The concentration of helicase needed will depend on the particular helicase used. See note 6.4 for details relating to experiments performed in Figs. 3 and 4. Prepare the helicase/substrate solution without helicase, chill on ice, and then add helicase 5 min before initiating the reaction. This incubation allows the enzyme to bind the substrate before beginning the reaction.
6. For the zero-second time point, mix 5 μL of the 2 × ATP/trap solution with the 10 μL of 2 × quenching solution already in the zero-second *time point tube* (orange in Fig. 2). Then add 5 μL of the enzyme/substrate solution and mix well.
7. For the heated controls, mix 5 μL of the 2 × ATP/trap solution with the 10 μL of 2 × quenching solution already in the heat *time point tube* (orange in Fig. 2). Then add 5 μL of the enzyme/substrate solution and heat at 95 °C for 10 min.
8. To begin an unwinding reaction, transfer 5 μL times the number of time points (excluding time zero and heated sample) plus an additional 5 μL of 2 × ATP/trap solution to a *reaction tube*. To initiate the reaction, add an equivalent volume of 2 × enzyme/substrate solution, and simultaneously start a timer counting up. Place the reaction tube at the appropriate temperature for the helicase of interest. See note 6.5 for

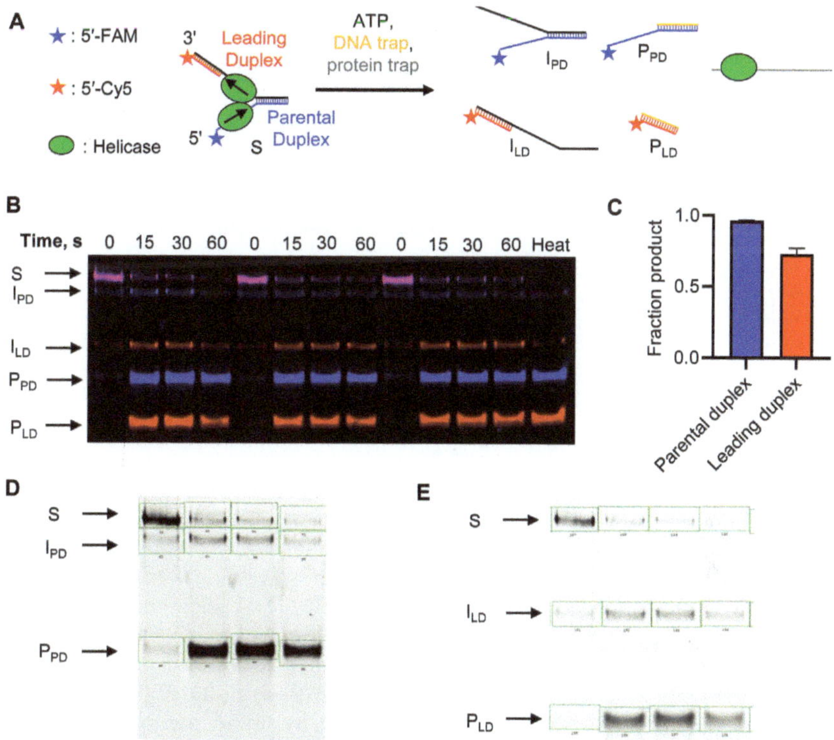

Fig. 3 Simultaneous visualization of multiple duplex unwinding. (A) Schematic of the helicase-catalyzed multiple duplex unwinding reaction. Helicase monomers (green) are bound to the ssDNA regions of the dual duplex unwinding substrate (S). Each helicase monomer will translocate 5′-to-3′ (as indicated by the arrow) and unwind the downstream parental duplex (blue) or the leading duplex (red). After ATP, DNA trapping strands (orange), and protein trap (gray) are added, reaction products (I_{PD}, I_{LD}, P_{PD}, and P_{LD}) are formed, and the helicase is prevented from rebinding to substrate by the protein trap. (B) Unwinding of the dual duplex substrate was performed in triplicate using Dda, and products were visualized after electrophoresis on a 20% native polyacrylamide gel. Multiple time points were collected to ensure completion of the unwinding reaction. (C) The fraction of product formed (either P_{PD} or P_{LD}) by unwinding of the duplex was quantified for each time point of each replicate. The mean fraction product formed was averaged to determine the quantity of product formed for each duplex. Plotted are means with standard deviations; $n = 3$. (D) Rectangles drawn for quantification of the FAM channel of a single replicate of the reaction in 3C are illustrated. (E) Rectangles drawn for quantification of the Cy5 channel of a single replicate of the reaction in 3C are illustrated.

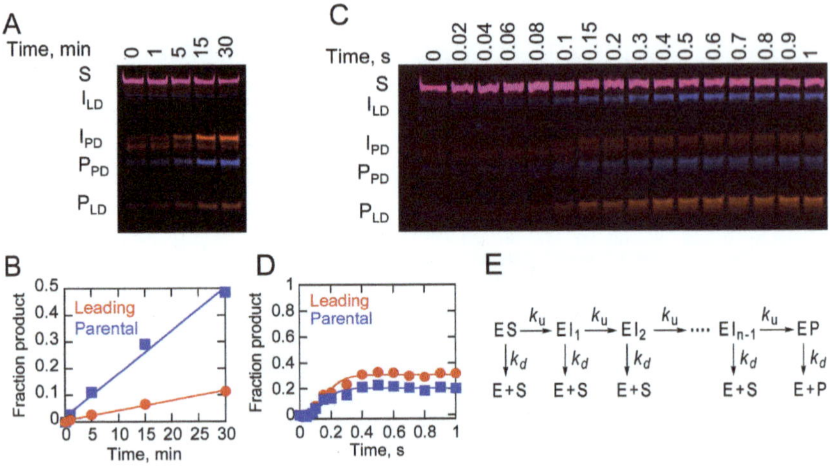

Fig. 4 Determining the rate of unwinding for multiple duplexes on the same substrate. (A) A multi-turnover unwinding reaction of a dual duplex substrate by human PIF1 helicase was performed (Su et al., 2019). (B) Product formation was quantified, and data was fit to a linear equation to determine that the rate of unwinding was greater for the parental duplex than the leading duplex (Su et al., 2019). (C) A single-turnover unwinding reaction of a dual duplex substrate by BaPif1 helicase was performed with a rapid chemical quench flow instrument (Su et al., 2019). (D) The fraction product formed for BaPif1-catalyzed unwinding of each duplex quantified (Su et al., 2019). (E) The rate constants of unwinding and dissociation were derived by fitting the data to a 5-step sequential mechanism. The rate constants for unwinding the leading and parental duplexes were the same, but the dissociation rate was higher for the parental duplex, leading to a reduced amplitude of product formation for the parental duplex (Byrd, Bell, & Raney, 2018; Su et al., 2019). Portions of two figures were combined with minor changes to the font and title. Additionally, data that was originally plotted in two separate plots is combined in panel B. *Adapted from Su, N., Byrd, A. K., Bharath, S. R., Yang, O., Jia, Y., Tang, X., et al. (2019). Structural basis for DNA unwinding at forked dsDNA by two coordinating Pif1 helicases.* Nature Communications *10(1), 1–11. doi:10.1038/s41467-019-13414-9. and Byrd, A. K., Bell, M. R., & Raney, K. D. (2018). Pif1 helicase unfolding of G-quadruplex DNA is highly dependent on sequence and reaction conditions.* Journal of Biological Chemistry, *293(46), 17792–17802. doi:10.1074/jbc.RA118.004499 under CC BY Creative Commons Attribution 4.0 International License (http://creativecommons.org/licenses/by/4.0/).*

temperature details relating to experiments performed in Figs. 3 and 4. The final reaction volume should be 10 μL per time point plus an additional 10 μL.

9. At the appropriate time, remove 10 μL of the reaction solution, and mix the reaction solution with the 10 μL of 2× quench solution in the appropriate *time point tube* to quench the unwinding reaction. The reaction solution should be added to the quench solution when

the timer reaches the time corresponding to the time point. Pipette the solution up three times after adding the reaction to the quench to ensure mixing.

10. Repeat step 9 for additional time points as appropriate.
11. Vortex the quenched reaction solutions in the *product tubes* and spin them down in a tabletop centrifuge.
12. Separate reaction products by gel electrophoresis as described in Section 5.1.3 *Separating samples by electrophoresis.*

6.1.2 Quantitation of product formation

1. Open the 2 channel fluorescent overlay image and confirm that the substrate, intermediate, and product migrate as expected based on your control gel from Section 5. Preparation of dual duplex DNA substrates. An example FAM and Cy5 channel fluorescent overlay of a multiple duplex unwinding reaction is in Fig. 3B.
2. Open the image from a single fluorescent channel. Analysis is performed on single color images, not the composite image of both fluorophores, so that quantitation of product formation for each duplex is unaffected by the signal from the other duplex. Fluorescence intensity of each duplex unwinding product can be measured using the software provided with your imager or ImageJ.
3. As observed in the gel to confirm substrate formation and migration patterns of the substrates and intermediates, three bands should be present in each fluorescence channel. An example FAM channel image of a single replicate from the gel in Fig. 3B is shown in Fig. 3D, and an example Cy5 channel image from the same replicate is shown in Fig. 3E.
4. Identify the substrate from the zero-second time point. Identify the product from the heated sample. The intermediate should migrate between the substrate and product.
5. Measure the density of the substrate, intermediate, and product in each lane. Many programs have built-in methods to do this automatically; however, manual quantitation is preferred to avoid missing lower intensity bands. To quantify manually, draw a rectangle around the substrate in the first lane and copy and paste for substrate, intermediate, and product in each lane. Copy and paste ensures that all rectangles are the same size, which reduces the effect of background. Example rectangles for quantitation are overlayed in green in Fig. 3D and E. Three extra rectangles are also pasted in regions with no bands for background correction.

6. Copy the data into a spreadsheet and average the three values for background density. Subtract this average background density from the density of each of the substrate, intermediate, and product band.
7. Calculate the fraction product in each lane by dividing the background corrected density of the product band by the sum of the background corrected densities of all three bands. The intermediate is counted as substrate because, in both cases, formation of the intermediate is due to unwinding of the duplex labeled with the fluorophore that is not scanned. For the FAM-labeled parental duplex, the intermediate I_{PD} has an unwound Cy5-labeled duplex, but the FAM-labeled duplex is intact. Likewise, for the Cy5-labeled leading duplex, the intermediate I_{LD} has an unwound FAM-labeled duplex, but the Cy5-labeled duplex is intact, as shown in Fig. 3A.
8. The fraction product for each duplex is corrected for any product present at time zero as follows:

$$\text{Fraction product} = \frac{\frac{P_t}{S_t + I_t + P_t} - \frac{P_0}{S_0 + I_0 + P_0}}{1 - \frac{P_0}{S_0 + I_0 + P_0}}.$$

   P_t is the background-corrected density of product (P_{PD} for FAM fluorescence; P_{LD} for Cy5 fluorescence) at a time point. P_0 is the background-corrected density of product at time zero. S_t is the background-corrected density of substrate at a time point. S_0 is the background-corrected density of substrate at time zero. I_t is the background-corrected density of intermediate with the other duplex unwound (I_{PD} for FAM fluorescence; I_{LD} for Cy5 fluorescence) at a time point. I_0 is the background-corrected density of intermediate with the other duplex unwound at time zero.
9. For experiments where the desired output is the quantity of product formed in a single cycle, the fraction product formed for each duplex is then averaged across each time point (except zero) and can be graphed as in Fig. 3C.
10. For experiments where a reaction time course has been collected, the shape of the reaction progress curve will determine the appropriate equation to use. For multi-turnover reactions, a linear function can be used to fit reactions that have not reached completion and have a linear appearance, as shown in Fig. 4A and B. For multi-turnover reactions that reach completion, a single exponential function is often appropriate. For single-turnover reactions, the reaction progress curve may be exponential or may be more complicated like the example in

Fig. 4C and D which exhibits a lag phase followed by a burst phase. This data was fit to a n-step sequential mechanism shown in Fig. 4E using KinTek Explorer (Johnson, Simpson, & Blom, 2009). Alternatively, the data can be fit with an equation describing an n-step mechanism (Ali et al., 1999; Ali & Lohman, 1997). In any case, the goodness of fit will need to be evaluated to determine whether the chosen function is appropriate.

6.2 Expected outcomes

As illustrated in Fig. 3, the fraction product formed from the parental duplex and the leading duplex (either P_{PD} or P_{LD}) can be assessed following electrophoresis. The zero-second time point serves as a negative control for substrate unwinding, and the heated sample serves as a positive control for duplex unwinding. As shown in Fig. 3C, quantification of the mean fraction product formed for each duplex allows for comparison of the processivity of the helicase for unwinding one duplex relative to the other.

As illustrated in Fig. 4, the product formation over time can be measured. For multi-turnover reactions, as in Fig. 4A, the rate of product formation for one duplex relative to the other can be determined. The data can fit with a linear function, as shown in Fig. 4B, or a single exponential function, depending on the shape of the curve. See Section *6.1.2 Quantitation of product formation* for more information on data fitting. As shown in Fig. 4C, a time course of product formation under single turnover conditions can be measured. This allows information about both the rate and processivity of product formation for each of the duplexes to be compared, as the amplitude of the product formation curve is affected by the processivity of the helicase, and the slope of the burst phase is affected by the unwinding rate of the helicase. Differences in the rate or amplitude of the product formation curves for the leading duplex and the parental duplex can be used to infer information about the helicase mechanism (Su et al., 2019).

6.3 Statistical approaches

Fraction product measurements of duplex unwinding are assumed to be normally distributed, so a student's *t*-test can be used to assess statistical significance between fraction product formation measurements of each duplex of interest. If assumptions of normality are violated, a non-parametric Mann-Whitney *U* test can be conducted instead.

6.4 Advantages

Measuring unwinding of two duplexes on the same substrate simultaneously provides an opportunity to measure the activity of enzymes bound to both strands of a forked DNA substrate. Since the activity of a helicase bound to the displaced strand is invisible in a traditional DNA unwinding assay, use of substrates with dual duplexes provides additional information about the unwinding mechanism. The use of fluorescently-labeled oligonucleotides reduces hazards created by use of ^{32}P-labeled oligonucleotides and allows for the monitoring of unwinding of more than one duplex at a time when each duplex is labeled with a different fluorophore. Additionally, the volume of purified enzyme required is small, and the required equipment is commonly available.

6.5 Limitations

The use of this ensemble assay does not elucidate any reaction intermediates or transient events that could be seen with single-molecule techniques.

6.6 Optimization and troubleshooting

6.6.1 Intermediates and/or products present in zero-second time point

The zero-second time point serves as a negative control for unwinding, as ATP hydrolysis is inhibited before addition of enzyme to the reaction. A high abundance of intermediate and product will prevent accurate quantitation.

6.6.2 Potential solution

Remake the substrate according to Section 5.1.1, ensuring that the concentrations of each component are equal. See also Section 5.3, "Presence of single-fluorophore-labelled species after substrate annealing along with annealed substrate band".

6.6.3 After initial rapid product formation, product formation slowly increases in a single-turnover reaction

If the product formation slowly increases after initially forming at a faster rate, it is likely that the concentration of the protein trap is not sufficiently high to ensure single-turnover conditions.

6.6.4 Potential solution

Increase the concentration of T_{50} protein trap to prevent rebinding of the helicase to the substrate to ensure single-turnover reaction conditions.

6.7 Notes

Note 6.1: The number of time points required depends on the desired output. For determining product amplitude, three time points should be collected in addition to time zero to ensure the reaction has reached completion. If only one time point is collected, it is difficult to determine whether changes in the quantity of product formed are due to changes in rate or processivity of product formation. If the quantity of product formed increases over time, the experiment can be repeated with longer time points where the fraction product formed is the same for each time point to determine the total quantity of product formed. Alternatively, the experiment can be performed across a broad time course to determine the rate of product formation. The number of time points required for a full time course depends on the helicase and substrate used. If product formation is linear or single exponential, 3 or 6 time points, respectively, are sufficient to define the curve. For more complex reaction progress curves with lag phases, more time points are required.

Note 6.2: T_{50} serves as a protein trap to prevent helicase that dissociates from the substrate from rebinding a substrate molecule and re-initiating unwinding (i.e., single-turnover conditions). The protein trap is added in at least 30-fold excess of binding sites relative to the helicase concentration. The protein trap can be omitted for multi-turnover reactions where the helicase dissociates and rebinds substrate throughout the reaction.

Note 6.3: The Parent16 and Lead16 are complementary to the FAM- and Cy5-labeled oligonucleotides. They are used to trap the unwound fluorescently labeled oligonucleotides to prevent reannealing. The annealing traps are added at a 20–30-fold excess, relative to the substrate.

Note 6.4: For the experiment in Figs. 3, 400 nM Dda was used in the 2 × solution, 2 μM hPIF1 was used in the 2 × solution for the experiments in Fig. 4A and B, and 4 μM BaPif1 was used in the 2 × solution for the experiments in Fig. 4C and D.

Note 6.5: For Dda, hPIF1, and BaPif1, reactions were performed at 25 °C.

7. Alternative methods/procedures

In addition to this procedure, there are other methods which allow monitoring of the unwinding of multiple duplexes simultaneously. This assay could be adapted for real-time measurement of unwinding of multiple duplexes on a single substrate by monitoring the increase in fluorescence upon unwinding of duplexes, each with a quenched fluorophore. Similar experiments using a plate reader or stopped flow spectrophotometer have been performed to monitor unwinding of a single duplex (Donmez & Patel, 2008; Houston & Kodadek, 1994; Marecki et al., 2019; Özeş et al., 2014).

Single molecule FRET (smFRET) microscopy with alternating laser excitation (ALEX) (Hohlbein, Craggs, & Cordes, 2014) allows simultaneous measurement of unwinding of multiple duplexes in a single molecule (Su et al., 2019). The main advantage of smFRET with ALEX is that by measuring unwinding of single molecules, intermediates in the unwinding process can be observed. For example, BaPif1 unwinding of a dual duplex substrate similar to those described here was monitored by smFRET with ALEX, and a mid-FRET intermediate was observed that was missed in the ensemble experiments (Su et al., 2019). This intermediate was observed for unwinding of both the leading and parental duplex (Su et al., 2019).

DNA footprinting and DNA protein-crosslinking can also be used to determine if helicases interact with both the translocase and displaced strand ssDNA arms of a forked duplex (Aarattuthodiyil et al., 2014; Windgassen et al., 2018; Windgassen & Keck, 2016), although these methods do not provide information about the activity of the helicases bound to the two strands.

References

Aarattuthodiyil, S., Byrd, A. K., & Raney, K. D. (2014). Simultaneous binding to the tracking strand, displaced strand and the duplex of a DNA fork enhances unwinding by Dda helicase. *Nucleic Acids Research*, *42*(18), 11707–11720. https://doi.org/10.1093/NAR/GKU845.

Ali, J. A., & Lohman, T. M. (1997). Kinetic measurement of the step size of DNA unwinding by Escherichia coli UvrD helicase. *Science*, *275*(5298), 377–380. https://doi.org/10.1126/SCIENCE.275.5298.377.

Ali, J. A., Maluf, N. K., & Lohman, T. M. (1999). An oligomeric form of E. coli UvrD is required for optimal helicase activity. *Journal of Molecular Biology*, *293*(4), 815–834. https://doi.org/10.1006/JMBI.1999.3185.

Amit, R., Gileadi, O., & Stavans, J. (2004). Direct observation of RuvAB-catalyzed branch migration of single Holliday junctions. *Proceedings of the National Academy of Sciences of the United States of America*, *101*(32), 11605–11610. https://doi.org/10.1073/PNAS.0404332101.

Belon, C. A., & Frick, D. N. (2008). Monitoring helicase activity with molecular beacons. *BioTechniques*, *45*(4), 433–442. https://doi.org/10.2144/000112834.

Bianco, P. R., Brewer, L. R., Corzett, M., Balhorn, R., Yeh, Y., Kowalczykowski, S. C., et al. (2001). Processive translocation and DNA unwinding by individual RecBCD enzyme molecules. *Nature*, *409*(6818), 374–378. https://doi.org/10.1038/35053131.

Bjomson, K. P., Amaratunga, M., Moore, K. J. M., & Lohman, T. M. (1994). Single-turnover kinetics of helicase-catalyzed DNA unwinding monitored continuously by fluorescence energy transfer. *Biochemistry*, *33*(47), 14306–14316. https://doi.org/10.1021/BI00251A044.

Brosh, R. M., & Matson, S. W. (2020). History of DNA helicases. In *Vol. 11. Genes* MDPI AG. https://doi.org/10.3390/genes11030255. Issue 3.

Byrd, A. K., Bell, M. R., & Raney, K. D. (2018). Pif1 helicase unfolding of G-quadruplex DNA is highly dependent on sequence and reaction conditions. *Journal of Biological Chemistry*, *293*(46), 17792–17802. https://doi.org/10.1074/jbc.RA118.004499.

Byrd, A. K., Malone, E. G., Hazeslip, L., Zafar, M. K., Harrison, D. K., Thompson, M. D., et al. (2021). A structural feature of Dda helicase which enhances displacement of streptavidin and trp repressor from DNA. *Protein Science*, *31*, 407–421. https://doi.org/10.1002/PRO.4232.

Byrd, A. K., Matlock, D. L., Bagchi, D., Aarattuthodiyil, S., Harrison, D., Croquette, V., et al. (2012). Dda helicase tightly couples translocation on single-stranded DNA to unwinding of duplex DNA: Dda is an optimally active helicase. *Journal of Molecular Biology*, *420*(3), 141. https://doi.org/10.1016/J.JMB.2012.04.007.

Byrd, A. K., & Raney, K. D. (2005). Increasing the length of the single-stranded overhang enhances unwinding of duplex DNA by bacteriophage T4 Dda helicase. *Biochemistry*, *44*(39), 12990–12997. https://doi.org/10.1021/bi050703z.

Carney, S. P., Ma, W., Whitley, K. D., Jia, H., Lohman, T. M., Luthey-Schulten, Z., et al. (2021). Kinetic and structural mechanism for DNA unwinding by a non-hexameric helicase. *Nature Communications*, *12*(1), 7015. https://doi.org/10.1038/S41467-021-27304-6.

Comstock, M. J., Whitley, K. D., Jia, H., Sokoloski, J., Lohman, T. M., Ha, T., et al. (2015). Direct observation of structure-function relationship in a nucleic acid—Processing enzyme. *Science*, *348*(6232), 352–354. https://doi.org/10.1126/SCIENCE.AAA0130.

Dawid, A., Croquette, V., Grigoriev, M., & Heslot, F. (2004). Single-molecule study of RuvAB-mediated Holliday-junction migration. *Proceedings of the National Academy of Sciences of the United States of America*, *101*(32), 11611–11616. https://doi.org/10.1073/PNAS.0404369101.

Dessinges, M. N., Lionnet, T., Xi, X. G., Bensimon, D., & Croquette, V. (2004). Single-molecule assay reveals strand switching and enhanced processivity of UvrD. *Proceedings of the National Academy of Sciences of the United States of America*, *101*(17), 6439–6444. https://doi.org/10.1073/PNAS.0306713101.

Dillingham, M. S., Wigley, D. B., & Webb, M. R. (1999). Demonstration of unidirectional single-stranded DNA translocation by PcrA helicase: Measurement of step size and translocation speed†. *Biochemistry*, *39*(1), 205–212. https://doi.org/10.1021/BI992105O.

Donmez, I., & Patel, S. S. (2008). Coupling of DNA unwinding to nucleotide hydrolysis in a ring-shaped helicase. *The EMBO Journal*, *27*(12), 1718–1726. https://doi.org/10.1038/EMBOJ.2008.100.

Eoff, R. L., & Raney, K. D. (2006). Intermediates revealed in the kinetic mechanism for DNA unwinding by a monomeric helicase. *Nature Structural & Molecular Biology*, *13*(3), 242–249. https://doi.org/10.1038/nsmb1055.

George, T., Wen, Q., Griffiths, R., Ganesh, A., Meuth, M., & Sanders, C. M. (2009). Human Pif1 helicase unwinds synthetic DNA structures resembling stalled DNA replication forks. *Nucleic Acids Research*, *37*(19), 6491–6502. https://doi.org/10.1093/NAR/GKP671.

Gupta, R., Sharma, S., Sommers, J. A., Jin, Z., Cantor, S. B., & Brosh, R. M. (2005). Analysis of the DNA substrate specificity of the human BACH1 helicase associated with breast cancer. *Journal of Biological Chemistry*, *280*(27), 25450–25460. https://doi.org/10.1074/JBC.M501995200.

He, X., Byrd, A. K., Yun, M. K., Pemble, C. W., IV, Harrison, D., Yeruva, L., et al. (2012). The T4 phage SF1B helicase Dda is structurally optimized to perform DNA strand separation. *Structure*, *20*(7), 1189–1200. https://doi.org/10.1016/j.str.2012.04.013.

Hohlbein, J., Craggs, T. D., & Cordes, T. (2014). Alternating-laser excitation: Single-molecule FRET and beyond. *Chemical Society Reviews, 43*(4), 1156–1171. https://doi.org/10.1039/C3CS60233H.

Houston, P., & Kodadek, T. (1994). Spectrophotometric assay for enzyme-mediated unwinding of double-stranded DNA. *Proceedings of the National Academy of Sciences of the United States of America, 91*(12), 5471–5474. https://doi.org/10.1073/PNAS.91.12.5471.

Jennings, T. A., Mackintosh, S. G., Harrison, M. K., Sikora, D., Sikora, B., Dave, B., et al. (2009). NS3 helicase from the hepatitis C virus can function as a monomer or oligomer depending on enzyme and substrate concentrations. *Journal of Biological Chemistry, 284*(8), 4806–4814. https://doi.org/10.1074/JBC.M805540200.

Jeong, Y. J., Levin, M. K., & Patel, S. S. (2004). The DNA-unwinding mechanism of the ring helicase of bacteriophage T7. *Proceedings of the National Academy of Sciences of the United States of America, 101*(19), 7264–7269. https://doi.org/10.1073/PNAS.0400372101.

Johnson, K. A., Simpson, Z. B., & Blom, T. (2009). Global kinetic explorer: A new computer program for dynamic simulation and fitting of kinetic data. *Analytical Biochemistry, 387*(1), 20–29. https://doi.org/10.1016/J.AB.2008.12.024.

Lahaye, A., Leterme, S., & Foury, F. (1993). PIF1 DNA helicase from Saccharomyces cerevisiae. Biochemical characterization of the enzyme. *Journal of Biological Chemistry, 268*(35), 26155–26161. https://doi.org/10.1016/S0021-9258(19)74294-X.

Levin, M. K., Wang, Y. H., & Patel, S. S. (2004). The functional interaction of the hepatitis C virus helicase molecules is responsible for unwinding processivity. *Journal of Biological Chemistry, 279*(25), 26005–26012. https://doi.org/10.1074/JBC.M403257200.

Lu, C., Le, S., Chen, J., Byrd, A. K., Rhodes, D., Raney, K. D., et al. (2019). Direct quantification of the translocation activities of Saccharomyces cerevisiae Pif1 helicase. *Nucleic Acids Research, 47*(14), 7494–7501. https://doi.org/10.1093/NAR/GKZ541.

Maluf, N. K., Fischer, C. J., & Lohman, T. M. (2003). A dimer of Escherichia coli UvrD is the active form of the helicase in vitro. *Journal of Molecular Biology, 325*(5), 913–935. https://doi.org/10.1016/S0022-2836(02)01277-9.

Marecki, J. C., Aarattuthodiyil, S., Byrd, A. K., Penthala, N. R., Crooks, P. A., & Raney, K. D. (2019). N-Naphthoyl-substituted indole thio-barbituric acid analogs inhibit the helicase activity of the hepatitis C virus NS3. *Bioorganic & Medicinal Chemistry Letters, 29*(3), 430–434. https://doi.org/10.1016/J.BMCL.2018.12.026.

Matson, S. W., & Kaiser-Rogers, K. A. (1990). DNA helicases. *Annual Review of Biochemistry, 59*(1), 289–329. https://doi.org/10.1146/ANNUREV.BI.59.070190.001445.

Matson, S. W., Tabor, S., & Richardson, C. C. (1983). The gene 4 protein of bacteriophage T7. Characterization of helicase activity. *Journal of Biological Chemistry, 258*(22), 14017–14024. https://doi.org/10.1016/S0021-9258(17)44018-X.

Morris, P. D., & Raney, K. D. (1999). DNA helicases displace streptavidin from biotin-labeled oligonucleotides. *Biochemistry, 38*(16), 5164–5171. https://doi.org/10.1021/BI9822269.

Myong, S., Bruno, M. M., Pyle, A. M., & Ha, T. (2007). Spring-loaded mechanism of DNA unwinding by hepatitis C virus NS3 helicase. *Science, 317*(5837), 513–516. https://doi.org/10.1126/SCIENCE.1144130.

Nanduri, B., Byrd, A. K., Eoff, R. L., Tackett, A. J., & Raney, K. D. (2002). Pre-steady-state DNA unwinding by bacteriophage T4 Dda helicase reveals a monomeric molecular motor. *Proceedings of the National Academy of Sciences of the United States of America, 99*(23), 14722–14727. https://doi.org/10.1073/pnas.232401899.

Özeş, A. R., Feoktistova, K., Avanzino, B. C., Baldwin, E. P., & Fraser, C. S. (2014). Real-time fluorescence assays to monitor duplex unwinding and ATPase activities of helicase proteins. *Nature Protocols, 9*(7), 1645. https://doi.org/10.1038/NPROT.2014.112.

Park, J., Myong, S., Niedziela-Majka, A., Lee, K. S., Yu, J., Lohman, T. M., et al. (2010). PcrA helicase dismantles RecA filaments by reeling in DNA in uniform steps. *Cell*, *142*(4), 544–555. https://doi.org/10.1016/J.CELL.2010.07.016.

Rajagopal, V., & Patel, S. S. (2008). Single Strand binding proteins increase the processivity of DNA unwinding by the hepatitis C virus helicase. *Journal of Molecular Biology*, *376*(1), 69–79. https://doi.org/10.1016/J.JMB.2007.10.070.

Ramanagoudr-Bhojappa, R., Chib, S., Byrd, A. K., Aarattuthodiyil, S., Pandey, M., Patel, S. S., et al. (2013). Yeast Pif1 helicase exhibits a one-base-pair stepping mechanism for unwinding duplex DNA. *Journal of Biological Chemistry*, *288*(22), 16185–16195. https://doi.org/10.1074/JBC.M113.470013.

Raney, K. D., & Benkovic, S. J. (1995). Bacteriophage T4 Dda helicase translocates in a unidirectional fashion on single-stranded DNA (*). *Journal of Biological Chemistry*, *270*(38), 22236–22242. https://doi.org/10.1074/JBC.270.38.22236.

Raney, K. D., Sowerst, L. C., Millar, D. P., Benkovic, S. J., Hacker, K., & Alberts, B. (1994). A fluorescence-based assay for monitoring helicase activity. *Proceedings of the National Academy of Sciences of the United States of America*, *91*, 6644–6648.

Rasnik, I., Jeong, Y. J., McKinney, S. A., Rajagopal, V., Patel, S. S., & Ha, T. (2008). Branch migration enzyme as a Brownian ratchet. *The EMBO Journal*, *27*(12), 1727–1735. https://doi.org/10.1038/EMBOJ.2008.106.

Saikrishnan, K., Powell, B., Cook, N. J., Webb, M. R., & Wigley, D. B. (2009). Mechanistic basis of 5′-3′ translocation in SF1B helicases. *Cell*, *137*(5), 849–859. https://doi.org/10.1016/j.cell.2009.03.036.

Sikora, B., Chen, Y., Lichti, C. F., Harrison, M. K., Jennings, T. A., Tang, Y., et al. (2008). Hepatitis C virus NS3 helicase forms oligomeric structures that exhibit optimal DNA unwinding activity in vitro. *Journal of Biological Chemistry*, *283*(17), 11516–11525. https://doi.org/10.1074/jbc.M708125200.

Su, N., Byrd, A. K., Bharath, S. R., Yang, O., Jia, Y., Tang, X., et al. (2019). Structural basis for DNA unwinding at forked dsDNA by two coordinating Pif1 helicases. *Nature Communications*, *10*(1), 1–11. https://doi.org/10.1038/s41467-019-13414-9.

Venkatesan, M., Silver, L. L., & Nossal, N. G. (1982). Bacteriophage T4 gene 41 protein, required for the synthesis of RNA primers, is also a DNA helicase. *Journal of Biological Chemistry*, *257*(20), 12426–12434. https://doi.org/10.1016/S0021-9258(18)33731-1.

Windgassen, T. A., & Keck, J. L. (2016). An aromatic-rich loop couples DNA binding and ATP hydrolysis in the PriA DNA helicase. *Nucleic Acids Research*, *44*(20), 9745–9757. https://doi.org/10.1093/NAR/GKW690.

Windgassen, T. A., Leroux, M., Satyshur, K. A., Sandler, S. J., & Keck, J. L. (2018). Structure-specific DNA replication-fork recognition directs helicase and replication restart activities of the PriA helicase. *Proceedings of the National Academy of Sciences of the United States of America*, *115*(39), E9075–E9084. https://doi.org/10.1073/pnas.1809842115.

Yodh, J. G., Stevens, B. C., Kanagaraj, R., Janscak, P., & Ha, T. (2009). BLM helicase measures DNA unwound before switching strands and hRPA promotes unwinding reinitiation. *The EMBO Journal*, *28*(4), 405–416. https://doi.org/10.1038/EMBOJ.2008.298.

Zhou, T., Ren, X., Adams, R. L., & Pyle, A. M. (2018). NS3 from hepatitis C virus strain JFH-1 is an unusually robust helicase that is primed to bind and unwind viral RNA. *Journal of Virology*, *92*(1), e01253-17. https://doi.org/10.1128/JVI.01253-17.

Zhou, X., Ren, W., Bharath, S. R., Tang, X., He, Y., Chen, C., et al. (2016). Structural and functional insights into the unwinding mechanism of Bacteroides sp Pif1. *Cell Reports*, *14*(8), 2030–2039. https://doi.org/10.1016/J.CELREP.2016.02.008.

Zhou, R., Zhang, J., Bochman, M. L., Zakian, V. A., & Ha, T. (2014). Periodic DNA patrolling underlies diverse functions of Pif1 on R-loops and G-rich DNA. *eLife*, *2014*(3), e02190. https://doi.org/10.7554/ELIFE.02190.001.

CHAPTER TWO

Alignment of helicases on single-stranded DNA increases activity

Deniz Ozaslan, Alicia K. Byrd, Binyam Belachew, and Kevin D. Raney*
Department of Biochemistry and Molecular Biology, University of Arkansas for Medical Sciences, Little Rock, AR, United States
*Corresponding author: e-mail address: akbyrd@uams.edu; raneykevind@uams.edu

Contents

Methods in Enzymology, Volume 672
ISSN 0076-6879
https://doi.org/10.1016/bs.mie.2022.03.066

Abstract

Helicases function in most biological processes that utilize RNA or DNA nucleic acids including replication, recombination, repair, transcription, splicing, and translation. They are motor proteins that bind ATP and then catalyze hydrolysis to release energy which is transduced for conformational changes. Different conformations correspond to different steps in a process that results in movement of the enzyme along the nucleic acid track in a unidirectional manner. Some helicases such as DEAD-box helicases do not translocate, but these enzymes transduce chemical energy from ATP hydrolysis to unwind secondary structure in DNA or RNA. Some helicases function as monomers while others assemble into defined structures, either dimers or higher order oligomers. Dda helicase from bacteriophage T4 and NS3 helicase domain from the hepatitis C virus are examples of monomeric helicases. These helicases can bind to single-stranded DNA in a manner that appears like train engines on a track. When monomeric helicases align on DNA, the activity of the enzymes increases. Helicase activity can include the rate of duplex unwinding and the total number of base pairs melted during a single binding event or processivity. Dda and NS3h are considered as having low processivity, unwinding fewer than 50 base pairs per binding event. Here, we report fusing two molecules of NS3h molecules together through genetically linking the C-terminus of one molecule to the N-terminus of a second NS3h molecule. We observed increased processivity relative to NS3h possibly arising from the increased probability that at least one of the helicases will completely unwind the DNA prior to dissociation. The dimeric enzyme also binds DNA more like the full-length NS3 helicase. Finally, the dimer can displace streptavidin from biotin-labeled oligonucleotide, whereas monomeric NS3h cannot.

1. Introduction

Helicase mechanisms have invited scrutiny for over 3 decades. A thorough account of the gradual development and understanding helicase mechanisms activity was recently published by Brosh and Matson (2020). Helicases contain critically important motifs that can be considered as the engine that produces energy through ATPase activity. Other helicase motifs serve as the transmission by conveying conformational changes that accompany ATP hydrolysis resulting in directionally biased movement along DNA. These motifs are located on two domains that are recA-like (Saikrishnan, Powell, Cook, Webb, & Wigley, 2009; Velankar, Soultanas, Dillingham, Subramanya, & Wigley, 1999). There are many proteins that share the helicase motifs, such as chromatin remodelers, illustrating that a fundamental aspect of these motifs is to drive movement along the DNA (Reyes, Marcum, & He, 2021). However as mentioned above, some helicases do not translocate, such as the DEAD-box helicases (Sharma, Putnam, & Jankowsky, 2017).

Helicases are classified as superfamilies based on sequence analysis (Gorbalenya & Koonin, 1993). Superfamily 1 (SF1) and superfamily 2 (SF2) make up the majority of helicases. These include well-studied helicases such as Rep (Makurath, Whitley, Nguyen, Lohman, & Chemla, 2019; Myong, Rasnik, Joo, Lohman, & Ha, 2005), UvrD (Ali & Lohman, 1997; Comstock et al., 2015; Lohman, Tomko, & Wu, 2008), and PcrA (Dillingham et al., 2008; Dillingham, Soultanas, Wiley, Webb, & Wigley, 2001; Park et al., 2010; Urrutia-Irazabal, Ault, Sobott, Savery, & Dillingham, 2021). These three helicases travel (translocate) along DNA in the 3′-to-5′ direction, further delineating the classification as SF1A helicases. A domain that is designated as domain 2B has been shown to play a major role in regulating the activity of Rep, UvrD, and PcrA. Domain 2B is not required for DNA helicase activity for Rep helicase (Cheng et al., 2002), while the conformation of domain 2B controls DNA unwinding activity (Makurath et al., 2019). Helicases that translocate in the 5′-to-3′ direction are designated as SF1B helicases. Examples include Dda from bacteriophage T4 (Eoff & Raney, 2010; He et al., 2012; Morris & Raney, 1999) and Pif1 (Buzovetsky et al., 2017; Byrd & Raney, 2017; Geronimo & Zakian, 2016), which is found in organisms ranging from bacteria to humans. These enzymes appear to lack domain 2B, but otherwise share most of the motifs found in SF1A helicases. At least one helicase, RecBCD, has SF1A and SF1B helicases as part of its structure, allowing it to move along both strands of DNA (Dillingham, Spies, & Kowalczykowski, 2003; Singleton, Dillingham, Gaudier, Kowalczykowski, & Wigley, 2004). Molecular mechanisms for DNA unwinding remain under discussion, although great progress has been made in understanding these molecular motors (Lohman & Fazio, 2018).

Fundamental properties that define helicases are the rate of translocation along single-stranded DNA, the rate of unwinding of duplex DNA, and the number of base pairs unwound or nucleotides traversed per binding event (processivity). Rates of translocation and DNA unwinding vary widely for SF1A and SF1B helicases as does the processivity. On the fast side is TraI which unwinds dsDNA at ~1100 bp/s and exhibits high processivity (Sikora, Eoff, Matson, & Raney, 2006). Helicases interact with proteins that can greatly alter processivity. For example, PcrA is not a processive helicase (Niedziela-Majka, Chesnik, Tomko, & Lohman, 2007), however it can bind to RepD and unwind DNA for thousands of base pairs (Chisty et al., 2013). Helicases that make up SF2 share many of the helicase motifs as SF1. Examples include the RecQ helicases and the Hepatitis C viral helicase NS3. The NS3 helicase has a protease domain that can be deleted from the enzyme leaving only the helicase domain which retains DNA unwinding activity. However, the full-length NS3 helicase has greater processivity.

SF1 and SF2 helicases can remodel protein-nucleic acid complexes by pushing proteins from DNA or by repositioning proteins along DNA (Sokoloski, Kozlov, Galletto, & Lohman, 2016). An assay that measures displacement of streptavidin from biotin-labeled oligonucleotides was developed in order to provide an orthogonal approach to measure directional translocation on ssDNA (Morris & Raney, 1999). Most helicases that have been examined in this assay can displace streptavidin in directionally biased manner. This result has been interpreted to favor a "power-stroke" mechanism, whereby a helicase actively disrupts proteins or nucleic acids in its path by producing force. NS3 helicase domain (NS3h) is one of the few helicases that translocates on ssDNA but does not appear to be able to displace streptavidin from biotin labeled oligonucleotides (Matlock et al., 2010).

Some SF1 and SF2 helicases can interact to form dimers or higher order oligomers with the oligomer generally displaying higher activity (Lohman et al., 2008). An additional type of interaction can occur when the ssDNA binding site is long enough to accommodate multiple SF1 or SF2 helicases. The model in Fig. 1 illustrates conditions for binding by one helicase molecule or three molecules whereby the enzymes align on the ssDNA. The ssDNA overhang can bind multiple molecules with the specific number depending on the length of the overhang and the binding

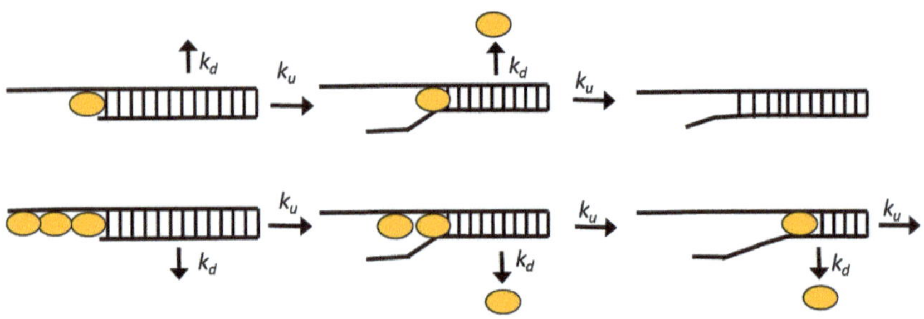

Fig. 1 Model for alignment of helicase molecules along single-stranded DNA. Multiple molecules assist in unwinding duplex or remodeling protein-DNA complexes. The alignment has been referred to as "trains" based on the idea of multiple engines moving along a track. The increase in the quantity of product formed has been referred to as functional cooperativity (Byrd & Raney, 2004; Levin, Wang, & Patel, 2004).

site size which is around 7–8 nucleotides for many SF1 and SF2 helicases. Longer overhangs can bind more than three molecules (Byrd & Raney, 2004, 2006; Ramanagoudr-Bhojappa et al., 2013).

If only one helicase molecule binds to the overhang, dissociation prior to completion of unwinding the duplex DNA will lead to hybridization the two strands (Levin et al., 2004). If multiple molecules bind to the ssDNA overhang, dissociation of one molecule leaves the trailing molecules bound and they can move along the nucleic acid to take the place of the dissociated lead molecule and prevent the substrate from hybridizing (Fig. 1). The model predicts a significant increase in processivity but not necessarily an increase in the rate of unwinding (Eoff & Raney, 2010). If the lead molecule encounters a chemical modification such as an abasic site or a neutral phosphonate deliberately placed to remove a negatively charged phosphodiester, then the lead molecule may stall or dissociate (Eoff & Raney, 2006; Eoff, Spurling, & Raney, 2005). However, the trailing molecules will collide with the stalled lead molecule and can interact with the lead molecule so that the lead molecule moves past the modification. Presumably, the trailing molecules generate force just as is likely to be the case when helicases displace streptavidin from biotin labeled oligonucleotides (Morris et al., 2002; Morris & Raney, 1999). Hence, the molecules that align on the DNA substrate move independently until the lead molecule collides with a structural or chemical block.

A model similar to that shown in Fig. 1 was suggested by the Patel lab for NS3 helicase domain (NS3h). The action of multiple molecules aligned on the DNA led to unwinding when the concentration of enzyme was in excess of the substrate concentration (Levin et al., 2004). Dda helicase has been shown to align on ssDNA and translocate with a 5′-to-3′ directional bias. When Dda encounters a block, such as streptavidin bound to a biotin-labeled oligonucleotide, the streptavidin dissociates (Byrd & Raney, 2004). More recently, RecQ helicase from *E. coli* was shown to function at a DNA fork as a result of assembly of multiple molecules (Rad, Forget, Baskin, & Kowalczykowski, 2015). In the case of NS3 and RecQ, a loosely held multimer of helicase molecules appears to bind to ssDNA. Not all SF1B or SF2 helicases align on ssDNA or ssRNA. The structure of a dimeric form of a Pif1-like helicase from *Bacteroides sp* bound to a DNA fork was recently reported (Su et al., 2019). Each monomeric unit was bound to one of the arms of the fork and each exhibited helicase activity. Hence, each unit of the dimer exhibited helicase activity on the strand to which it was bound.

2. Preparation of an artificial dimeric helicase, NS3h-fusion

We sought to test the idea of an increase in helicase processivity due simply to increased probability of a single trailing molecule taking the place of a lead molecule when the lead dissociates. Our strategy was to link two helicase molecules together at the genetic level. Consideration of the structure of NS3h indicated that the C-terminus and the N-terminus were accessible without altering the structure (Fig. 2). We refer to the protein as NS3 fusion dimer (NS3h-fusion).

The linker length was varied by designing three different peptides that would easily reach from the C-terminus of molecule 1 to the N-terminus of molecule 2. The amino acid linker was made up of a serine-glycine mix which would impart little or no structural constraints and remain soluble. A detailed, stepwise description of considerations to take into account when linking proteins together at the genetic level has recently been published, so a general procedure is provided here (Gran-Scheuch, Aalbers, Woudstra, Parra, & Fraaije, 2021) (Fig. 3).

2.1 NS3h fusion dimer cloning and insertion of the peptide linker

The region of the NS3 gene encoding the helicase domain was amplified by PCR from HCV replicon plasmid (HCVrep1b-I377/NS3-3′UTR) using primers containing *Sac*II digestion sites (forward primer 5′-GGTTGCG AAGGCGCCGCGGTGGAGTGGACTTTGTACC-3′ and reverse primer

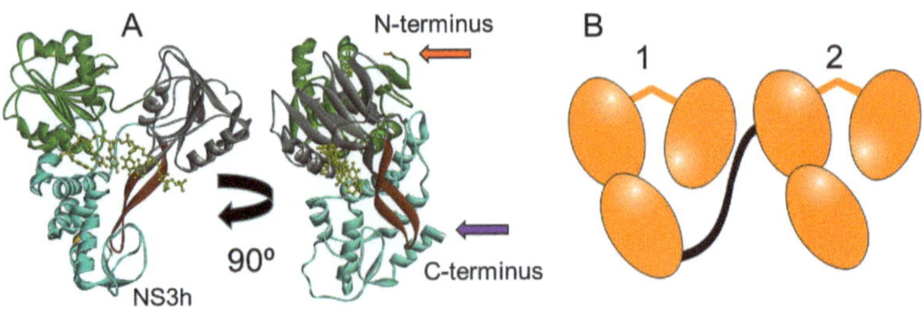

Fig. 2 (A) The X-ray crystal structure of the Hepatitis NS3 helicase domain (NS3h) (Gu & Rice, 2016). Two RecA-like domains (green and gray, also referred to as helicase domains), contain the helicase motifs that allow for coupling of ATP hydrolysis to movement of the helicase along ssDNA (yellow). The N-terminus and C-terminus are indicated. (B) Diagram of the NS3 helicase domain fused to a second molecule of NS3h through a peptide linker. The linker reaches from the C-terminus of molecule 1 to the N-terminus of molecule 2.

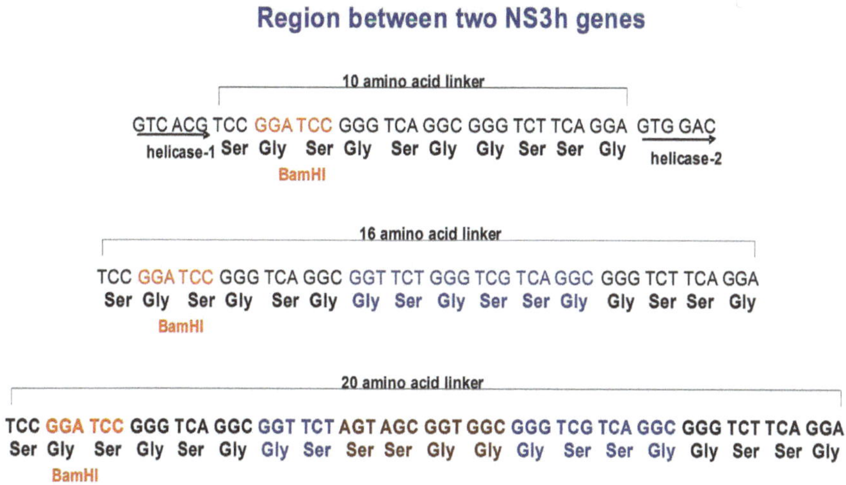

Fig. 3 Construction of the NS3h-fusion dimer NS3H-FUSION. Oligonucleotide primers were designed to encode a linker made up of serine–glycine residues in order to have solubility and flexibility. The primers were used to place the linker between two molecules of NS3h. Primers used for cloning are shown in Table 1. The cloning procedure followed standard protocols described in Section 2.1.

5′-CACCCAGGTGCTCCGCGGCCTCCGGACGTGACGACCTC-3). The PCR product was ligated into pET26b-Ub-NS3h at the unique SacII site at the 3′ end of the NS3h coding region, creating a single open reading frame encoding an NS3h dimer containing a small linker region (TCCGGAGG CCGCGGTGGA) between the two monomers. The resulting pET26-Ub-NS3h-NSh sequence was verified at DNA Sequencing Core Facility, UAMS (Little Rock, AR).

Three consequent mutagenesis reactions were carried out using QuickChange Site-Directed Mutagenesis Kit (Stratagene, La Jolla, CA) to increase the linker length between the two NS3h encoding sequences. The primers and their sequences are shown in Table 1. The length of the linker was increased to 10 amino acids, 16 amino acids and 20 amino acids at the end of the first, second and third insertion reactions, respectively. The first reaction was designed to create a unique BamHI site instead of SacII site between the two NS3h genes allowing for selective subcloning for future experiments.

2.2 NS3h and NS3h-fusion dimer protein expression and purification

Escherichia coli BL21 (DE3) cells were co-transformed with the plasmid encoding either NS3h or the NS3h fusion dimer-ubiquitin construct

Table 1 Oligonucleotides for cloning linkers for fusion of two molecules of NS3h.

Linker length	Primers	
10 amino acids	Forward	5′-GTC GTC ACG TCC GGA TCC GGG TCA GGC GGG TCT TCA GGA GTG GAC TTT GTA CCC-3′
	Reverse	5′-GGG TAC AAA GTC CAC TCC TGA AGA CCC GCC TGA CCC GGA TCC GGA CGT GAC GAC-3′
16 amino acids	Forward	5′-CGT CCG GAT CCG GGT CAG GCG GTT CTG GGT CGT CAG GCG GGT CTT CAG GAG TGG AC-3′
	Reverse	5′-GTC CAC TCC TGA AGA CCC GCC TGA CGA CCC AGA ACC GCC TGA CCC GGA TCC GGA CG-3′
20 amino acids	Forward	5′-GGG TCA GGC GGT TCT AGT AGC GGT GGC GGG TCG TCA GGC GGG-3′
	Reverse	5′-CCC GCC TGA CGA CCC GCC ACC GCT ACT AGA ACC GCC TGA CCC-3′

(pET26-Ub-NS3h or pET26-Ub-NS3h-NS3h with either one of the linkers) and a plasmid encoding the ubiquitin protease according to Novagen specifications. 200 mL of NZCYM media (pH 7.6) containing 30 μg/mL kanamycin and 34 μg/mL chloramphenicol was inoculated with a single colony of the transformed *E. coli* BL21 (DE3) cells and incubated in a 30 °C shaker/incubator at 250 rpm overnight (~16 h).

15 mL of the overnight culture was transferred into 500 mL NZCYM media (pH 7.6) containing 30 μg/mL kanamycin and 34 μg/mL chloramphenicol. The cells were swirled to mix after the addition of 500 mL NZCYM media (pH 7.6) containing 30 μg/mL kanamycin and 34 μg/mL chloramphenicol and then induced by adding IPTG to a final concentration of 0.5 mM and dextrose to 0.2%. Induction proceeded for 4–6 h at 15 °C. Cells were harvested by centrifugation at 6000 rpm for 10 min in a Beckman JLA 10.500 rotor.

5 mL of cold lysis buffer (25 mM Tris, 500 mM NaCl, 1 mM EDTA, pH 8.0, 5 mM βME and 10% glycerol) was added per gram of cells and stirred until the solution was homogenous. Lysis was achieved by adding 2 mM

PMSF, 4 μg pepstatin A and 100 μL lysozyme per gram of cells. The cells were then passed through microfluidizer for one cycle at 18,000 psi. Lysate was centrifuged at 20,000 rpm and 4 °C for 20 min in a Beckman JA 25.50 rotor. The supernatant was further clarified by ultracentrifugation at 45,000 rpm and 4 °C for 2 h in a Beckman Type 50.2 Ti rotor. NS3h or NS3h fusion dimer in the supernatant was purified by column chromatography and for each step of the chromatography, fractions containing the protein were identified by SDS-PAGE. The supernatant was passed through a 40 mL Macro-Prep High Q strong anion exchange column equilibrated in lysis buffer. NS3h and NS3h fusion dimer proteins do not bind to this column under these conditions. Protein was then precipitated by addition of ammonium sulfate to 55% saturation. After centrifugation at 20,000 rpm and 4 °C for 1 h, the pellet was reconstituted in resuspension buffer (25 mM Tris, 500 mM NaCl, 1 mM EDTA, pH 8.0, 5 mM βME, 1 M $(NH_4)_2SO_4$ and 10% glycerol). The suspension was then loaded onto a column containing 25 mL of Macro-Prep methyl HIC resin and eluted with a linear gradient of 2–0 M ammonium sulfate in HIC buffer (25 mM Tris, 50 mM NaCl, 1 mM EDTA, pH 8.0, 2 mM βME and 10% glycerol). Fractions containing NS3h or NS3h fusion protein were identified, collected and dialyzed against low salt buffer (25 mM Tris, 10 mM NaCl, 1 mM EDTA, pH 8.0, 5 mM βME and 10% glycerol). The dialyzed solution was then loaded onto a column containing 5 mL of poly(U)-Sepharose 4B and eluted with a linear gradient of 0.01–1 M NaCl in low salt buffer. Appropriate fractions were collected, dialyzed against the storage buffer (25 mM HEPES, 50 mM NaCl, 1 mM EDTA, pH 8.0, 5 mM βME and 20% glycerol), concentrated by ultrafiltration if needed, frozen in liquid nitrogen and stored at −80 °C. Concentrations of the proteins were determined by using a Bradford assay.

3. Characterization of the NS3h-fushion by DNA binding

To determine whether the linker disrupted the structure of NS3h, DNA binding was measured by using a fluorescence polarization experiment (Lundblad, Laurance, & Goodman, 1996; Raines, 2015) (Fig. 4). This experiment measures the change in polarization of the fluorescence from a FAM labeled oligonucleotide as protein binds to the DNA (usually a 15mer or 20mer made up of only thymidines). Low binding of NS3h-fusion compared to NS3h alone might indicate a misfolded protein. For example, if an amino acid plays a key role within or near one of the helicase motifs, then the variant helicase is not likely to display full activity. For the experimental design here,

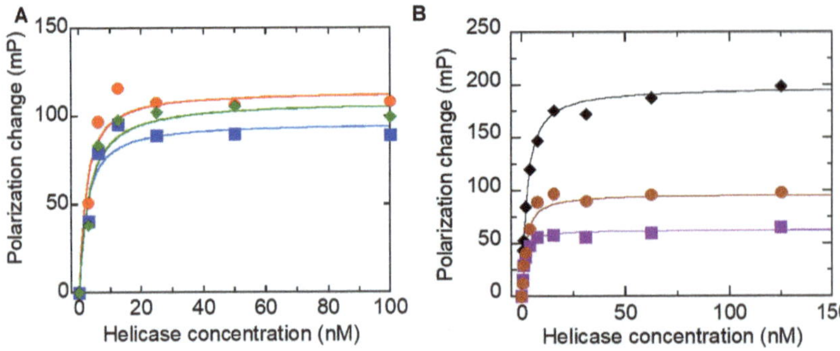

NS3h-fusion 10AA linker	NS3h-fusion 16AA linker	NS3h-fusion 20AA linker
1.8 ± .7 nM	2.0 ± .1 nM	2.6 ± .9 nM

NS3	NS3h-fusion	NS3h
2.0 ± .02 nM	1.5 ± 0.3 nM	0.9 ± .4 nM

Fig. 4 (A) Binding of NS3h-fusions to a 15mer (U)15. The binding assay was performed as described in experimental methods. Increasing concentration of protein was added to 2 nM fluorescein labeled (U)15. Protein was incubated at 37 °C for 20 min. Fluorescence polarization was determined by using a Perkin Elmer plate reader. The K_d value was determined by fitting the data to Eq. (1).

the addition of the linker is not expected to disrupt the inherent ATPase activity. Additional ways to characterize the basic structural folding is to measure DNA-dependent ATP hydrolysis. This assay is more time consuming and requires more reagents, but it can provide more details if needed. For example, if DNA unwinding is found to be reduced, then it is important to determine ATPase activity to know if the DNA unwinding activity is limited by reduced ATPase.

3.1 Equipment

- 96 well assay plate (Corning Inc.)
- PerkinElmer 1420 multilabel counter victor ^{3}V plate reader or equivalent
- Wallac 1420 Workstation software
- KaleidaGraph Software

3.2 Buffers and reagents

- Binding buffer (25 mM MOPS, 50 mM NaCl, 0.1 mM EDTA, 1 mM BME, 0.1 mg/mL BSA)
- 2 × Protein concentration in binding buffer
- FAM labeled oligonucleotide in binding buffer

3.3 Procedure

1. Prepare 2 × stock of fluorescein labeled oligo in binding buffer
2. Label small Eppendorf tubes for 100, 50, 25, 12.5, 6.25, 3.125, and 1.5625 nM enzyme as well as for the reference (no enzyme)
3. In the 100 nM tube, prepare 220 μL 200 nM enzyme by diluting with binding buffer
4. Transfer 110 μL binding buffer to all the remaining tubes
5. Make serial dilutions of the protein solution by transferring 110 μL to the next tube containing 110 μL binding buffer and mixing well. Repeat for each tube (excluding the reference), and discard 110 μL from the last one
6. Add 110 μL 2 × fluorescein-labeled oligo to each of the protein and reference tubes, for a final oligo concentration of x and a final protein concentration matching the tube label
7. Transfer 200 μL binding buffer to the far left well of a black 96-well plate to serve as the blank
8. Load 200 μL from each tube—the reference next to the blank and enzyme concentration increasing toward the right
9. When all samples have been loaded, cover the plate and wrap it in aluminum foil
10. Incubate for 30 min at 37 °C
11. Read polarization using PerkinElmer 1420 multilabel counter victor ^{3}V plate reader on FP fluorescein 0.2 s setting
12. Use KaleidaGraph (Synergy Software) to plot enzyme concentration versus signal change
13. Use the quadratic equation to fit the data and obtain dissociation constant values (K_d) (Fig. 4).

Quadratic equation:

$$y = (\Delta mp/[DNA]) * \left((Kd + [DNA]_T + [E]_T)\right.$$
$$\left. -sqrt\left((KD + [DNA] + [E]_T)\hat{}2 - 4[E]_T[DNA]_T\right)\right)) \qquad (1)$$

3.4 Notes

The fluorescence binding assay provides reproducible results for binding of helicases to DNA. The length of DNA is important to consider. If one molecule of helicase binds to one strand of fluorescein labeled DNA, then a 1:1 binding event ensures that the fit of the data to the quadratic equation provides an appropriate measure of the K_d value. A positive control should be included for first time measurements. Other approaches to measure binding are microscale thermophoresis for low nM dissociation constants (Plach, Grasser, & Schubert, 2017) or isothermal calorimetry (Johnson, 2021) if binding is expected to be in the high nanomolar or low micromolar range. A discussion of methods was provided by Li-Blatter, Zweifel, and Sharpe (2021).

4. DNA unwinding using the rapid quench-flow instrument from Kintek Corp

Helicase-catalyzed DNA unwinding was performed with the three NS3h-fusion helicases to determine whether a particular linker was superior in supporting activity. Single-turnover conditions were chosen because they provide a direct measure of rates for unwinding. Methods in which the substrate is in excess of the helicase can be useful but may not provide a measure of DNA unwinding due to the need for trapping strands that can sequester a helicase which reduces the amount of enzyme actually involved in DNA unwinding. To measure rapid rates of DNA unwinding, the rapid quench-flow method was applied so that very fast reaction rates could be measured. Several descriptions of using pre-steady state methods for studying DNA polymerase activity or DNA endonuclease activity have been published (Liu, Gong, Yu, Taylor, & Johnson, 2019; Su & Guengerich, 2016).

We provide an overview and then specific stepwise use of the RQ3 from Kintek corp. Helicase can be incubated with nucleic acid in the absence of ATP, then the reaction is initiated by rapid mixing of these two samples. Small volumes (~1 mL) are adequate to produce data for one time course. Samples are introduced into the instrument by a 1-mL syringe that fit into a port within the instrument. Substrate is delivered into the mixing chamber by carefully and slowly pressing the syringe so that as little as 15 μL of sample is readied for a reaction to be initiated. When the instrument is loaded with the helicase and substrate, the two solutions are pushed through a mixer under conditions of turbulent flow. The rapid flow rates of helicase solution and substrate solution ensures that complete mixing will occur. When the reaction time is complete, the reaction mixture is rapidly mixed with a quench solution. Thus, the reagents are rapidly mixed to start a reaction and then quenched equally fast. The instrument can measure 2 ms reaction times or rates up to 1000/s.

4.1 Equipment

- KinTek rapid chemical quench-flow instrument
- Typhoon Trio phosphorimager
- ImageQuant software

4.2 Reagents

- Unwinding reaction buffer (25 mM MOPS, pH 6.5, 50 mM KCl, 0.1 mM EDTA, 2 mM BME, and 0.1 mg/mL BSA)
- Quench buffer (200 mM EDTA, and 0.7% SDS)
- ^{32}P radiolabeled DNA substrate, 20 nM
- Q = quench solution
- L = DNA substrate (left syringe)
- R = ATP (right syringe)

4.3 Overall procedure for rapid quench-flow experiment

1. Incubate radiolabeled partial duplex DNA and NS3 at 37 °C for 30 min in a reaction buffer containing 25 mM MOPS, pH 6.5, 50 mM KCl, 0.1 mM EDTA, 2 mM BME, and 0.1 mg/mL BSA
2. Simultaneously, in another reaction tube, preincubate 5 mM ATP, 10 mM $MgCl_2$, 60 nM annealing trap. For a single cycle reaction, the second reaction tube also contained 0.5 mM T_{50} in nucleotide as a protein trap
3. The contents of the solution containing the enzyme and the solution containing ATP are drawn into 1-mL syringes then 15 μL of each are loaded into the reaction lines
4. The reaction is initiated when the contents of the two reaction lines are rapidly mixed by the instrument. The reaction is terminated at different time points with a quench buffer containing 200 mM EDTA, and 0.7% SDS
5. Samples are separated by gel electrophoresis
6. The gel image is visualized using a Typhoon Trio phosphorimager
7. Band intensity was quantified using ImageQuant software

4.4 A stepwise, detailed method to measure DNA unwinding using the rapid chemical quench-flow

1. Plan your reactions
   a. Prepare 10 mL Quench. For NS3Fl unwinding, use 0.7% SDS and 200 mM EDTA
   b. Label tubes: L, R (L—DNA; R—ATP; all 1.5 mL Eppendorf tube).

c. Also, label 1.5 mL Eppendorf tubes and 0.65 mL tubes for each time points except the heated blank. Pre-incubate the L and R solution for 30 mins
d. In the meantime, Prepare the Rapid Quench
   i. Fill the reservoirs
   ii. Set the platform position
   iii. Flush the inlet lines (if you know that someone else had used the instrument before you).
   iv. Flush the reaction loop

2. Using Rapid Quench Instrument
   a. Make sure the water lines are connected. Plug in the circulator power cord, switch on the power and cooling, and set the temperature while flashing
   b. Plug in the keyboard and turn on the Quench-Flow controller
      i. At the system error message, type "RUN" and press return
   c. Fill the reservoirs.
      i. Using a 10-mL syringe filled with H_2O, fill the right syringe (in the "Load" position) and then the Left and finally the middle (Q) syringe. Load all three of the drive syringes while keeping the loading valves all in the horizontal position until you are done. This will prevent solution, especially the quench solution, from backing up into any of the other syringes. Switch to the "Fire" position and disconnect the syringe. Make sure there are no air bubbles present. To remove air bubbles from the drive syringes while loading, work the solution back and forth rapidly, pausing to allow air bubbles to rise in the external syringe and then again forcing solution rapidly up into the drive syringe and back out again. After turning the valves to "Fire" position, bring the drive plate down to contact the drive syringes and force a small amount of buffer out of the exit line. Leave the L and R syringes on "Fire" and change the Q to "Load" position
   d. Set the platform position
      i. Connect the middle tubing to the vacuum assembly and switch the bottom knobs to "Fire."
      ii. On the keyboard, press 2 (Change Position).
      iii. Press + to move down or − to move up; press Start to move (stops when the button is released) far enough to fill the tubes to the loop (just above the 5-mL mark on the syringes). After turning the valves to "Fire" position, bring the drive plate down

to contact the drive syringes and force a small amount of buffer out of the exit line. Leave the L and R syringes on "Fire" and change the Q to "Load" position

e. Flush the inlet lines
   i. With Q still in the "Load" position (also L and R on Load position), switch the L inlet to "Load" and the bottom R to "Fire" (sample load valves).
   ii. Reconnect to the vacuum assembly and turn it on
   iii. Connect a 10-mL H_2O syringe to the L inlet and pass the entire volume through the line
   iv. Repeat with MeOH. Then use a KimWipe to dry the inlet connector and draw air through
   v. Disconnect the vacuum and turn it off
   vi. Switch the L inlet to "Fire" and the R to "Load"; reconnect to the vacuum assembly and turn it on
   vii. Flush the R inlet with H_2O followed by MeOH and air
   viii. Disconnect from the vacuum and turn it off

f. The Load-Fire-Flush cycle:
   i. Attach a 1 mL syringe containing your samples to each of the load lines
   ii. For each time point
      1. Turn the valve to Load position and load the sample one at a time (when one side is on Load the other has to be on Fire). After loading each sample, turn the valve to the fire position so that now the valve is pointing upward and connected to the drive syringe. In other words, With Q still on "Load", switch the L inlet to "Load" and the R inlet to "Fire." Attach the 1-mL syringe containing L to the L inlet. Load the left side of the reaction just to fill the tube. Switch the L inlet to "Fire" and the R inlet to "Load", and repeat with R
      2. Make sure all the Knobs are on "Fire." On the keyboard press 0 (Cancel), then 1 (to take time points). Enter the time for the point and confirm that the correct loop is selected. Place the middle tubing into the appropriate 1.5-mL tube, double-check that all knobs are in the "Fire" position, and press Start. The reaction mixture is collected in the 1.5 mL tube and placed on ice until all the reaction times are completed. It is useful to hold the exit line in a position that allows the reaction mix to

run down the inside wall of the tube to ensure that it does not splash out of the 1.5 mL tube

3. Flush the loop
   a. Attach the vacuum line to the exit line from the quench flow but do not yet turn the vacuum
   b. Turn each of the sample valves to the FLUSH position, then immerse the flush lines into water, and then turn on the vacuum. Do this with the flush lines in Methanol and then air dry
   c. Disconnect the vacuum line, change the sample valves and Q to fire, and then start the next time point

g. Clean up the instrument
   i. Flush the loop after the last time point
   ii. Turn all knobs to fire and raise the platform (on the keyboard, press 0, return and then press 2 to go back to change Position).
   iii. Switch the R syringe to Load and press the drive to remove the water into an empty 10 mL syringe. Repeat for L and Q syringes
   iv. Fill the Q with water; switch the Q and inlet valves (sample inlet valves) to Fire
   v. Flush water through; switch Q back to load and drain the syringe
   vi. For sample inlets
      1. With L on load and R on Fire, pull the syringe plunger to withdraw samples to empty the line
      2. Flush the inlet line with water, phosphate buffer, methanol, and dry air
      3. Repeat with R inlet
      4. Flush the reaction loop
      5. Turn off the circulator and QF controller, and unplug the keyboard
      6. Empty the vacuum waste container and H_2O/MeOH beakers into the sink
      7. Spray the flask with NoCount, and Check the area with the Geiger counter

Two Things you MUST NOT do when using the rapid quench

1. Do not use the vacuum while the sample valve is on Load
2. You should always be careful that the valves are in the correct position before pressing the start button

$$ssDNA = amp^*(1 - (1 + k_u{}^*t)^* \exp(k_u{}^*t)) \quad (2)$$

amp is the amplitude and k_u equals the rate associated with each step of the unwinding reaction.

The results in panel B compare full length NS3, NS3-fusion, and NS3h. NS3h-fusion unwound the 30 bp duplex at a similar rate as full length NS3 and NS3h. However, the quantity of product from NS3-fusion was greater than NS3h indicating that it has higher processivity than NS3h.

4.5 Notes

The rapid quench-flow produces highly reproducible data and the Kintek instrument is robust and can last for decades with proper handling. Data in Fig. 5 illustrate that the NS3h-fusion behaves well, and each linker produced very similar results. The 10AA linker was selected for the DNA footprinting experiment because it produced similar results as the other linkers and the preparation produced more total protein.

4.6 Safety considerations and standards

The major risk here is that DNA or RNA substrates are typically radiolabeled with ^{32}P. This allows very low concentration of sometimes expensive DNA or RNA synthetic oligonucleotides. More researchers are using fluorescently labeled substrates to avoid the need for ^{32}P. The fluorophores are not as sensitive, but they are sensitive enough for many experiments. If fluorophores are utilized, then the quantum yield must be the same in substrate as in the product.

4.7 Alternative methods/procedures

The repetitive nature of conducting one time point and then repeating the process for the subsequent time points can be difficult if multiple time courses are obtained in 1 day. It is not a technique that every lab member will gravitate toward. However, the Quench-Flow instrument from Kintek produces very reproducible data and can measure rates that other methods cannot. An alternative approach for ensemble experiments is stopped-flow fluorescence spectroscopy. This method is generally easier to perform, but the data are sometimes difficult to interpret. The quantity of product formed over time is not always known because relating a fluorescence change to a change in concentration can lead to uncertainty.

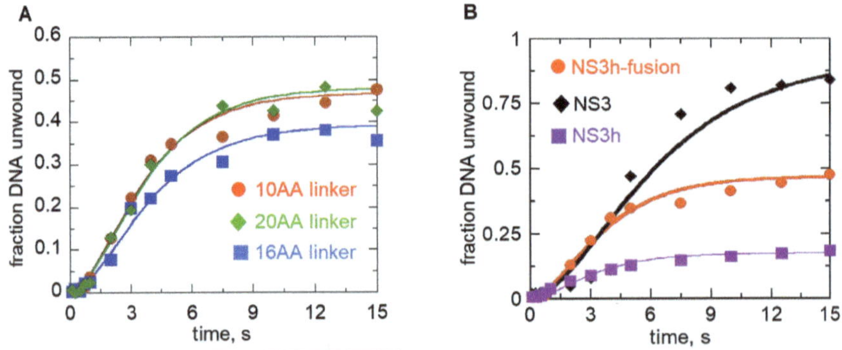

	NS3h-fusion 10AA linker	NS3h-fusion 16AA linker	NS3h-fusion 20AA linker
amp	0.47 ± .01	0.39 ± .01	0.48 ± .02
k_u	0.49 ± .04 s^{-1}	0.47 ± .04 s^{-1}	0.49 ± .04s^{-1}

	NS3	NS3h-fusion	NS3h
amp	0.92 ± .06	0.47 ± .01	0.17 ± .04
k_u	0.29 ± .03 s^{-1}	0.51 ± .04 s^{-1}	0.58 ±.01s^{-1}

Fig. 5 (A) Helicase-catalyzed DNA unwinding of a 30 bp substrate with a 15 nt ssDNA overhang. Each NS3h-fusion helicase was incubated with the DNA substrate for 5 min on ice. Helicase, 500 nM (in helicases motors) and substrate was 2 nM. The reaction was initiated upon rapid mixing with ATP, 5 mM. The mixing was performed by using a KinTek RQF-3 Quench-Flow instrument (Kintek, Snow Shoe, PA). The multi-phase appearance of ssDNA was analyzed by fitting the data to Eq. (2), a three-step mechanism by using Kaleidagraph. The three NS3h-fusion proteins unwound the DNA substrate with similar rate constants and amplitudes as shown in the table below the graph. (B) DNA unwinding catalyzed by NS3 ◆, NS3h-fusion ●, and NS3h ■. Data were analyzed by fitting to the stepping equation (3-steps) (Lucius, Maluf, Fischer, & Lohman, 2003). Results are shown in the table below the graph.

5. DNA footprinting to compare DNA binding of NS3H-FUSION to NS3h and NS3 full length helicase

The unwinding reaction indicated that the NS3h-fusion exhibited greater unwinding processivity than NS3h but lower activity compared to full length NS3. This result is consistent with the idea that the fusion dimer may behave more like the full length NS3 rather than NS3h. The ability for the three helicases to bind to ssDNA was investigated by DNA footprinting. Potassium permanganate reacts with thymidine residues and subsequent treatment of the DNA with pyridine causes a cascade of reactions that results in cleavage of the phosphodiester backbone. The DNA used in this experiment was 24 thymidine in length while the binding site size of NS3 is 7–8. The DNA band intensities produced a footprinting pattern resulting from NS3 binding (Fig. 6, panel A). Quantitation of the bands

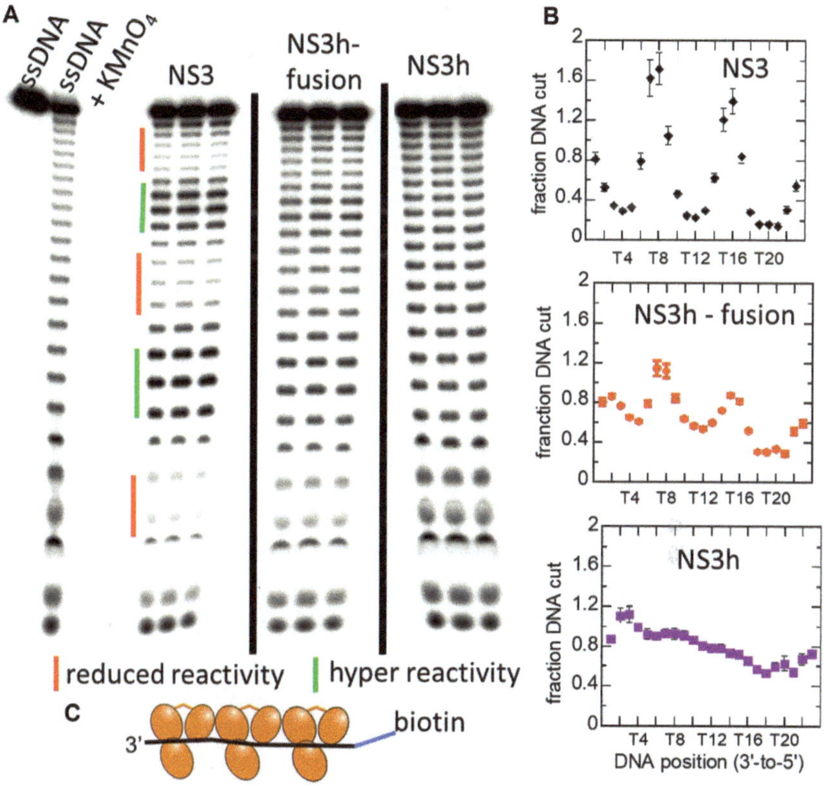

Fig. 6 DNA footprinting to characterize the interaction of multiple helicase molecules aligning on single-stranded DNA. (A) NS3 (500 nM) was incubated with ssDNA (biotinylated T-30mer,10 nM) for 5 min at room temperature, followed by treatment with $KMnO_4$ (30 mM). The DNA was isolated by capture with magnetic streptavidin beads. (B) After washing with buffer, DNA was treated as described in the methods section, then loaded onto a denaturing polyacrylamide and separated by electrophoresis. The DNA was visualized by using a phosphorimager and bands were quantified by using ImageQuant software. NS3 gave rise to a pattern of hyper-reactivity with regions of protection as was reported previously (Raney et al., 2012). (C) The number of nucleotides between the hyper reactive peaks was 8, which is the binding site size for NS3. This indicates that binding of multiple molecules occurred in a uniform structural pattern. NS3h-fusion exhibited a similar footprinting pattern as NS3, but the peak heights were less pronounced than NS3. NS3h did not exhibit the pattern of protection and hyper reactivity.

and determining the fraction of DNA cut relative to the control lane clearly show three regions of protection at positions T4, T12, and T20 (Fig. 6, panel B, top graph). In between the regions of protection are two peaks corresponding to hyper-reactivity centered at positions T8 and T16

(Raney et al., 2012). In contrast, NS3h binding to the T24-mer resulted in little or no pattern of protection (Fig. 6, panel B, bottom graph). The NS3h-fusion gave rise to a pattern that was much the same as full length NS3 with three protected regions interspersed with two hyper-reactive regions at positions 8 and 16 (Fig. 6, panel B, middle graph). Therefore, the linker appears to play a role in stabilizing the artificial dimeric helicase onto the DNA. This can explain the increase in processivity for DNA unwinding. Furthermore, stabilizing the helicase is due to the linker holding the two molecules in close proximity which is consistent with the model for functional cooperativity (Fig. 6, panel C).

6. Streptavidin displacement assay indicates NS3h-fusion can increase dissociation of streptavidin from a biotin-labeled oligonucleotide

This assay is used to determine the directional bias in translocation if not already known, and whether a helicase can increase dissociation of macromolecules in its path (Byrd & Raney, 2004; Morris & Raney, 1999). Biotin-labeled oligonucleotides with streptavidin bound to either the 5′-end or the 3′-end serves as the substrate. NS3 translocates 3′-to-5′, and so the biotin is place near the 5′-end. We found previously that Dda helicase aligned on ssDNA and the assembly functioned together to increase dissociation of streptavidin from biotin labeled oligonucleotides. However, NS3h was not able to displace streptavidin under any conditions (Matlock et al., 2010). The streptavidin displacement assay has been described previously in methods manuscripts (Brüning, Howard, & McGlynn, 2016; Morris, Tackett, & Raney, 2001; Shin & Kelman, 2006), so the description below is an overview of the experimental protocol. The assay includes a biotin label place at or near the end of an oligonucleotide. For a 3′-to-5′ helicase, the biotin is placed on the 5′-end. Streptavidin is added to form a blockade at the end of the oligonucleotide. In the presence of ATP and magnesium, helicases that translocate collide with the streptavidin and cause it to dissociate much faster than its normal dissociation rate. The reaction can be observed by native gel electrophoresis which readily separates streptavidin-bound DNA from free DNA (Fig. 7, panel A).

6.1 General description of streptavidin displacement assay

A 5′ radiolabeled 30mer with biotin one nucleotide from the 5′ end was utilized in this experiment (5′-T-BiodT-T_{30}-3′). All concentrations listed are

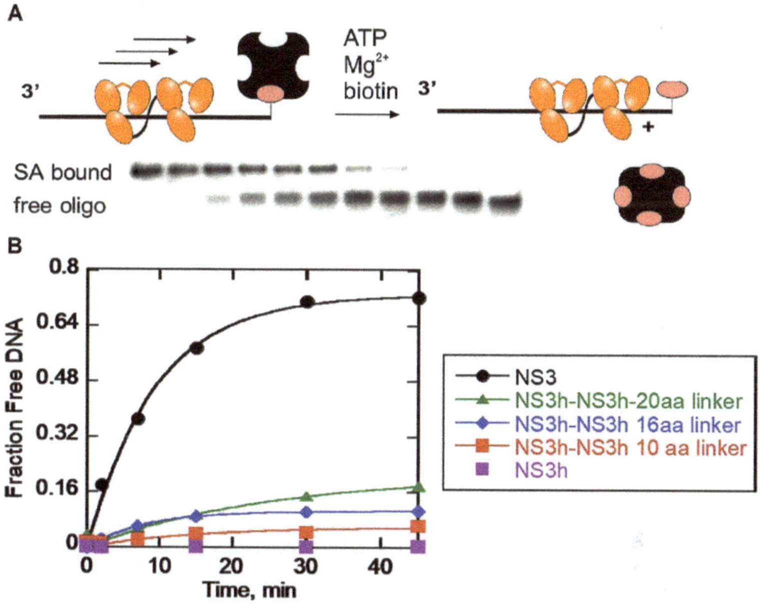

	NS3	20AA	16AA	10AA	NS3h
amp	0.73	0.21	0.11	0.06	None

Fig. 7 Streptavidin displacement assay. (A). Diagram of a ssDNA oligonucleotide with biotin attached on the 5′-end and streptavidin bound to the biotin. NS3h-fusion is depicted as moving in the 3′-to-5′ direction as the reaction is initiated upon addition of ATP (5 mM), Mg2+ (10 mM), and biotin (1 μM). When NS3h-fusion reaches the 5′-end, streptavidin is displaced and rapidly binds to the free biotin. The reaction can be visualized by separating the streptavidin-bound DNA from the free DNA by native gel electrophoresis as shown in the gel image. (B) The radioactive DNA in each band can be visualized by a phosphorimager and quantified by using ImageQuant software. The fraction of free DNA is plotted over time and the data were fit to a single exponential. The amplitude (amp) for each helicase-catalyzed reaction is listed in the table below the graph.

final concentrations after addition of all reaction components. 10 nM oligonucleotide was incubated at 37 °C with 2 μM streptavidin in reaction buffer (25 mM MOPS pH 7.0, 10 mM NaCl, 2 mM β-mercaptoethanol, 0.1 mg/mL bovine serum albumin). After 3 min, 500 nM full length NS3, NS3h-fusion, or NS3h was added, and the tube was incubated at 37 °C for 5 min before beginning the reaction. In a separate tube, 5 mM ATP, 10 mM $MgCl_2$, and 10 μM biotin to trap displaced streptavidin were mixed in reaction buffer. Also included with the ATP was 10 U/mL pyruvate kinase/lactate dehydrogenase and 4 mM phosphoenol pyruvate as an ATP regenerating system. The ATP tube was also incubated at 37 °C for 5 min. During the 5 min incubation, the 0 s timepoint was prepared by mixing 5 μL of the ATP solution with 10 μL of quench, followed by 5 μL of the enzyme-substrate solution. Concentrations of quench solution before adding to the reaction are (0.6% SDS, 200 mM EDTA pH 8.0, 50 μM poly(dT) in nucleotides, 5 mg/mL bromophenol blue, 5 mg/mL xylene cyanol, 5% glycerol). The reactions were initiated by mixing equal quantities of the enzyme-substrate and ATP solutions and incubating at 37 °C. At increasing times after ATP addition, 10 μL of sample was removed and mixed with 10 μL of helicase quench. Samples were then separated on a 15% native polyacrylamide-TBE gel. Gels were exposed to a phosphor screen overnight and imaged using a Typhoon Trio Phosphorimager. The percentage of free oligonucleotide vs. streptavidin bound oligonucleotide was calculated for each time point using ImageQuantTL software.

The results from the streptavidin displacement assay are shown in Fig. 7. Full-length NS3 was readily able to displace streptavidin but NS3h gave rise to no measurable product (Fig. 7, panel B, compare black circles to purple squares). NS3h-fusion containing 10AA, 16AA, or 20AA linker displaced streptavidin from the biotin-labeled oligo (Fig. 7, blue diamonds, green triangles, and red squares). Hence, NS3h-fusion exhibited the ability to "push" on a block placed in its path.

7. Conclusions and summary

The model for functional cooperativity can apply to active helicases as well as passive helicases (Manosas, Xi, Bensimon, & Croquette, 2010). Dda classifies as an active helicase because it unwinds DNA and translocates along ssDNA at similar rates. NS3h translocates on DNA faster than it unwinds dsDNA (Matlock et al., 2010), thereby placing it into the category as passive. However, linking two molecules together appears to have a synergistic impact on activity. The rate of unwinding dsDNA was similar, but the processivity of NS3h-fusion was increased. The DNA footprinting results indicated that NS3h-fusion interacts with the ssDNA in a defined manner, similar to full-length NS3. The pattern of protection for NS3h-fusion was not as distinct as full length NS3, possibly because only two of the molecules bound to the ssDNA contained a linker. The streptavidin displacement assay illustrated that NS3h-fusion could increase dissociation of streptavidin from the biotin-oligonucleotide, whereas NS3h could not. This is possibly due to the rapid dissociation of trailing molecules during translocation or DNA unwinding by NS3h. For NS3h-fusion, the trailing molecule can dissociate, but likely re-binds very quickly, thereby increasing processivity. The significance of the DNA footprinting pattern remains to be explored, but the similar results between full length NS3 and NS3h-fusion is consistent with a similar mode of binding. The binding pattern may reflect the synergistic binding which may also explain why NS3h-fusion can displace streptavidin whereas NS3h does not. Overall, linking two molecules of helicase provided results that were consistent with the idea that some helicases can align on ssDNA and exhibit higher activity as described in Fig. 1 and has been described as functional cooperativity.

Acknowledgments

This work was funded by NIH R01 GM098922 and R35 GM122601 (K.D.R). and the UAMS Research Council. The UAMS DNA Sequencing Core is supported by the UAMS Center for Microbial Pathogenesis and Host Inflammatory Responses (National Institutes of Health Grant P20 GM103625), and the UAMS Translational Research Institute (National Institutes of Health Grant UL1TR000039). The UAMS Digital Microscopy Core facility is supported through user fees and university funding. We thank Bartek Sikora and Melody K. Harrison for expert technical assistance.

References

Ali, J. A., & Lohman, T. M. (1997). Kinetic measurement of the step size of DNA unwinding by Escherichia coli UvrD helicase. *Science*, *275*(5298), 377–380. https://doi.org/10.1126/science.275.5298.377.

Brosh, R. M., & Matson, S. W. (2020). History of DNA helicases. *Genes*, *11*(3). https://doi.org/10.3390/genes11030255. MDPI AG.

Brüning, J. G., Howard, J. A. L., & McGlynn, P. (2016). Use of streptavidin bound to biotinylated DNA structures as model substrates for analysis of nucleoprotein complex disruption by helicases. *Methods (San Diego, Calif.)*, *108*, 48–55. https://doi.org/10.1016/J.YMETH.2016.03.017.

Buzovetsky, O., Kwon, Y., Pham, N. T., Kim, C., Ira, G., Sung, P., et al. (2017). Role of the Pif1-PCNA complex in pol δ-dependent strand displacement DNA synthesis and break-induced replication. *Cell Reports*, *21*(7), 1707–1714. https://doi.org/10.1016/j.celrep.2017.10.079.

Byrd, A. K., & Raney, K. D. (2004). Protein displacement by an assembly of helicase molecules aligned along single-stranded DNA. *Nature Structural and Molecular Biology*, *11*(6), 531–538. https://doi.org/10.1038/nsmb774.

Byrd, A. K., & Raney, K. D. (2006). Displacement of a DNA binding protein by Dda helicase. *Nucleic Acids Research*, *34*(10), 3020–3029. https://doi.org/10.1093/nar/gkl369.

Byrd, A. K., & Raney, K. D. (2017). Structure and function of Pif1 helicase. *Biochemical Society Transactions*, *45*(5), 1159–1171. Portland Press Ltd https://doi.org/10.1042/BST20170096.

Cheng, W., Brendza, K. M., Gauss, G. H., Korolev, S., Waksman, G., & Lohman, T. M. (2002). The 2B domain of the Escherichia coli rep protein is not required for DNA helicase activity. *Proceedings of the National Academy of Sciences of the United States of America*, *99*(25), 16006–16011. https://doi.org/10.1073/pnas.242479399.

Chisty, L. T., Toseland, C. P., Fili, N., Mashanov, G. I., Dillingham, M. S., Molloy, J. E., et al. (2013). Monomeric PcrA helicase processively unwinds plasmid lengths of DNA in the presence of the initiator protein RepD. *Nucleic Acids Research*, *41*(9), 5010–5023. https://doi.org/10.1093/nar/gkt194.

Comstock, M. J., Whitley, K. D., Jia, H., Sokoloski, J., Lohman, T. M., Ha, T., et al. (2015). Direct observation of structure-function relationship in a nucleic acid -processing enzyme. *Science*, *348*(6232), 352–354. https://doi.org/10.1126/science.aaa0130.

Dillingham, M. S., Soultanas, P., Wiley, P., Webb, M. R., & Wigley, D. B. (2001). Defining the roles of individual residues in the single-stranded DNA binding site of PcrA helicase. *Proceedings of the National Academy of Sciences of the United States of America*, *98*(15), 8381–8387. https://doi.org/10.1073/pnas.131009598.

Dillingham, M. S., Spies, M., & Kowalczykowski, S. C. (2003). RecBCD enzyme is a bipolar DNA helicase. *Nature*, *423*(6942), 893–897. https://doi.org/10.1038/nature01673.

Dillingham, M. S., Tibbles, K. L., Hunter, J. L., Bell, J. C., Kowalczykowski, S. C., & Webb, M. R. (2008). Fluorescent single-stranded DNA binding protein as a probe for sensitive, real-time assays of helicase activity. *Biophysical Journal*, *95*(7), 3330–3339. https://doi.org/10.1529/biophysj.108.133512.

Eoff, R. L., & Raney, K. D. (2006). Intermediates revealed in the kinetic mechanism for DNA unwinding by a monomeric helicase. *Nature Structural and Molecular Biology*, *13*(3), 242–249. https://doi.org/10.1038/nsmb1055.

Eoff, R. L., & Raney, K. D. (2010). Kinetic mechanism for DNA unwinding by multiple molecules of Dda helicase aligned on DNA. *Biochemistry*, *49*(21), 4543–4553. https://doi.org/10.1021/bi100061v.

Eoff, R. L., Spurling, T. L., & Raney, K. D. (2005). Chemically modified DNA substrates implicate the importance of electrostatic interactions for DNA unwinding by Dda helicase. *Biochemistry*, *44*(2), 666–674. https://doi.org/10.1021/bi0484926.

Geronimo, C. L., & Zakian, V. A. (2016). Getting it done at the ends: Pif1 family DNA helicases and telomeres. *DNA Repair*, *44*, 151–158. Elsevier B.V https://doi.org/10.1016/j.dnarep.2016.05.021.

Gorbalenya, A. E., & Koonin, E. V. (1993). Helicases: Amino acid sequence comparisons and structure-function relationships. *Current Opinion in Structural Biology*, *3*(3), 419–429. https://doi.org/10.1016/S0959-440X(05)80116-2.

Gran-Scheuch, A., Aalbers, F., Woudstra, Y., Parra, L., & Fraaije, M. W. (2021). Optimizing the linker length for fusing an alcohol dehydrogenase with a cyclohexanone monooxygenase. *Methods in Enzymology*, *647*, 107–143. https://doi.org/10.1016/BS.MIE.2020.09.008.

Gu, M., & Rice, C. M. (2016). The spring α-helix coordinates multiple modes of HCV (hepatitis C virus) NS3 helicase action. *The Journal of Biological Chemistry*, *291*(28), 14499–14509. https://doi.org/10.1074/JBC.M115.704379.

He, X., Byrd, A. K., Yun, M. K., Pemble, C. W., IV, Harrison, D., Yeruva, L., et al. (2012). The T4 phage SF1B helicase Dda is structurally optimized to perform DNA strand separation. *Structure*, *20*(7), 1189–1200. https://doi.org/10.1016/j.str.2012.04.013.

Johnson, C. M. (2021). Isothermal titration calorimetry. *Methods in Molecular Biology*, *2263*, 135–159. https://doi.org/10.1007/978-1-0716-1197-5_5.

Levin, M. K., Wang, Y. H., & Patel, S. S. (2004). The functional interaction of the hepatitis C virus helicase molecules is responsible for unwinding processivity. *Journal of Biological Chemistry*, *279*(25), 26005–26012. https://doi.org/10.1074/jbc.M403257200.

Li-Blatter, X., Zweifel, L., & Sharpe, T. (2021). A familiar protein-ligand interaction revisited with multiple methods. *Methods in Molecular Biology (Clifton, N.J.)*, *2263*, 47–79. https://doi.org/10.1007/978-1-0716-1197-5_2.

Liu, M. S., Gong, S., Yu, H. H., Taylor, D. W., & Johnson, K. A. (2019). Kinetic characterization of Cas9 enzymes. *Methods in Enzymology*, *616*, 289–311. https://doi.org/10.1016/bs.mie.2018.10.022.

Lohman, T. M., & Fazio, N. T. (2018). How does a helicase unwind DNA? Insights from RecBCD helicase. *BioEssays*, *40*(6). https://doi.org/10.1002/bies.201800009.

Lohman, T. M., Tomko, E. J., & Wu, C. G. (2008). Non-hexameric DNA helicases and translocases: Mechanisms and regulation. *Nature Reviews Molecular Cell Biology*, *9*(5), 391–401. Nature Publishing Group https://doi.org/10.1038/nrm2394.

Lucius, A. L., Maluf, N. K., Fischer, C. J., & Lohman, T. M. (2003). General methods for analysis of sequential "n-step" kinetic mechanisms: Application to single turnover kinetics of helicase-catalyzed DNA unwinding. *Biophysical Journal*, *85*(4), 2224–2239. https://doi.org/10.1016/S0006-3495(03)74648-7.

Lundblad, J. R., Laurance, M., & Goodman, R. H. (1996). Fluorescence polarization analysis of protein-DNA and protein-protein interactions. *Molecular Endocrinology (Baltimore, Md.)*, *10*(6), 607–612. https://doi.org/10.1210/MEND.10.6.8776720.

Makurath, M. A., Whitley, K. D., Nguyen, B., Lohman, T. M., & Chemla, Y. R. (2019). Regulation of rep helicase unwinding by an auto-inhibitory subdomain. *Nucleic Acids Research*, *47*(5), 2523–2532. https://doi.org/10.1093/nar/gkz023.

Manosas, M., Xi, X. G., Bensimon, D., & Croquette, V. (2010). Active and passive mechanisms of helicases. *Nucleic Acids Research*, *38*(16), 5518–5526. https://doi.org/10.1093/nar/gkq273.

Matlock, D. L., Yeruva, L., Byrd, A. K., Mackintosh, S. G., Langsten, C., Brown, C., et al. (2010). Investigation of translocation, DNA unwinding, and protein displacement by NS3h, the helicase domain from the hepatitis C virus helicase. *Biochemistry*, *49*(10), 2097–2109. https://doi.org/10.1021/bi901977k.

Morris, P. D., Byrd, A. K., Tackett, A. J., Cameron, C. E., Tanega, P., Ott, R., et al. (2002). Hepatitis C virus NS3 and simian virus 40 T antigen helicases displace streptavidin from 5′-biotinylated oligonucleotides but not from 3′-biotinylated oligonucleotides: Evidence for directional bias in translocation on single-stranded DNA. *Biochemistry*, *41*(7), 2372–2378. https://doi.org/10.1021/BI012058B.

Morris, P. D., & Raney, K. D. (1999). DNA helicases displace streptavidin from biotin-labeled oligonucleotides. *Biochemistry*, *38*(16), 5164–5171. https://doi.org/10.1021/bi9822269.

Morris, P. D., Tackett, A. J., & Raney, K. D. (2001). Biotin-streptavidin-labeled oligonucleotides as probes of helicase mechanisms. *Methods*, *23*(2), 149–159. https://doi.org/10.1006/meth.2000.1116.

Myong, S., Rasnik, I., Joo, C., Lohman, T. M., & Ha, T. (2005). Repetitive shuttling of a motor protein on DNA. *Nature*, *437*(7063), 1321–1325. https://doi.org/10.1038/nature04049.

Niedziela-Majka, A., Chesnik, M. A., Tomko, E. J., & Lohman, T. M. (2007). Bacillus stearothermophilus PcrA monomer is a single-stranded DNA translocase but not a processive helicase in vitro. *The Journal of Biological Chemistry*, *282*(37), 27076–27085. https://doi.org/10.1074/JBC.M704399200.

Park, J., Myong, S., Niedziela-Majka, A., Lee, K. S., Yu, J., Lohman, T. M., et al. (2010). PcrA helicase dismantles RecA filaments by reeling in DNA in uniform steps. *Cell*, *142*(4), 544–555. https://doi.org/10.1016/j.cell.2010.07.016.

Plach, M. G., Grasser, K., & Schubert, T. (2017). MicroScale thermophoresis as a tool to study protein-peptide interactions in the context of large eukaryotic protein complexes. *Bio-Protocol*, 7(23), e2632. https://doi.org/10.21769/BioProtoc.2632.

Rad, B., Forget, A. L., Baskin, R. J., & Kowalczykowski, S. C. (2015). Single-molecule visualization of RecQ helicase reveals DNA melting, nucleation, and assembly are required for processive DNA unwinding. *Proceedings of the National Academy of Sciences of the United States of America*, *112*(50), E6852–E6861. https://doi.org/10.1073/pnas.1518028112.

Raines, R. T. (2015). Fluorescence polarization assay to quantify protein-protein interactions: An update. *Methods in Molecular Biology (Clifton, N.J.)*, *1278*, 323–327. https://doi.org/10.1007/978-1-4939-2425-7_19.

Ramanagoudr-Bhojappa, R., Chib, S., Byrd, A. K., Aarattuthodiyil, S., Pandey, M., Patel, S. S., et al. (2013). Yeast Pif1 helicase exhibits a one-base-pair stepping mechanism for unwinding duplex DNA. *Journal of Biological Chemistry*, *288*(22), 16185–16195. https://doi.org/10.1074/jbc.M113.470013.

Raney, V. M., Reynolds, K. A., Harrison, M. K., Harrison, D. K., Cameron, C. E., & Raney, K. D. (2012). Binding by the hepatitis C virus NS3 helicase partially melts duplex DNA. *Biochemistry*, *51*(38), 7596–7607. https://doi.org/10.1021/bi300654v.

Reyes, A. A., Marcum, R. D., & He, Y. (2021). Structure and function of chromatin remodelers. *Journal of Molecular Biology*, *433*(14). https://doi.org/10.1016/J.JMB.2021.166929.

Saikrishnan, K., Powell, B., Cook, N. J., Webb, M. R., & Wigley, D. B. (2009). Mechanistic basis of 5′-3′ translocation in SF1B helicases. *Cell*, *137*(5), 849–859. https://doi.org/10.1016/j.cell.2009.03.036.

Sharma, D., Putnam, A. A., & Jankowsky, E. (2017). Biochemical differences and similarities between the DEAD-box helicase Orthologs DDX3X and Ded1p. *Journal of Molecular Biology*, *429*(23), 3730–3742. https://doi.org/10.1016/j.jmb.2017.10.008.

Shin, J. H., & Kelman, Z. (2006). DNA unwinding assay using streptavidin-bound oligonucleotides. *BMC Molecular Biology*, 7. https://doi.org/10.1186/1471-2199-7-43.

Sikora, B., Eoff, R. L., Matson, S. W., & Raney, K. D. (2006). DNA unwinding by Escherichia coli DNA helicase I (TraI) provides evidence for a processive monomeric molecular motor. *The Journal of Biological Chemistry*, *281*(47), 36110–36116. https://doi.org/10.1074/JBC.M604412200.

Singleton, M. R., Dillingham, M. S., Gaudier, M., Kowalczykowski, S. C., & Wigley, D. B. (2004). Crystal structure of RecBCD enzyme reveals a machine for processing DNA breaks. *Nature*, *432*(7014), 187–193. https://doi.org/10.1038/NATURE02988.

Sokoloski, J. E., Kozlov, A. G., Galletto, R., & Lohman, T. M. (2016). Chemo-mechanical pushing of proteins along single-stranded DNA. *Proceedings of the National Academy of Sciences of the United States of America*, *113*(22), 6194–6199. https://doi.org/10.1073/pnas.1602878113.

Su, N., Byrd, A. K., Bharath, S. R., Yang, O., Jia, Y., Tang, X., et al. (2019). Structural basis for DNA unwinding at forked dsDNA by two coordinating Pif1 helicases. *Nature Communications*, *10*(1), 5375. https://doi.org/10.1038/s41467-019-13414-9.

Su, Y., & Guengerich, F. P. (2016). Pre-steady-state kinetic analysis of single-nucleotide incorporation by DNA polymerases. *Current Protocols in Nucleic Acid Chemistry*, *2016*, 7.23.1–7.23.10. https://doi.org/10.1002/cpnc.2.

Urrutia-Irazabal, I., Ault, J. R., Sobott, F., Savery, N. J., & Dillingham, M. S. (2021). Analysis of the pcra-rna polymerase complex reveals a helicase interaction motif and a role for pcra/uvrd helicase in the suppression of r-loops. *eLife*, *10*. https://doi.org/10.7554/eLife.68829.

Velankar, S. S., Soultanas, P., Dillingham, M. S., Subramanya, H. S., & Wigley, D. B. (1999). Crystal structures of complexes of PcrA DNA helicase with a DNA substrate indicate an inchworm mechanism. *Cell*, *97*, 75–84.

CHAPTER THREE

Use of an unnatural amino acid to map helicase/DNA interfaces via photoactivated crosslinking

Alexander T. Duckworth and James L. Keck*
Department of Biomolecular Chemistry, University of Wisconsin-Madison School of Medicine and Public Health, Madison, WI, United States
*Corresponding author: e-mail address: jlkeck@wisc.edu

Contents

Abstract

Formation of protein/nucleic acid complexes is essential for life. From DNA replication and repair to transcription and translation, myriad different proteins bind nucleic acids to execute their essential cellular functions. Our understanding of the mechanisms

Methods in Enzymology, Volume 672
ISSN 0076-6879
https://doi.org/10.1016/bs.mie.2022.02.019

underlying recognition and processing of nucleic acids can be greatly informed by mapping protein domains and residues that form interfaces with their DNA or RNA targets. Here we describe a crosslinking protocol in which the unnatural amino acid *p*-benzoyl-*l*-phenylalanine (Bpa) integrated at selected sites within the PriA DNA helicase is used to map surfaces of the protein that interact with specific positions in a synthetic DNA replication fork in vitro.

1. Introduction

Nearly all biological functions rely on formation of protein/nucleic acid complexes. Genome maintenance, transcription, and translation are among the essential processes that are controlled directly by proteins binding to nucleic acids. A multitude of different nucleic acid binding proteins have evolved to mediate these essential functions. Many such proteins associate with their targets using sequence- and/or structure-specific DNA or RNA binding strategies. Sequence-specific contacts generally rely on interactions between the protein and bases within the targeted DNA or RNA. Structure-specific interactions (or interactions that are neither sequence- nor structure-specific) often utilize electrostatic binding to the phosphodiester backbone. Furthermore, proteins can recognize specific nucleic acid structures, such as single-strand (ss) or double-strand (ds) DNA or RNA, hairpins, bent DNA or RNA, fork structures, ss/ds junctions, tRNA, guanine quadruplexes, and more.

The wide variety of protein/nucleic acid interactions have necessitated the development of many different complementary experimental methods to study complex formation. These span biochemical (e.g., gel mobility, activity, single molecule, crosslinking), structural (e.g. X-ray crystallography, cryo-electron microscopy, NMR), and in vivo (e.g. chromatin immunoprecipitation, fluorescently-tagged proteins) approaches. In this chapter, we describe a biochemical crosslinking strategy that can provide distance information to define protein/DNA contacts in structure-specific DNA-binding proteins. This methodology has allowed mapping of helicase/DNA interactions (Windgassen & Keck, 2016; Windgassen, Leroux, Sandler, & Keck, 2019; Windgassen, Leroux, Satyshur, Sandler, & Keck, 2018), providing information that is useful on its own or in conjunction with other structural data.

Helicases, the subject of this volume, are enzymes that convert the chemical energy of ATP hydrolysis to the mechanical energy needed to unwind nucleic acids. These enzymes comprise critical genome maintenance factors in all cells. For example, DNA replication requires unwinding of the genomic DNA so that each strand can be copied. Helicases also function in DNA recombination and repair processes. DNA unwinding is required to bypass unusual DNA structures such as hairpins or guanine quadruplexes that could otherwise hinder DNA replication. Helicases often contain positively charged surfaces that allow them to bind the negatively charged phosphodiester backbone of DNA. These basic surfaces also facilitate translocation along nucleic acids.

The *E. coli* PriA DNA helicase mediates re-initiation of DNA replication, or "replication restart," at locations distant from the bacterial chromosomal origin sequence (Windgassen, Wessel, Bhattacharyya, & Keck, 2018). To initiate canonical replication in bacteria such as *E. coli*, the DNA replication machinery (replisome) is loaded at *oriC*, the chromosomal origin of replication (O'Shea & Berger, 2014; Ozaki & Katayama, 2009). This origin sequence is recognized by loader proteins that promote the assembly of the DNA replisome. Once two copies of the replisome are loaded, they begin synthesizing DNA bidirectionally around the circular chromosome until they reach the termination sequence. However, it is estimated that each replisome in *E. coli* synthesizes only a fraction of the chromosome before encountering an impediment that causes it to dissociate (Cox et al., 2000; Mangiameli, Merrikh, Wiggins, & Merrikh, 2017; Mok & Marians, 1987; Sandler, 2000). Dissociation of the replisome leaves an abandoned replication fork that would be lethal if replication is not re-initiated. Since replisome failure can occur anywhere within the genome, replication restart must initiate via a mechanism that recognizes abandoned replication fork structures rather than a particular sequence. PriA is the major structure-specific helicase in bacteria that recognizes DNA replication fork structures to allow the replisome to be reloaded (reviewed in Windgassen, Wessel, et al., 2018).

In the protocol described in this chapter, we outline a crosslinking technique that has been used to map interactions between specific surfaces of PriA and a synthetic DNA replication fork. The approach identified interactions between individual residues of PriA and different positions in the replication fork, allowing development of a model of the PriA/DNA replication fork complex (Windgassen et al., 2019; Windgassen & Keck, 2016; Windgassen, Leroux, et al., 2018). The binding mechanism derived from these studies suggests how PriA is able to discriminate between DNA replication forks and other DNA structures present in cells, ultimately restricting its replication restart activity to bona fide replication forks. Our approach is broadly applicable to other helicase/DNA complexes.

The overall goal of a crosslinking experiment is to define inter- or intra-molecular interaction points in a given molecular complex. In vitro, this can be accomplished by conjugating a crosslinking agent to one molecule and identifying the crosslinking location on the interacting partner. For protein complexes, the latter step can be achieved using techniques such as mass spectrometry, which can identify crosslinked peptides (e.g. Nguyen, Sabat, & Sussman, 2018). However, protein/nucleic acid crosslinking presents a challenge for mass spectrometry methods. Instead, the method described here relies on incorporation of a crosslinker into a fixed position within a protein, crosslinking to DNA, then primer extension to determine the crosslinking site on the DNA.

The method uses an unnatural amino acid with a benzophenone side chain as a crosslinking agent. Benzophenones are photoactivatable, undergoing homolysis under UV light (Fig. 1) (Dorman & Prestwich, 1994). The light-generated radical can then react with any carbon-hydrogen bond in close proximity. Benzophenones are thus versatile in that they can react with

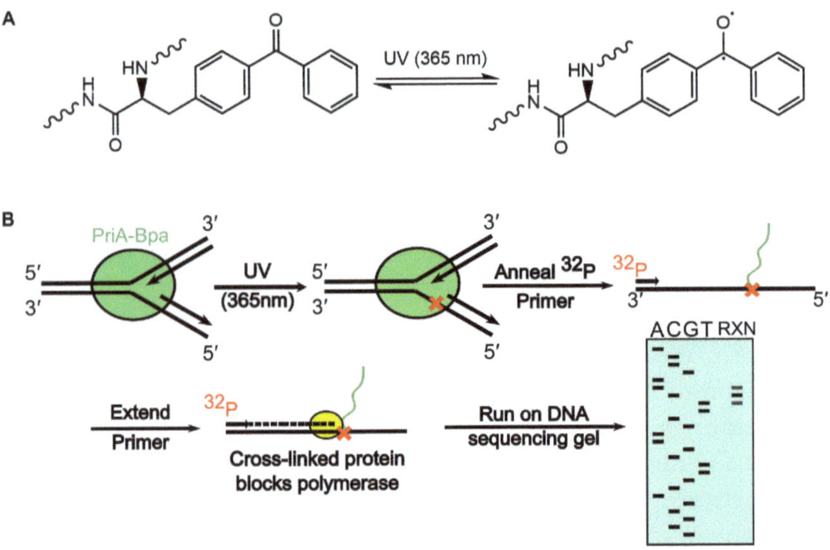

Fig. 1 Schematic of primer extension protocol. (A) Structure and homolysis of *p*-benzoyl-*l*-phenylalanine (Bpa). (B) PriA-Bpa variants are incubated with synthetic DNA forks and crosslinked via exposure to 365 nm UV light. Once crosslinked, the samples are heat-denatured, and a radiolabeled primer is annealed. Extension of this primer via a polymerase proceeds until it is physically obstructed by the covalently attached protein. This radiolabeled DNA fragment is run on a DNA sequencing gel to map the location of the crosslink with single-nucleotide precision.

proteins or nucleic acids, with a moderate preference for thymine nucleobases (Cuquerella, Lhiaubet-Vallet, Cadet, & Miranda, 2012). Benzophenones have been derivatized with thiol-reactive groups for conjugation to cysteine residues (Guo et al., 2005), and incorporated into amino acids (Kauer, Erickson-Viitanen, Wolfe, & DeGrado, 1986). *p*-benzoyl-*l*-phenylalanine (Bpa) is an amino acid derivative of phenylalanine that contains a benzophenone active group (Fig. 1A). It can be incorporated into any position in a protein using existing unnatural amino acid incorporation protocols (Chin, Martin, King, Wang, & Schultz, 2002; Ryu & Schultz, 2006; Xie & Schultz, 2005). Exposure of Bpa-containing proteins to UV light induces crosslinking to carbon-hydrogen bonds within ~10 Å of the Bpa α-carbon (Dorman & Prestwich, 1994; Wilkins, Daggett, & Cropp, 2008; Wittelsberger, Mierke, & Rosenblatt, 2008).

We have successfully incorporated Bpa into several positions in *E. coli* PriA and used Bpa variants to map surfaces that bind to different positions in a synthetic DNA fork substrate (Windgassen et al., 2019; Windgassen & Keck, 2016; Windgassen, Leroux, et al., 2018). Denaturing gel mobility shift assays are used to determine specific oligonucleotide linkages for each Bpa variant and primer extension experiments are used to map the precise locations of the crosslinks (Fig. 1B). Development of this method was inspired by an earlier crosslinking study of *E. coli* RNA polymerase/DNA interactions (Winkelman et al., 2015).

2. Design of Bpa incorporation sites

The crosslinking approach described here has been used to define the surfaces of PriA that bind to specific elements within a synthetic replication fork. Replication forks have three distinct regions with which PriA domains can interact: the parental duplex, the leading strand, and the lagging strand (Figs. 1 and 2). As *E. coli* PriA is a relatively large helicase (82 kDa, 732 amino acids) containing multiple functional domains (Bhattacharyya et al., 2014), we sought to create enough Bpa variants to sample the PriA surface. Bpa sites were chosen using two criteria: (1) we selected surface exposed residues that we predicted could accommodate the relatively large Bpa side chain and (2) we chose residues that were poorly conserved to minimize the likelihood of replacing a residue that is important for DNA binding. Conservation status was analyzed via a multiple sequence alignment generated through the ConSurf server (Landau et al., 2005). Bpa site selection in PriA was aided greatly by structures of the protein (Bhattacharyya et al., 2014;

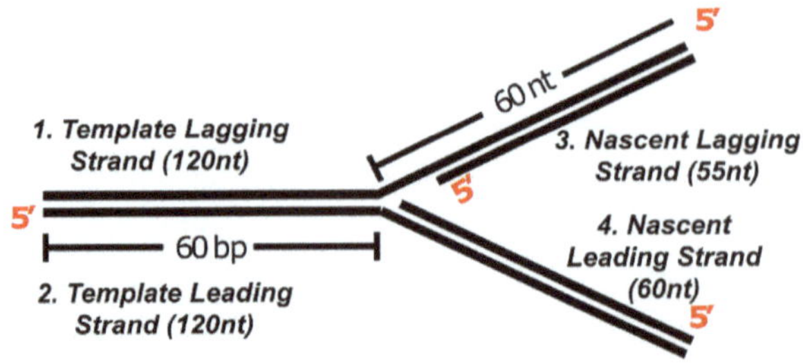

Fig. 2 Schematic of the synthetic DNA replication fork used in this protocol. *With permission from Windgassen, T. A., & Keck, J. L. (2016). An aromatic-rich loop couples DNA binding and ATP hydrolysis in the PriA DNA helicase.* Nucleic Acids Research, 44, *9745–9757.*

Windgassen, Leroux, et al., 2018). We used site-directed mutagenesis to replace the native codons at these locations with the *E. coli* amber stop codon (TAG). As described in the next section, this codon is recognized by the orthogonal tRNA-Bpa machinery to insert Bpa during translation of the mutant PriA gene. In total, we have successfully replaced ~15 individual PriA residues with Bpa.

3. Expression/incorporation of Bpa into PriA

The goal of this portion of the protocol is to incorporate Bpa via amber stop codon suppression. Briefly, the Shultz lab has developed an orthogonal aminoacyl tRNA synthetase/tRNA pair that can aminoacylate an amber-suppressor tRNA with Bpa (Chin et al., 2002; Ryu & Schultz, 2006). This tRNA recognizes the amber stop codon (TAG) in *E. coli*. Thus, Bpa is incorporated at TAG sequences via normal translation. Overexpression plasmids carrying *priA* genes with TAG codons at individual positions selected for Bpa incorporation were expressed in the Bpa overexpression system as described below.

3.1 Equipment

- 37 °C incubator and shaker
- Centrifuge with 1 L and 50 mL rotors
- ÄKTA pure 25 L Fast Protein Liquid Chromatography System (Cytiva)
- 5 mL HisTrap FF crude FPLC column (Cytiva)
- 20 mL HiPrep sulfopropyl sepharose fast flow (SPFF) FPLC column (Cytiva)
- 120 mL HiPrep S300 FPLC column (Cytiva)
- Vivaspin 20 10 kDa Spin Concentrators (Fisher)

3.2 Reagents

- BL21-AI One Shot Chemically Competent *E. coli* (ThermoFisher)
- pET-15b protein overexpression plasmid containing *priA* gene with a single amber codon at the selected Bpa site (N-terminal hexahistidine tag, Amp^R)
- pEVOL-pBpF (plasmid containing arabinose-inducible tRNA/tRNA synthetase pair, Cam^R, Addgene catalog number: 31190) (Chin et al., 2002)
- Luria Broth (LB) media
- H-p-Bz-Phe-OH (Bpa, BACHEM, 100 mM in 1 M NaOH)
- Lysis Buffer (10 mM HEPES-HCl pH 7.0, 10% glycerol, 500 mM NaCl, 100 mM glucose, 20 mM imidazole, 1 mM β-mercaptoethanol, 1 mM phenylmethylsulfonyl fluoride, 1 mM benzamidine, 1 EDTA-free protease inhibitor tablet)
- Ni Elution Buffer (10 mM HEPES-HCl pH 7.0, 10% glycerol, 500 mM NaCl, 300 mM imidazole, 1 mM β-mercaptoethanol)
- Dilution Buffer (10 mM HEPES-HCl pH 7.0, 10% glycerol, 1 mM β-mercaptoethanol)
- Buffer A (10 mM HEPES-HCl pH 7.0, 10% glycerol, 100 mM NaCl, 10 mM dithiothreitol)
- Buffer B (10 mM HEPES-HCl pH 7.0, 10% glycerol, 1000 mM NaCl, 10 mM dithiothreitol)
- Storage Buffer (10 mM HEPES-HCl pH 7.0, 50% glycerol, 500 mM NaCl, 10 mM dithiothreitol)

3.3 Protocol

3.3.1 *Expression of Bpa-containing PriA*

1. Co-transform BL21(Arabinose inducible) *E. coli* cells with a PriA-Bpa plasmid (pET-15b vectors containing an IPTG-inducible N-terminal hexahistidine tagged PriA variant, Amp^R) and pEVOL-pBpF (Cam^R)
2. Isolate transformants and grow in 50 mL of LB overnight at 37 °C
3. Add 1 mM Bpa and antibiotics to 1 L LB flasks and inoculate with 20 mL of the overnight culture
4. Grow at 37 °C to an OD_{600} of ~0.6
5. Induce protein expression with 1 mM IPTG and 0.5% arabinose
6. Incubate cultures overnight at 18 °C
7. Harvest cells by centrifuging at 7000 RPM for 12 min and store at −80 °C

3.3.2 Purification of PriA-Bpa variants

1. Thaw cell pellet on ice and resuspend in 40 mL Lysis Buffer
2. Lyse the cells by sonication on ice
3. Centrifuge for 30 min at 30,000 RPM at 4 °C to separate soluble and insoluble material

   Note: The following steps were all done using an ÄKTA Pure FPLC system at 4°C. We disconnected the UV lamp to prevent inadvertent Bpa activation.
4. Load the soluble supernatant directly onto a 5 mL HisTrap FF crude FPLC column equilibrated in Wash Buffer (Lysis Buffer without protease inhibitors)
5. Wash HisTrap column with 50 mL Wash Buffer
6. Elute bound protein with 30 mL Ni Elution Buffer
7. Dilute the eluate with 70 mL Dilution Buffer (100 mL final volume), then filter through a 0.22 μm filter to remove aggregated protein
8. Load the diluted eluate directly onto a 20 mL HiPrep sulfopropyl sepharose fast flow (SPFF) FPLC column equilibrated in Buffer A
9. Wash the SPFF column with 40 mL Buffer A
10. Elute the protein with a 160 mL gradient from 0% to 100% Buffer B (100 mM to 1000 mM NaCl)
    a. Collect 3 mL fractions
11. Identify fractions containing PriA by SDS-PAGE, combine fractions, and concentrate to 2 mL using a spin concentrator
12. Load the 2 mL solution onto a 120 mL HiPrep S300 FPLC column equilibrated in Buffer B
13. Elute the protein using 180 mL Buffer B
    a. Collect 3 mL fractions
14. Identify fractions containing PriA by SDS-PAGE, combine fractions, and concentrate to 2 mL using a spin concentrator
15. Dialyze against 500 mL Storage Buffer
16. Store purified PriA at −20 °C

3.4 Notes

1. We make the Bpa stock fresh right before use. It will turn yellow over time and may not be as effective if used after this point
2. Bpa is stable under ambient light, but care should be taken to minimize exposure to UV radiation by wrapping containers in aluminum foil or using opaque containers

3. In our system, arabinose induces the expression of the T7 polymerase from the BL21(AI) genome as well the orthogonal tRNA/tRNA synthetase pair from the pEVOL-pB-pF plasmid. This prevents issues associated with unwanted suppression of amber stop codons during cell growth. IPTG will trigger dissociation of the LacI repressor from the PriA-encoding plasmid, allowing the *priA* mutant gene to be expressed
4. If possible, we recommend any purification tags be placed on the C-terminus of the protein of interest. This will prevent purification of proteins that are prematurely truncated due to normal translation termination at amber codons
5. Diluting the HisTrap eluate with Dilution Buffer is necessary to reduce the NaCl concentration to levels that allow PriA to bind the cation exchange column in the next step. The NaCl concentration is reduced from 500 mM to 150 mM in this step
6. As Bpa is an amino acid that is only slightly larger than phenylalanine, its incorporation into proteins often does not greatly impact the structure when exposed on the surface of the protein. Thus, Bpa-containing PriA variants can generally be purified via the same method as the wild-type protein
7. After purification, we generally perform DNA-binding experiments to confirm that Bpa integration has not disrupted DNA-binding affinity

4. Design/construction of DNA replication forks

We used a derivative of a synthetic DNA replication fork described by the Marians group (Heller & Marians, 2005) in our crosslinking studies. This fork contains 60 base-pairs (bp) of dsDNA representing the parental template, 60 bp of dsDNA representing the leading strand duplex, and 60 bp representing the lagging strand duplex (note: there is a 5 nucleotide (nt) ssDNA gap between the lagging strand and fork junction) (Fig. 2). This structure was created using four oligonucleotides (Table 1). Two 120 nt oligonucleotides representing the parental template strands were designed to have 60 nt of complementarity and 60 nt of non-complementarity. Thus, the complementary regions of each oligonucleotide will anneal together and leave 60 nt of ssDNA. The nascent leading and lagging strand oligonucleotides can then be annealed to the two regions of non-complementarity in the parental strands (Fig. 2 and Table 1). Each strand can be individually radiolabeled with ^{32}P and annealed to the others to create radiolabeled replication forks. As noted in the Introduction, benzophenones have a moderate preference for crosslinking to thymine bases (Cuquerella et al., 2012). Thus,

Table 1 Oligonucleotide sequences.

Oligo name	Oligo sequence (5′ - 3′)	Strand in Fig. 2
oTW140 (Parental leading strand template)	GCA AGC CTT CTA CAG GTC GAC CGT CCA TGG CGA CTC GAG ACC GCA ATA CGG ATA AGG GCT GAG CAC GCC GAC GAA CAT TCA CCA CGC CAG ACC ACG TAG TGC TAT AAG CTT CGC TGA CGT	2
oTW141 (Parental lagging strand template)	AGC GTA ACG GAT CGA CCG CTT TGA CTA TCT ACG TCC GAG GCT CGC GCC GCA GAC TCA TTT AGC CCT TAT CCG TAT TGC GGT CTC GAG TCG CCA TGG ACG GTC GAC CTG TAG AAG GCT TGC	1
oTW168 (Nascent leading strand)	ACG TCA GCG AAG CTT ATA GCA CTA CGT GGT CTG GCG TGG TGA ATG TTC GTC GGC GTG CTC	4
oTW143 (Nascent lagging strand)	AGT CTG CGG CGC GAG CCT CGG ACG TAG ATA GTC AAA GCG GTC GAT CCG TTA CGC T	3
oTW144 (Primer for oTW141)	GCA AGC CTT CTA CAG GTC G	
oTW206 (Primer for oTW140)	ACG TCA GCG AAG CTT ATA GC	
oTW146 (Primer for oTW143)	AGC GTA ACG GAT CGA C	

From Windgassen, T. A., & Keck, J. L. (2016). An aromatic-rich loop couples DNA binding and ATP hydrolysis in the PriA DNA helicase. *Nucleic Acids Research, 44*, 9745–9757.

we recommend doing appropriate controls to examine the possible sequence dependence of any observed crosslinks (see Windgassen & Keck, 2016).

4.1 Equipment

- Heat block
- PROTEAN II xi Cell large format gel electrophoresis rig (16x16cm, BioRad)
- Storage phosphor screen (Molecular Dynamics)
- Typhoon FLA 9000 Gel Scanner (Cytiva) capable of phosphorimaging
- Agarose electrophoresis rig.

4.2 Reagents

- DNA oligonucleotides (Table 1)
- T4 Polynucleotide Kinase (20U/μL, New England Biolabs) and 10× Kinase buffer (New England Biolabs)

- ATP-[γ-^{32}P] (PerkinElmer)
- 5 × annealing buffer (50 mM Tris–HCl pH 7, 35 mM MgCl2, 1000 mM NaCl)
- 6 × DNA loading buffer (New England Biolabs)
- 10 × TBE (Fisher)

4.3 Protocol

4.3.1 *Labeling oligonucleotides with ^{32}P for DNA fork generation (written for labeling the lagging template strand)*

1. In a 50 μL reaction, combine 1 × kinase buffer, 534 nM oTW141, 8 μL ATP-[γ-^{32}P], and 25 units of T4 polynucleotide kinase
2. Using a heat block, incubate at 37 °C for 1 h and then 70 °C for 20 min

4.3.2 *Annealing radiolabeled forks*

1. In a 24.5 μL reaction volume, combine 653 nM oTW140, 784 nM oTW143, 784 nM oTW168, 262 nM ^{32}P-oTW141, and 1 × annealing buffer
2. Using a heat block, incubate at 95 °C for 10 min and then 70 °C for 1 h
3. Remove the heat block containing the samples from the heater and slowly cool on the benchtop for ~3 h
4. Store the labeled DNA forks at −20 °C overnight

4.3.3 *Purification of ^{32}P-labeled DNA forks*

1. Add 1 × DNA loading dye to the sample
2. Resolve through a large format native 7.5% TBE-PAGE gel (175 V for 3 h)
3. Running buffer is 1 × TBE
4. Image the gel via autoradiography and a Typhoon gel imager
5. Locate, excise, and place relevant bands in a dialysis bag containing 400 μL 0.5 × TBE
6. Place the dialysis bag in an agarose electrophoresis rig containing 500 mL 0.5 × TBE and electroelute for 1 h at 100 V
7. Remove the dialysis bag and dialyze against 20 mM Tris–HCl pH 7 and 5 mM MgCl2 for 45 min at room temperature
8. Store the purified replication fork (should be ~16 nM) at −20 °C

4.4 Notes

1. As the oligonucleotides we use to synthesize these forks are long (55–120 nt), we have them PAGE purified to remove sequences that are not full-length

2. To locate the labeled forks in the gel, we place filter paper on either side of the gel and spot DNA loading dye spiked with ATP-[γ-^{32}P] onto it in a manner that resembles a ladder. When the gel is imaged, this creates visible radioactive reference points that can be used to locate the radiolabeled forks

5. Determining strand-specific PriA-Bpa crosslinking

The goal of this experiment is to determine which oligonucleotide in the four-strand synthetic replication fork is crosslinked to PriA. This will help determine which primer to use for the primer extension experiment. For this, we create four variants of the DNA fork where each different oligonucleotide was radiolabeled separately. Incubating with PriA-Bpa and exposing to UV will produce covalent crosslinks that can be observed via a bandshift in a native TBE-PAGE gel (note: the protein in the samples is denatured with SDS while the replication fork remains assembled). Heat denaturing the different radiolabeled forks allows for the determination of which strands are crosslinked to Bpa (Fig. 3).

5.1 Equipment

- 365 nM UV lamp (Spectroline)
- PROTEAN II xi Cell large format gel electrophoresis rig (16 x 16 cm, BioRad)
- Vacuum gel dryer
- Storage phosphor screen (Molecular Dynamics)
- Typhoon FLA 9000 Gel Scanner (Cytiva) capable of phosphorimaging

5.2 Reagents

- DNA oligonucleotides (Table 1)
- T4 Polynucleotide Kinase (20 U/μL, New England Biolabs) and 10 × Kinase buffer (New England Biolabs)
- ATP-[γ-^{32}P] (PerkinElmer)
- 5 × annealing buffer (50 mM Tris–HCl pH 7, 35 mM $MgCl_2$, 1000 mM NaCl)
- 2 × SDS loading buffer (100 mM Tris–HCl pH 6.8, 4% sodium dodecyl sulfate, 0.2% bromophenol blue, 20% glycerol, 200 mM dithiothreitol)
- 10 × TBE (Fisher)

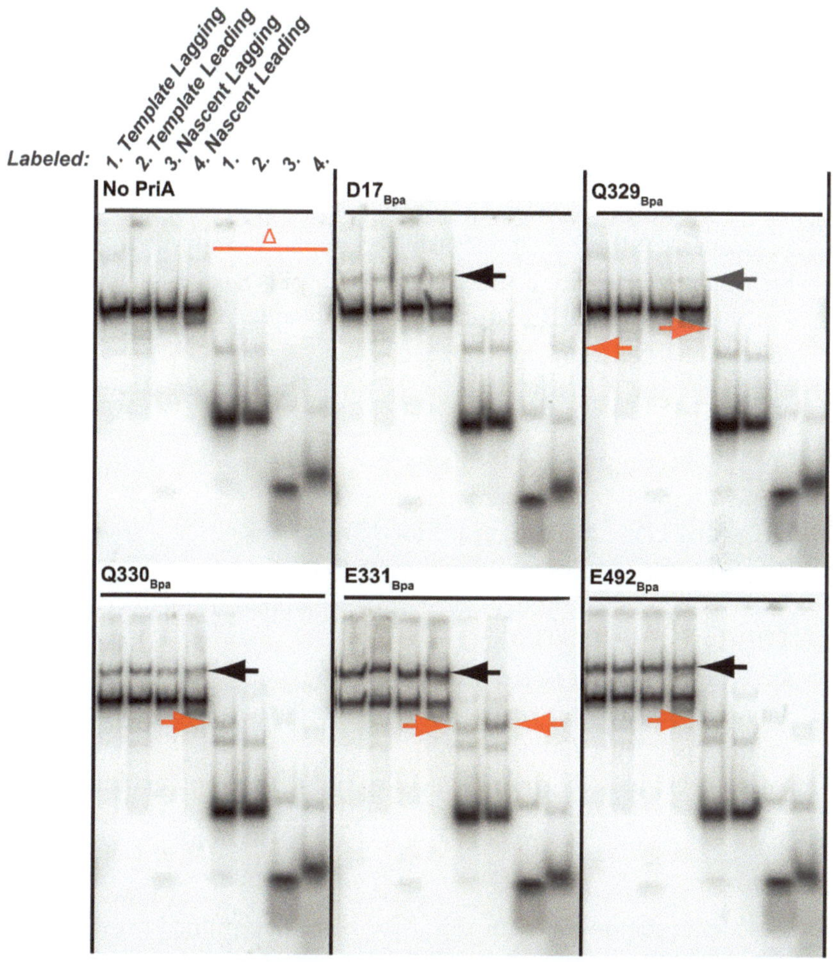

Fig. 3 Differential radiolabeling to determine strand-specificity of PriA-Bpa crosslinks. One of the four oligonucleotides is radiolabeled and annealed to the remaining three unlabeled oligonucleotides to form a replication fork structure. By labeling one of the four strands in separate annealing reactions, we produce four synthetic DNA replication forks each with a different strand radiolabeled. Heat denaturing (right four lanes shown for each PriA-Bpa variant) allows tracking of individual strands. In the left four lanes of each gel, PriA-Bpa is incubated with the four differentially radiolabeled DNA forks and exposed to UV. The presence of bandshifts (black arrows) in a native PAGE gel (note that the protein is denatured with SDS) indicates formation of a covalent crosslink. The right four lanes of each gel resolve the same reactions after heat denaturation. The presence of bands with slowed mobility in these lanes (red arrows) indicates crosslinks to the individual DNA strand that was radiolabeled. *With permission from Windgassen, T. A., & Keck, J. L. (2016). An aromatic-rich loop couples DNA binding and ATP hydrolysis in the PriA DNA helicase.* Nucleic Acids Research, 44, *9745–9757.*

5.3 Protocol

1. Radiolabel each oligo (oTW140, oTW141, oTW168, oTW143) via the method described above
2. Anneal the three remaining, unlabeled oligonucleotides to each radio-labeled oligonucleotide with excess non-labeled strands and purify as described above
3. Incubate 1 nM of each radiolabeled replication fork with 3.5 nM PriA-Bpa variant in 50 mM Tris–HCl, pH 8.0, 0.1 mg/mL BSA, 2 mM dithiothreitol, 5 mM ethylenediaminetetracetic acid, 6% glycerol, and 50 mM NaCl on ice
4. Illuminate the reactions with a 365 nm UV lamp for 5 min at room temperature
5. Put the reactions on ice for 2 min
6. Repeat steps 4–5 two additional times
7. Do a final UV exposure on ice for 15 min
8. Boil relevant samples for 10 min at 95 °C
9. Add 1 × SDS loading dye and resolved samples through a native 5% TBE-PAGE gel (100 V for 2.5 h)
   a. Running buffer is 1 × TBE
10. Transfer gels to filter paper, cover with plastic wrap, and dry with a vacuum gel dryer (80 °C for 1 h)
11. Image dried gels via autoradiography and a Typhoon gel imager

5.4 Notes

1. The concentrations used in this assay were determined previously via electrophoretic mobility shift. Only one shifted species is observed with 3.5 nM PriA and 1 nM DNA fork. This is ideal as multiple binding events may confound crosslinking results.
2. It may be tedious to perform this experiment on a large panel of Bpa variants, especially because any given Bpa variant may not produce any crosslinks. Thus, we recommend first doing an experiment to determine which of the variants will in fact crosslink to the DNA. For this, we incubate each of the Bpa variants with a radiolabeled DNA replication fork and expose to UV as above (and a no-UV control). If UV-induced crosslinking occurs for a given Bpa variant, a bandshift will appear on a native TBE-PAGE gel (the protein is denatured with SDS) relative to

the no-UV control. From here we begin to dissect the mechanism of crosslinking via the strand-specific and primer extension experiments.

6. Primer extension to determine interaction points

This protocol allows for the identification of the PriA-Bpa/DNA replication fork crosslinking site at single-nucleotide resolution. Briefly, PriA-Bpa is covalently crosslinked to an unlabeled DNA replication fork, the DNA and protein are denatured, and a radiolabeled primer is annealed to the crosslinked oligonucleotide. Extension of the radiolabeled primer with a polymerase will proceed until it is physically blocked by the crosslinked protein. This produces a radiolabeled piece of DNA of a length that corresponds to the location of the crosslinked PriA-Bpa. Resolution of the extended primer on a sequencing gel along with a DNA ladder maps the location of the crosslink on the DNA (Fig. 4).

6.1 Equipment

- Heat block
- Microspin G-25 Columns (Sigma)
- 365 nM UV lamp (Spectroline)
- Sequenase Cycle Sequencing Kit (Thermo)
- T100 Thermal Cycler (BioRad)
- DNA sequencing gel rig
- Vacuum gel dryer
- Storage phosphor screen (Molecular Dynamics)
- Typhoon FLA 9000 Gel Scanner (Cytiva) capable of phosphorimaging

6.2 Reagents

- DNA oligos (Table 1), unlabeled DNA fork
- ATP-[γ-^{32}P] (PerkinElmer)
- T4 Polynucleotide Kinase (20 U/μL, New England Biolabs) and 10 × Kinase buffer (New England Biolabs)
- Taq DNA polymerase (5 U/μL, New England Biolabs) and 10 × Taq Buffer (New England Biolabs)
- 10 × TBE (Fisher)
- 2 × denaturing urea buffer (8 M urea, 0.5 × TBE, bromophenol blue and xylene cyanol)

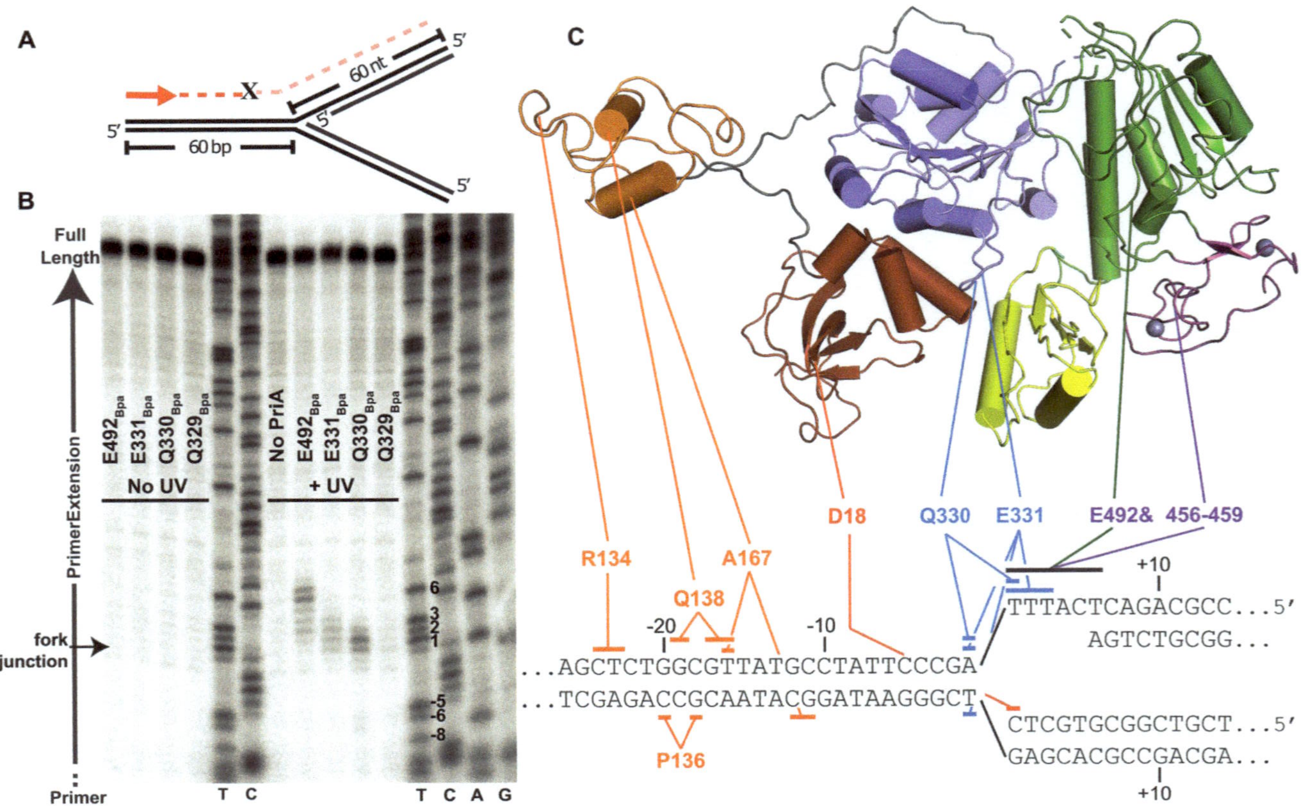

Fig. 4 See figure legend on opposite page.

6.3 Protocol

6.3.1 Labeling primers with ^{32}P for primer extension (written for PriA-Bpa variants that crosslink with the lagging strand template strand, as determined in Section 5)

1. In a 10 μL reaction volume, combine 1 × kinase buffer, 10 μM oTW144, 5 μL ATP-[γ-^{32}P], and 10 units of T4 polynucleotide kinase
2. Incubate the mixture at 37 °C for 1 h, then 70 °C for 20 min
3. Dilute the labeled primer to 2 μM and purify via a G-25 spin column
4. Store the purified oligonucleotide at −20 °C or proceed to next step

6.3.2 Setting up crosslinking reaction

1. Set up crosslinking reactions as described in Section 5.3 (steps 3–7) except with unlabeled DNA forks and with 2 nM DNA fork and 7 nM PriA-Bpa

6.3.3 Primer extension reaction/sequencing ladder setup

1. In a 12.5 μL reaction volume, combine 2.25 μL of each crosslinking reaction, 1.25 units Taq DNA Polymerase, 1 × Taq buffer, 200 μM of each dNTP, 5% DMSO, 2 M betaine and 320 nM ^{32}P-labeled primer
2. Generate the sequencing ladder with the Thermo Sequenase Cycle Sequencing Kit using 229 nM labeled oTW144 and 7.7 nM unlabeled oTW141
3. Place both the primer extension and sequencing ladder reactions in a thermal cycler and cycle via the following conditions:
   a. 95 °C for 30 s
   b. 95 °C for 30 s
   c. 48 °C for 30 s
   d. 68 °C for 15 s
   e. Repeat b–d 20 times

Fig. 4 Primer extension mapping of crosslink locations. (A) Schematic of the synthetic DNA replication fork showing the annealing location of the radiolabeled primer (red arrow) used to map crosslinks on the template lagging strand. The dashed red line depicts primer extension by a polymerase. (B) Primer extension reactions resolved on a DNA sequencing gel. Bands that appear in the +UV lanes are indicative of the polymerase colliding with the covalently crosslinked PriA and terminating polymerization. The location of these bands can be compared to the sequencing ladder to determine the location of the crosslink on the original oligonucleotide. (C) The crystal structure of *Ec*PriA (Windgassen, Leroux, et al., 2018, PDB: 6DGD) and a schematic of the DNA fork are shown. Colored lines link Bpa incorporation sites with their respective crosslinking site on the DNA fork. (A) and (B) are modified versions of figures that appear in Windgassen & Keck (2016). *These modified figures are used with permission.*

4. Add 1 × denaturing urea buffer to each sample
   a. The ladder reactions usually need to be diluted 4 × with denaturing urea buffer
5. Store samples at −20 °C or proceed to next step

6.3.4 Pouring/running DNA sequencing gel

1. In a 40 mL reaction volume, combine 8 M urea, 1 × TBE, and 12% bis/acrylamide
2. Dissolve mixture at 37 °C with frequent mixing
3. Add 100 μL 10% APS and 30 μL TEMED and load mixture into 35 cm glass plates with 0.25 mm spacers
4. Set up gel rig and pre-run sequencing gel for 1 h at 30 W
   a. Running buffer is 1 × TBE
5. Boil the primer extension and sequencing ladder reactions at 95 °C for 5 min and load into the gel
   a. We generally load 15 μL
6. Run the gel at 40 W for 1 h, and subsequently at 35 W for 2 h
7. Transfer gels to filter paper, cover with plastic wrap, and dry with a vacuum gel dryer (80 °C for 1 h)
8. Image dried gels via autoradiography and a Typhoon gel imager

6.4 Notes

1. This method is best done with fresh ATP-[γ-^{32}P]. Primer extension blockage products can be faint and sensitivity is quickly lost when ATP-[γ-^{32}P] has gone through multiple half-lives.
2. Transferring the sequencing gel to the filter paper for drying can be difficult due to the gel sticking to the glass rather than the filter paper. We recommend silanizing the glass plates prior to pouring the gel. It may take some trial and error to find a method to transfer the gel without ripping it.

7. Summary

The method outline above allows mapping of protein-nucleic acid interaction points via the benzophenone-derivative Bpa. Bpa can be readily incorporated into proteins via established methodology and, depending on the location of its incorporation, may cause negligible changes to local structure. Once incorporated, Bpa is a versatile crosslinker with a short distance restraint (~10 Å from the α-carbon). After exposure to UV, Bpa homolysis allows for indiscriminate crosslinking to any carbon-hydrogen bond. Thus, Bpa-containing proteins can be crosslinked not only to nucleic acids, as above, but also to proteins. Mass spectrometry techniques exist to determine where protein-protein Bpa crosslinks occur (e.g. Nguyen et al., 2018) however, nucleic acid crosslinking locations cannot be identified via mass spectrometry. Thus, we present a primer extension protocol that allows for the mapping of the crosslink locations. Overall, crosslinking with Bpa coupled to a primer extension assay provides a robust method of mapping protein-nucleic acid interaction points.

References

Bhattacharyya, B., George, N. P., Thurmes, T. M., Zhou, R., Jani, N., Wessel, S. R., et al. (2014). Structural mechanisms of PriA-mediated DNA replication restart. *Proceedings of the National Academy of Sciences of the United States of America*, *111*, 1373–1378.

Chin, J. W., Martin, A. B., King, D. S., Wang, L., & Schultz, P. G. (2002). Addition of a photocrosslinking amino acid to the genetic code of Escherichia coli. *Proceedings of the National Academy of Sciences of the United States of America*, *99*, 11020–11024.

Cox, M. M., Goodman, M. F., Kreuzer, K. N., Sherratt, D. J., Sandler, S. J., & Marians, K. J. (2000). The importance of repairing stalled replication forks. *Nature*, *404*, 37–41.

Cuquerella, M. C., Lhiaubet-Vallet, V., Cadet, J., & Miranda, M. A. (2012). Benzophenone photosensitized DNA damage. *Accounts of Chemical Research*, *45*, 1558–1570.

Dorman, G., & Prestwich, G. D. (1994). Benzophenone photophores in biochemistry. *Biochemistry*, *33*, 5661–5673.

Guo, L.-W., Hajipour, A. R., Gavala, M. L., Arbabian, M., Martemyanov, K. A., Arshavsky, V. Y., et al. (2005). Sulfhydryl-reactive, cleavable, and radioiodinatable benzophenone photoprobes for study of protein−protein interaction. *Bioconjugate Chemistry*, *16*, 685–693.

Heller, R. C., & Marians, K. J. (2005). Unwinding of the nascent lagging strand by rep and PriA enables the direct restart of stalled replication forks. *Journal of Biological Chemistry*, *280*, 34143–34151.

Kauer, J. C., Erickson-Viitanen, S., Wolfe, H. R., & DeGrado, W. F. (1986). P-benzoyl-L-phenylalanine, a new photoreactive amino acid. Photolabeling of calmodulin with a synthetic calmodulin-binding peptide. *Journal of Biological Chemistry*, *261*, 10695–10700.

Landau, M., Mayrose, I., Rosenberg, Y., Glaser, F., Martz, E., Pupko, T., et al. (2005). ConSurf2005: The projection of evolutionary conservation scores of residues on protein structures. *Nucleic Acids Research*, *33*, W299–W302.

Mangiameli, S. M., Merrikh, C. N., Wiggins, P. A., & Merrikh, H. (2017). Transcription leads to pervasive replisome instability in bacteria. *eLife*, *6*, e19848.

Mok, M., & Marians, K. (1987). The Escherichia coli preprimosome and DNA B helicase can form replication forks that move at the same rate. *Journal of Biological Chemistry*, *262*, 16644–16654.

Nguyen, T. T., Sabat, G., & Sussman, M. R. (2018). In vivo cross-linking supports a head-to-tail mechanism for regulation of the plant plasma membrane P-type H+-ATPase. *Journal of Biological Chemistry*, *293*, 17095–17106.

O'Shea, V. L., & Berger, J. M. (2014). Loading strategies of ring-shaped nucleic acid translocases and helicases. *Current Opinion in Structural Biology*, *25*, 16–24.

Ozaki, S., & Katayama, T. (2009). DnaA structure, function, and dynamics in the initiation at the chromosomal origin. *Plasmid, 62*, 71–82.
Ryu, Y., & Schultz, P. G. (2006). Efficient incorporation of unnatural amino acids into proteins in Escherichia coli. *Nature Methods, 3*, 263–265.
Sandler, S. J. (2000). Multiple genetic pathways for restarting DNA replication forks in Escherichia coli K-12. *Genetics, 155*, 487–497.
Wilkins, B. J., Daggett, K. A., & Cropp, T. A. (2008). Peptide mass fingerprinting using isotopically encoded photo-crosslinking amino acids. *Molecular BioSystems, 4*, 934.
Windgassen, T. A., & Keck, J. L. (2016). An aromatic-rich loop couples DNA binding and ATP hydrolysis in the PriA DNA helicase. *Nucleic Acids Research, 44*, 9745–9757.
Windgassen, T. A., Leroux, M., Sandler, S. J., & Keck, J. L. (2019). Function of a strand-separation pin element in the PriA DNA replication restart helicase. *Journal of Biological Chemistry, 294*, 2801–2814.
Windgassen, T. A., Leroux, M., Satyshur, K. A., Sandler, S. J., & Keck, J. L. (2018). Structure-specific DNA replication-fork recognition directs helicase and replication restart activities of the PriA helicase. *Proceedings of the National Academy of Sciences of the United States of America, 115*, 9075–9084.
Windgassen, T. A., Wessel, S. R., Bhattacharyya, B., & Keck, J. L. (2018). Mechanisms of bacterial DNA replication restart. *Nucleic Acids Research, 46*, 504–519.
Winkelman, J. T., Winkelman, B. T., Boyce, J., Maloney, M. F., Chen, A. Y., Ross, W., et al. (2015). Crosslink mapping at amino acid-base resolution reveals the path of scrunched DNA in initial transcribing complexes. *Molecular Cell, 59*, 768–780.
Wittelsberger, A., Mierke, D. F., & Rosenblatt, M. (2008). Mapping ligand–receptor interfaces: Approaching the resolution limit of benzophenone-based photoaffinity scanning. *Chemical Biology & Drug Design, 71*, 380–383.
Xie, J., & Schultz, P. G. (2005). An expanding genetic code. *Methods, 36*, 227–238.

CHAPTER FOUR

Quantitative methods to study helicase, DNA polymerase, and exonuclease coupling during DNA replication

Anupam Singh and Smita S. Patel*
Department of Biochemistry and Molecular Biology, Rutgers University, Robert Wood Johnson Medical School, Piscataway, NJ, United States
*Corresponding author: e-mail address: patelss@rutgers.edu

Contents

Methods in Enzymology, Volume 672
ISSN 0076-6879
https://doi.org/10.1016/bs.mie.2022.03.011

Abstract

Genome replication is accomplished by highly regulated activities of enzymes in a multi-protein complex called the replisome. Two major enzymes, DNA polymerase and helicase, catalyze continuous DNA synthesis on the leading strand of the parental DNA duplex while the lagging strand is synthesized discontinuously. The helicase and DNA polymerase on their own are catalytically inefficient and weak motors for unwinding/replicating double-stranded DNA. However, when a helicase and DNA polymerase are functionally and physically coupled, they catalyze fast and highly processive leading strand DNA synthesis. DNA polymerase has a 3′-5′ exonuclease activity, which removes nucleotides misincorporated in the nascent DNA. DNA synthesis kinetics, processivity, and accuracy are governed by the interplay of the helicase, DNA polymerase, and exonuclease activities within the replisome. This chapter describes quantitative biochemical and biophysical methods to study the coupling of these three critical activities during DNA replication. The methods include real-time quantitation of kinetics of DNA unwinding-synthesis by a coupled helicase-DNA polymerase complex, a 2-aminopurine fluorescence-based assay to map the precise positions of helicase and DNA polymerase with respect to the replication fork junction, and a radiometric assay to study the coupling of DNA polymerase, exonuclease, and helicase activities during processive leading strand DNA synthesis. These methods are presented here with bacteriophage T7 replication proteins as an example but can be applied to other systems with appropriate modifications.

Abbreviations

2-AP 2-aminopurine
nt nucleotide
NTP nucleoside triphosphate
SSB single-strand DNA binding protein
TLC thin layer chromatography

1. Introduction

The general mechanism of DNA replication is conserved in prokaryotes and eukaryotes (Li & O'Donnell, 2019; Yang, Seidman, Rupp, & Gao, 2019). Minimally, DNA replication requires four basic enzymatic activities: a DNA helicase to catalyze the unwinding of the parental double-stranded (ds) DNA into single-stranded (ss) DNA, a primase to make RNA primers *de novo* from nucleoside triphosphate (NTP) substrates, DNA polymerase to elongate the primer and copy the parental DNA strands in the 5'–3' direction using dNTPs as substrates, and a 3'–5' exonuclease to proofread mistakes made by the polymerase (Benkovic, Valentine, & Salinas, 2001; Burgers & Kunkel, 2017; Hamdan & Richardson, 2009). Due to the antiparallel arrangement of dsDNA strands and the fixed directionality of DNA synthesis by the polymerase, one of the parental strands, called the leading strand, is copied continuously by the action of the helicase and leading strand polymerase (Ogawa & Okazaki, 1980). The complementary lagging strand is copied as Okazaki fragments discontinuously through frequent priming events by the primase and elongation by the lagging strand polymerase (Duderstadt et al., 2016; Ogawa & Okazaki, 1980; Okazaki, Okazaki, Sakabe, Sugimoto, & Sugino, 1968; Pandey et al., 2009). The hexameric ring-shaped helicases are the central organizers of the DNA replication machinery (Donmez & Patel, 2006; O'Donnell & Li, 2018; Yang et al., 2019). The replicative helicase interacts with the leading and lagging strand DNA polymerases, primase, and accessory proteins such as processivity factor and single-strand DNA binding protein (SSB) (Benkovic & Spiering, 2017; Burgers & Kunkel, 2017; Hamdan & Richardson, 2009; Kose, Xie, Cameron, Strycharska, & Yardimci, 2020; Li & O'Donnell, 2019; Manosas, Spiering, Ding, Croquette, & Benkovic, 2012; Nandakumar, Pandey, & Patel, 2015). Functional and physical interactions with the partner proteins are essential in regulating the helicase and the interacting proteins. This chapter discusses ensemble biochemical and biophysical methods to study the functional coupling between replicative helicase, polymerase, and exonuclease activities during leading strand DNA synthesis.

As motor proteins, helicases utilize NTP hydrolysis to translocate unidirectionally on nucleic acid substrates (Abdelhaleem, 2010; Bourgeois, Mortreux, & Auboeuf, 2016; Patel & Picha, 2000). Their ability to translocate along ssDNA is critical for assisting the DNA polymerase in catalyzing strand displacement leading strand DNA synthesis. Nearly all replicative helicases are poor at unwinding duplex DNA on their own. The rates and processivities of the replicative helicases in isolation are often lower than those observed within the replisomes (Delagoutte & von Hippel, 2002, 2003; Patel, Pandey, & Nandakumar, 2011). Biochemical studies show that bacteriophage T7 and T4 helicases are mostly passive unwindases, and they rely on thermal melting of the base-pairs at the fork junction to separate the strands of the duplex DNA (Johnson, Bai, Smith, Patel, & Wang, 2007; Lionnet, Spiering, Benkovic, Bensimon, & Croquette, 2007; Manosas, Xi, Bensimon, & Croquette, 2010). Fluorescence, optical, and magnetic tweezers-based single-molecule techniques reveal frequent helicase "slippage" and rezipping events limiting their rates and processivity (Manosas, Spiering, Ding, Croquette, & Benkovic, 2012; Schlierf, Wang, Chen, & Ha, 2019; Sun et al., 2011, 2018). Replicative DNA polymerases are efficient at catalyzing DNA synthesis on ssDNA templates but cannot catalyze processive strand displacement synthesis on dsDNA templates (Manosas et al., 2012; Nandakumar et al., 2015; Pandey & Patel, 2014; Yuan & McHenry, 2009). How do the slow and inefficient helicase and DNA polymerase accomplish the task of catalyzing fast and highly processive leading strand synthesis during replication? The answer lies in the physical and functional coupling of the helicase and DNA polymerase activities, which mutually benefits the two proteins.

Prokaryotic and eukaryotic replicative helicases have an opposite polarity of translocation on the DNA, giving rise to different leading strand replisome architectures (Fig. 1). The RecA-type prokaryotic (bacteria and bacteriophages) and mitochondrial replicative helicases translocate and unwind DNA in the 5' to 3' direction. This class of helicases bind on the opposite lagging strand to the DNA polymerase bound on the leading strand. In contrast, AAA^{+}-type eukaryotic replicative helicases (and eukaryotic viruses) unwind DNA in the 3' to 5' direction and bind on the same leading strand as the DNA polymerase. Despite these differences, recent structures of phage T7 and eukaryotic replisomes show conserved mechanisms of interactions between the helicase and leading strand DNA polymerase. In both cases, the C-terminal domain of the helicase, harboring the NTPase and helicase activities, are involved in interactions with the leading strand DNA polymerase (Li & O'Donnell, 2019; Yang et al., 2019). In the case of the T7 replisome, the helicase subunits are physically coupled to the leading strand DNA polymerase via their C-terminal tail regions (Foster et al., 2019; Gao et al., 2019). This physical coupling provides stable positioning of the helicase and polymerase at the fork junction, which is critical

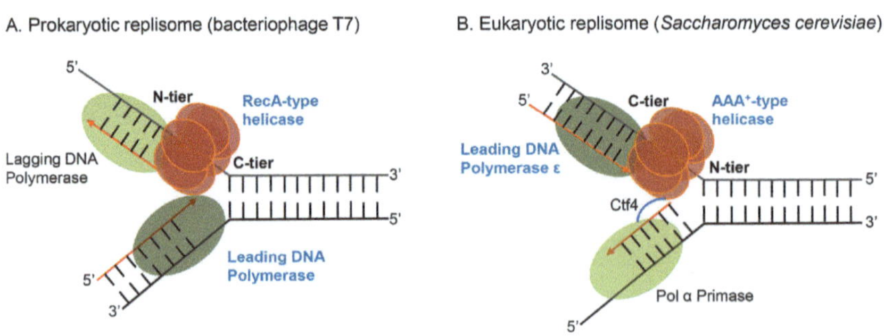

Fig. 1 Architectures of prokaryotic (A) and eukaryotic (B) replisomes show similar modular structures with C tier of replicative helicase interacting with leading strand DNA polymerase and N tier with lagging strand DNA polymerase or primase. The polarity of the helicase rings is opposite in the two cases.

for efficient strand displacement synthesis of the leading strand without slippages and fork reannealing. Often replication hurdles such as reduced nucleotide levels, lesions on the DNA, protein blocks, or secondary structures uncouple helicase-polymerase, prompting slippage and junction base-pair reannealing. These events result in the fraying of the primer-end and excessive excision of the nascent DNA by the exonuclease activity (Singh et al., 2020). Measurements of the polymerase to exonuclease activity ratio provide a way to assess helicase-polymerase coupling.

The poor unwinding and synthesis activities of the replicative helicase and DNA polymerase, respectively, in isolation, and the mutual dependency ensure that inappropriate unwinding or copying of DNA regions does not occur in an unregulated manner. The partnering of the replicative helicase with the leading strand DNA polymerase increases the replication rate and processivity, eliminates nucleotide sequence dependency, and prevents rezipping of the unwound DNA (Delagoutte & von Hippel, 2002; Georgescu et al., 2014; Kim, Dallmann, McHenry, & Marians, 1996; Manosas, Spiering, Ding, Bensimon, et al., 2012; Manosas, Spiering, Ding, Croquette, & Benkovic, 2012; Pandey & Patel, 2014; Stano et al., 2005; Sun et al., 2018). Here, we describe three *in vitro* ensemble methods to study the functional coupling between helicase, DNA polymerase, and exonuclease activities on synthetic replication fork substrates mimicking leading strand DNA synthesis. The methods developed using bacteriophage T7 proteins apply to other systems with appropriate modifications.

A fluorescence-based method to quantify the single turnover kinetics of DNA unwinding-synthesis by the coupled T7 helicase-DNA polymerase complex is detailed (Nandakumar et al., 2015). As the isolated helicase is poor at unwinding the fork DNA substrate, and the T7 DNA polymerase (1:1 complex of T7 gp5 and its processivity factor, *Escherichia coli* thioredoxin) cannot synthesize beyond 5–7 base-pairs on its own, the separation of the two strands by the coupled helicase-DNA polymerase complex essentially provides the rate of unwinding-synthesis. We also describe a 2-aminopurine (2-AP) fluorescence-based method that provides structural information on helicase and DNA polymerase positions with respect to the replication fork junction. This is a powerful technique that can be applied to quantify the contributions made by individual proteins in the unwinding of the fork junction (Nandakumar et al., 2015). Finally, we describe a radiometric, thin layer chromatography (TLC)-based method to study DNA polymerase and exonuclease coupling during processive leading strand synthesis on a minicircle DNA substrate (Singh et al., 2020).

This method provides valuable information on the polymerase-exonuclease active-site switching and its dependence on helicase-DNA polymerase coupling as it allows simultaneous measurement of polymerase and exonuclease activities in the same reaction. The methods described here are fast, less tedious than gel-based, and can be adapted to design high-throughput assays.

2. Real-time quantitation of unwinding-synthesis by coupled helicase-DNA polymerase

A fork DNA substrate is designed with 5' and 3' single-stranded tails and a 40 bp duplex DNA region, mimicking the DNA replication fork (Fig. 2A). Based on several biochemical trials, we chose 35 nucleotide long 5' tail for optimal assembly and unwinding of the fork DNA by T7 helicase. The trials consisted of changing the length of the 5' tail and measuring the unwinding yield and rate. Best yields/rates were obtained when the 5' tail was > 25 nt. Similarly, on the 3' tail of the fork DNA, a 24-mer primer was annealed to assemble the T7 DNA polymerase. The crystal structure of T7 DNA polymerase on the DNA shows interactions with ~21 bp of dsDNA (Doublie, Tabor, Long, Richardson, & Ellenberger, 1998). At the blunt end of the fork DNA duplex, a fluorophore-quencher pair was positioned; fluorescein was attached at the 3' end and BHQ-1 quencher at the 5' end (Fig. 2A). The proximity of the quencher to fluorescein keeps the fluorescence intensity low. The fluorophore-quencher typically works well, and FRET pairs are not necessary.

The DNA unwinding-synthesis reactions on the short fork DNA substrates occur in the milliseconds time range; hence, reactions are carried out in a rapid mixing stopped-flow device (Fig. 2B). A stopped-flow apparatus consists of two motor-powered syringes, a rapid mixing chamber, a flow cell connected to a fluorescence detector, and a stop-syringe for controlling the flow of the reaction in the flow cell. The temperature of the syringes, mixing chamber, and flow cell are precisely controlled with the help of a circulating water bath. A pre-assembled leading strand complex containing the fork DNA, helicase, dTTP, and DNA polymerase is filled in one syringe. T7 helicase binds stably to the fork substrate as a hexamer only in the presence of dTTP, and we leave out Mg^{2+} from the buffer to prevent the unwinding of the DNA fork (Ahnert & Patel, 1997; Picha & Patel, 1998). Reactions are initiated with the addition of Mg^{2+}, 100-fold excess of dT_{90} ssDNA over the fork DNA substrate, and the rest of the three dNTPs (dCTP, dGTP and dATP, referred to as dVTPs hereafter) from the

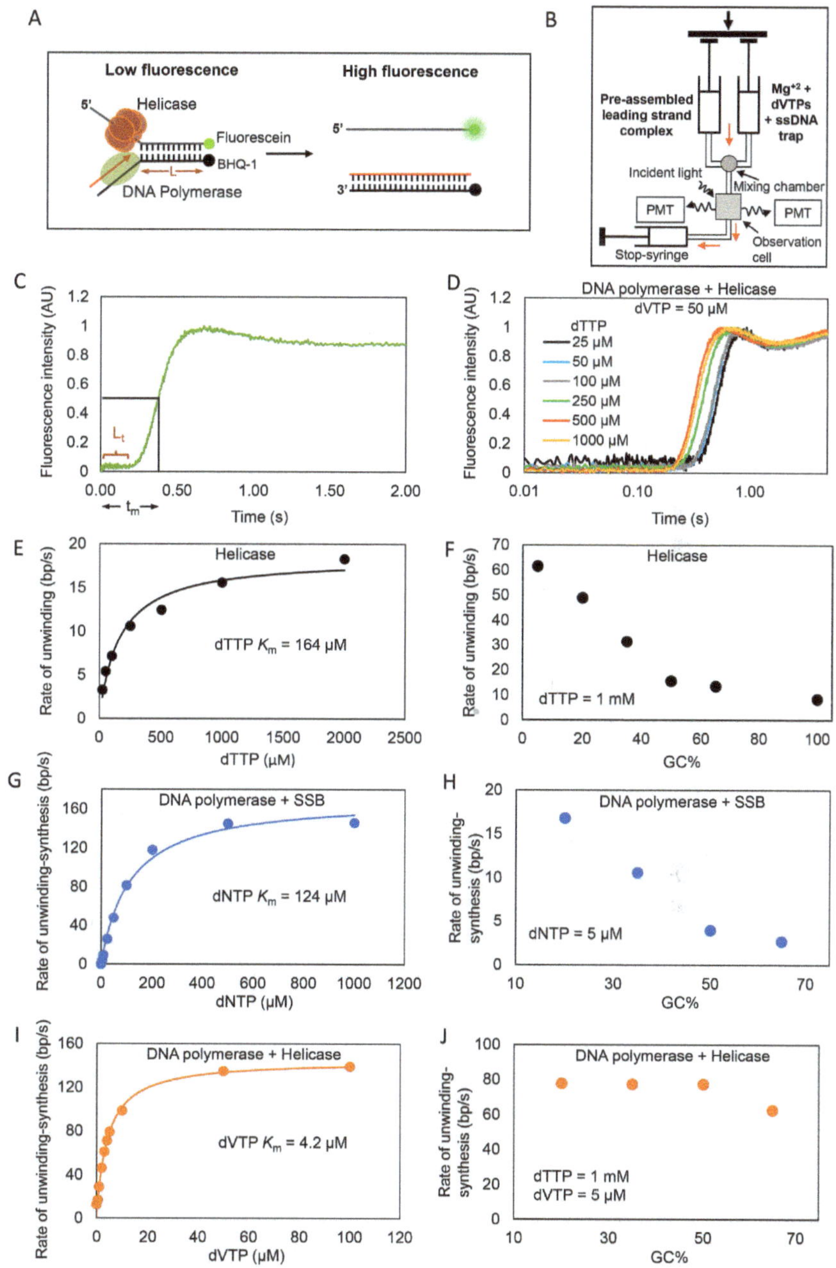

Fig. 2 Real-time measurement of unwinding-synthesis by coupled helicase-DNA polymerase complex (Nandakumar et al., 2015). (A) Schematic representation of the experimental design used to determine rates of unwinding-synthesis by T7 helicase and DNA polymerase complex. The leading strand complex is assembled on a primed fork DNA

(Continued)

other syringe. The dT_{90} ssDNA traps the free or dissociated helicase molecules from rebinding to the fork DNA, ensuring single turnover conditions for rate measurements.

The unwinding-synthesis reaction is measured in real-time on the stopped-flow apparatus by following the increase in fluorescence intensity of fluorescein with time (1–2 ms onward). A typical time course shows an initial lag phase, representing the helicase-DNA polymerase complex's time to reach the blunt-end of the fork DNA. The lag phase is followed by a steep, exponential increase in fluorescence signal and a plateau representing the complete separation of the two strands (Fig. 2C). The transient dip in fluorescence intensity at the reaction end is due to the helicase going over the fluorescein probe. The lag time before the steep fluorescence increase should directly correlate with the length of the DNA duplex used. With minor modifications, the method can be used to measure rates of DNA unwinding and strand-displacement synthesis by isolated helicases and DNA polymerases, respectively (Nandakumar et al., 2015; Ozes, Feoktistova, Avanzino, Baldwin, & Fraser, 2014).

Fig. 2—Cont'd substrate with "L" bp long DNA duplex. Fluorescence signal by the fluorescein probe increases in response to the separation of the two strands of the duplex DNA. (B) A schematic representation of the experiment in the stopped-flow rapid mixing device (KinTek Corp). (C) A representative time course shows the increase in fluorescence from leading strand synthesis reaction by T7 helicase and DNA polymerase. The initial lag (L_t) reflects the time the helicase-DNA polymerase complex takes to reach the blunt-end of the fork DNA. The time corresponding to the mid-point of fluorescence transition, t_m, can be used to obtain the rate of unwinding-synthesis reaction. (D) Fluorescence time traces from leading strand synthesis reactions by T7 helicase and DNA polymerase conducted at different dTTP concentrations and fixed 50 μM dVTPs. (E) Rate constants of fork DNA (50% GC) unwinding by T7 helicase at increasing dTTP concentrations. The solid line shows the fit to a hyperbola. (F) Dependence of the helicase unwinding rate constant on the GC-content of the dsDNA region of the fork DNA substrate at a fixed 1 mM dTTP concentration. (G) Rate constants of fork DNA (50% GC) unwinding-synthesis reactions by DNA polymerase and *E. coli* SSB as a function of dNTP concentration. The solid line shows the fit to a hyperbola. (H) Dependence of unwinding-synthesis rates by DNA polymerase and SSB complex on the GC-content of the dsDNA region of the fork DNA substrate. All reactions were performed with 5 μM dNTPs. (I) Rate constants of fork DNA (50% GC) unwinding-synthesis reactions by the coupled helicase-DNA polymerase complex as a function of dVTP concentration at a fixed dTTP concentration of 1 mM. The solid line shows the fit to a hyperbola. (J) Rate constants of unwinding-synthesis reactions by coupled helicase-DNA polymerase complex on fork DNA substrates with varying GC content conducted with 1 mM dTTP and 5 μM dVTP.

The desired parameter is the rate of DNA unwinding-synthesis, which is expressed in base-pairs unwound-synthesized per second (bp/s). The average unwinding-synthesis rate can be obtained by fitting the fluorescence time traces into a stepping-mechanism model using a publicly available *gfit* program, described previously (Levin, Hingorani, Holmes, Patel, & Carson, 2009; Pandey, Levin, & Patel, 2010), or using other software such as the KinTek Explorer (Chib, Byrd, & Raney, 2016; Johnson, 2009; Johnson, Simpson, & Blom, 2009). The average rate of unwinding-synthesis can also be estimated from the time required to reach the midpoint of fluorescence transition (t_m) and multiplying $1/t_m$ and dsDNA length (Fig. 2C). One can measure the rate of unwinding-synthesis at single-nucleotide resolution by conducting leading strand synthesis reactions using a fork DNA substrate with a fluorescently labeled or radiolabeled DNA primer in a quenched-flow rapid mixing apparatus and resolving the DNA products by sequencing PAGE (Pandey & Patel, 2014). Although the technique provides high-esolution information, it is not high-throughput and requires substantial time and effort compared to the stopped-flow method described above.

Factors such as dTTP concentration, dVTP concentration, and GC content of the duplex region of DNA substrates will affect the observed rate of unwinding-synthesis reactions. For example, the kinetics of unwinding-synthesis by T7 helicase-DNA polymerase complex depends on dTTP concentration, requiring at least 500 μM dTTP to produce optimum unwinding-synthesis (Fig. 2D). Measuring the rates of helicase unwinding, DNA polymerase strand displacement synthesis, and unwinding-synthesis by the helicase-DNA polymerase complex as a function of these parameters provides mechanistic insights into the coupling of the two proteins. T7 helicase in isolation has poor unwinding activity (Fig. 2E), and the rates depend on the GC content of the duplex DNA (rate of unwinding decreases from 62 bp/s for 5% GC fork to 8.3 bp/s for 100% GC fork) (Fig. 2F). Likewise, T7 DNA polymerase on its own is unable to extend the DNA primer beyond a few base-pairs on a fork DNA substrate. *E. coli* SSB or T7 gp2.5 (T7 SSB) supports strand displacement synthesis (Fig. 2G), but the rate depends on the GC content (Fig. 2H) (Nandakumar et al., 2015). The GC dependency indicates that helicase and DNA polymerase on their own are primarily passive motors. When T7 helicase and DNA polymerase are combined, they synergistically enhance each other's activity to catalyze unwinding-synthesis at a fast rate of ~140 bp/s (Fig. 2I). Moreover, the coupled action of the two proteins eliminates the dependency of the rate of unwinding-synthesis on GC content (Fig. 2J), which is crucial for the accurate replication of various regions of the genome. Although SSB supports DNA polymerase's strand displacement synthesis, high dNTP concentrations must be added to reach the maximal rate of ~140 bp/s (dNTP K_m with SSB is 124 μM) (Fig. 2G). On the other hand, the helicase-coupled strand displacement synthesis occurs at a much lower dNTP concentration (dVTP K_m with helicase is ~4 μM), which appears to be more physiologically relevant (Fig. 2I). These assays are highly convenient for measuring the dNTP K_m values, which provide physiologically meaningful mechanistic information.

2.1 Equipment

- Stopped Flow instrument equipped with fluorescence detection capabilities and KinTek software (KinTek Corp, Snow Shoe, PA).
- Water baths to maintain the stopped-flow syringes, mixing chamber, flow cell, and lamp at desired temperatures.

2.2 Buffers and reagents

- Buffer A: 50 mM Tris–HCl, pH 7.6, 40 mM NaCl and 10% glycerol
- Ethylenediaminetetraacetic acid (EDTA)
- Dithiothreitol (DTT)
- Fork DNA substrates prepared by annealing appropriate DNA oligos
- A dT_{90} or similar ssDNA
- Magnesium chloride solution
- dNTPs
- Purified T7 helicase and T7 DNA polymerase proteins (Donlin, Patel, & Johnson, 1991; Pandey et al., 2009).

2.3 Procedure

1. Set the temperature of the stopped-flow reaction chamber to 18 °C.
2. Set up the experiment with an excitation wavelength of 480 nm and fluorescence emission with a long pass 515 nm cut-off filter.
3. Prepare "Mixture A" by mixing Buffer A, 1.5 mM EDTA, 5 mM DTT, 20 nM fork DNA substrate, 40 nM T7 helicase, 40 nM T7 DNA polymerase, and 2000 μM dTTP.
4. Mix $MgCl_2$, 100 μM dVTPs, and 2 μM dT_{90} ssDNA in reaction buffer A to prepare "Mixture B."
5. Pre-incubate Mixtures A and B separately at 18 °C for 10 min.
6. Load the Mixtures A and B in separate loading syringes.

7. Trigger the rapid mixing of A and B in the mixing chamber and measure the fluorescence change starting in the millisecond range to up to 5 s. Change the recording time depending upon the time it takes to complete the reaction. Measure at least three time courses for each reaction and use the averaged trace to quantitate the rates.

2.4 Analysis

1. The average rate of unwinding-synthesis (bp/s) can be obtained by fitting the time traces into a stepping-mechanism model using the publicly available *gfit* program, described previously (Levin et al., 2009; Pandey et al., 2010). The data fitting into a multistep unwinding-synthesis reaction can also be carried out using the KinTek Explorer software to obtain the stepping rate (Chib et al., 2016; Johnson, 2009; Johnson et al., 2009).
2. A rough estimate of the average unwinding-synthesis rate constant (k) on a fork DNA with a duplex length of L bp can be obtained from the time required to reach the midpoint of fluorescence transition (t_m) (Fig. 2C):

$$k = 1/t_m \times L \tag{1}$$

We have previously shown that the rate constants determined by the two methods are comparable (Nandakumar & Patel, 2016).

3. Average rate constants from the reactions are plotted as a function of dNTP concentration and are fitted to hyperbolic trend to obtain dNTP K_m and k_{cat} using the following equation:

$$k = k_{cat}.[\text{dNTP}]/(K_m + [\text{dNTP}]) \tag{2}$$

2.5 Notes

1. To make the fork DNA substrate, DNA primer, template DNA and fluorescein labeled non-template DNA flap are mixed in a ratio of 1:1:0.9 in an annealing buffer consisting of 1X TE and 100 mM NaCl. A slightly lesser concentration of fluorophore-labeled oligonucleotide is used to keep the baseline fluorescence intensity low. DNA annealing is carried out by heating the mixed strands at 95 °C for 2 min on a heat-block and allowing the heat block to gradually cool down to room temperature by turning the heat-block off. Proper annealing of the fork DNA substrate should be confirmed by running the annealed DNA on a native PAGE. The fluorescein-labeled DNA

strand and the duplex DNA can be visualized from the fluorescein fluorescence or stained with SYBR Green to view all strands.

2. The 5' and 3' tail lengths in the fork DNA should be optimized for the helicase-DNA polymerase complex based on their known footprints on the DNA.
3. Kinetics of nucleotide incorporation by T7 replisome is extremely fast which makes accurate rate measurements difficult at physiologically relevant temperatures. Therefore, 18 °C is used to slow down the reaction rates and to make the kinetics measurable. The measurement temperatures can be changed for other replication systems based on the kinetics of the replication proteins, deadtime of the stopped-flow apparatus, etc.
4. In Fig. 2A, fluorescein was attached at the 3' end and BHQ-1 quencher at the 5' end on the opposite strand. A sequence of 3–4 dG residues in place of BHQ-1 also efficiently quenches the fluorescence of fluorescein and is a cost-effective alternative to BHQ-1.
5. Fluorescence increase with gentle slope may suggest problems in replisome assembly, variable rates of unwinding-synthesis by heterogeneous complexes, and interrupted DNA synthesis.
6. Experiments should be carried out with and without the dT trap. We have experimentally confirmed that dT_{90} ssDNA does not affect the T7 replication complex. However, it can deleteriously affect the kinetic measurements of other replication systems by dissociating the preassembled replisomes. Thus, an experimental control without a dT ssDNA trap should be conducted.

3. Aminopurine-based method to study cooperative base-pair melting at replication fork junction

It is essential to know the exact position of the helicase and DNA polymerase relative to each other and the replication fork junction in the assembled leading strand replisome complex. This allows a proper understanding of the contributions made by the helicase and DNA polymerase in unwinding the junction base-pairs. 2-AP is a fluorescent analog of adenine and guanine that base-pairs with thymine and cytosine (Law, Eritja, Goodman, & Breslauer, 1996; Ward, Reich, & Stryer, 1969). The 2-AP fluorescence is quenched when it is base-paired and has stacking interactions with the neighboring bases. This property is used to monitor enzyme-catalyzed DNA unwinding reaction and to study the interaction of helicase with the junction base-pairs in the fork DNA (Jose, Datta, Johnson, & von Hippel, 2009; Reha-Krantz, 2009). By strategically incorporating a 2-AP residue at different positions around the junction base-pair of the fork DNA, one can map the precise position of the helicase and DNA polymerase with respect to the 2-AP base-pair. Furthermore, the method reveals the roles of individual helicase and DNA polymerase proteins and their coupled action in unwinding the junction base-pairs.

A DNA fork substrate can be designed with a single 2-AP residue on the lagging strand base-paired to a dT on the leading strand at the fork junction (junction 2-AP) (Fig. 3A). The position of the primer-template junction from the fork junction can be changed by shortening the primer and creating 1–3 nt gaps between the primer-end and the junction base-pair (Fig. 3A). This reveals the position of the DNA polymerase active site with respect to the junction base-pair. A different type of DNA fork can be designed with an internal 2-AP probe at a position downstream to the junction base-pair (internal 2-AP) (Fig. 3A). Such probe position reveals the ability of helicase and DNA polymerase to melt the internal base-pairs upon binding to the fork DNA. Fork DNA substrates are prepared by annealing appropriate oligonucleotides.

Two fluorescence intensity measurements are made, one with fork DNA alone and the other with fork DNA incubated with DNA polymerase, helicase, or their combination (Fig. 3B and C). A parallel experiment is carried out with an unlabeled replication fork substrate to correct for the protein fluorescence changes resulting from the DNA binding. Optimal protein concentration that produces the maximum fluorescence intensity change must be determined empirically by carrying out experiments are various protein concentrations. The 2-AP fluorescence change in response to protein binding is expressed as "fold change." When a "no gap" fork DNA is used, addition of T7 DNA polymerase to the fork DNA increases the fluorescence intensity by twofold over 2-AP fork DNA alone, demonstrating that the DNA polymerase unwinds the junction base-pair to accommodate itself at the primer-end (Fig. 3D). Interestingly, the increase in fluorescence intensity in response to DNA polymerase binding is also observed with the "1-nt gap" fork DNA, slightly with 2-nt gap but not on 3-nt gap substrate (Fig. 3D). This indicates that the DNA polymerase unwinds two base-pairs at the junction to create a 2-nt gap between the primer-end and junction base-pair to optimally accommodate itself on the replication fork.

Similar experiments with T7 helicase were conducted with dTMP-PCP, a non-hydrolysable analog of dTTP, needed to stabilize the helicase hexamer on the fork DNA without unwinding the fork. Other analogs such as dTDP.BeF_3 or dTDP.AlF_3 can also be used. T7 helicase also increases the fluorescence intensity of 2-AP at the junction base-pair upon binding to the

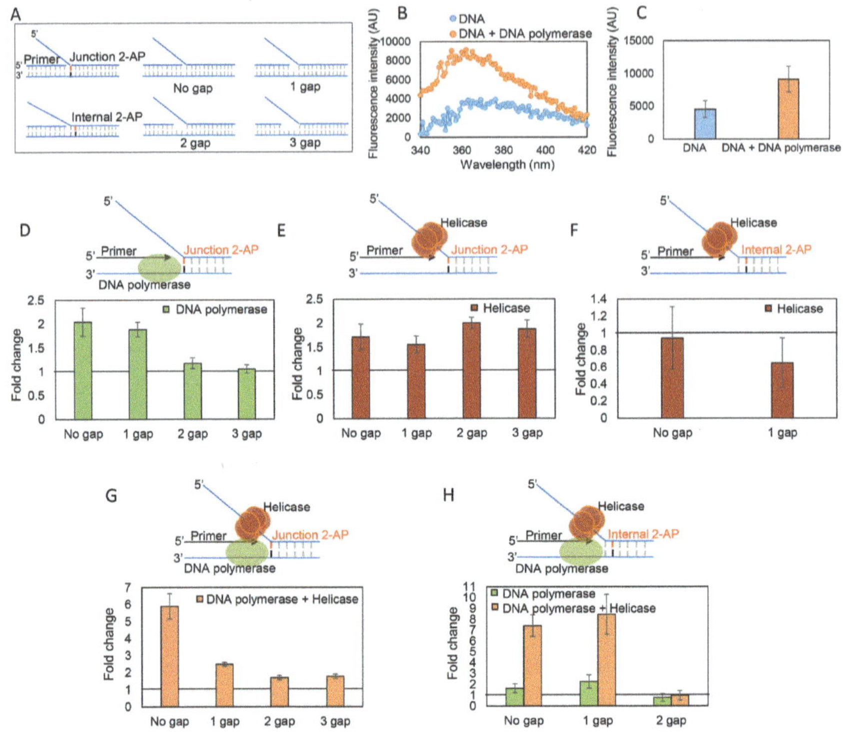

Fig. 3 2-AP fluorescence assay to determine the relative contribution of the T7 helicase and T7 DNA polymerase to the unwinding of fork DNA duplex region (Nandakumar et al., 2015). (A) Schematic of 2-AP labeled fork DNA substrates. A single 2-AP probe (red) is incorporated in the lagging strand at the fork junction base-pair (Junction 2-AP) or the base-pair at 1-nt downstream of the fork junction (Internal 2-AP). The DNA primer length is changed to introduce a 0 to 3-nt gap between the primer-end and the fork junction. (B) Fluorescence intensities (in arbitrary units) of 2-AP containing replication fork DNA with and without T7 DNA polymerase. (C) Representative 2-AP fluorescence signals measured at 370 nm in the presence and absence of T7 DNA polymerase. (D) Fold change in "junction" 2-AP fluorescence upon binding of T7 DNA polymerase to the fork DNA relative to free DNA (shown as black line). A significant increase in 2-AP fluorescence is only observed when the DNA fork has no gap and 1-nt gap between primer-end and junction base-pair. (E) Fold change in "junction" 2-AP fluorescence in response to binding of T7 helicase to the fork DNA. The increase in fluorescence signal is independent of the gap size. (F) Fold change in "internal" 2-AP fluorescence in response to binding of T7 helicase to the fork DNA. No increase in 2-AP fluorescence is observed. (G) Fold change in "junction" 2-AP fluorescence in response to T7 helicase and DNA polymerase binding to the fork DNA. The enhanced 2-AP fluorescence is observed with no gap and 1-nt gap fork DNA substrates. (H) Fold change in "internal" 2-AP fluorescence in response to binding of T7 DNA polymerase alone or the T7 helicase-DNA polymerase complex to the fork DNA. Enhanced 2-AP fluorescence was observed with no gap and 1-nt gap fork DNA substrates.

fork DNA, like the DNA polymerase (Fig. 3E). However, this increase is observed in all gap-sized forks but only when the 2-AP probe is at the junction base-pair and not at an internal position (Fig. 3E and F) (Nandakumar et al., 2015). This indicates that T7 helicase binds at the fork junction and can melt only the junction base-pair.

We observe the greatest increase in 2-AP fluorescence when both helicase and DNA polymerase are added to the fork DNA (Fig. 3G). The enhanced melting of the 2-AP at the junction base-pair is observed in the no-gap substrate, and the change is much less when the gap size is longer than 1-nt (Fig. 3G). Enhanced unwinding by helicase-DNA polymerase is apparent when the 2-AP is placed internally one nucleotide downstream of the fork junction on the no-gap and 1-nt gap substrate and not seen in the 2-nt gap substrate (Fig. 3H). These experiments indicate that helicase and DNA polymerase unwind one to two base-pairs of the fork junction to accommodate them at the replication fork of the leading strand. Such studies could be instrumental in designing fork DNA scaffolds for structural studies of the replisome. For example, the replisome architecture was confirmed by the cryoEM structure of the T7 leading strand complex (Gao et al., 2019). The agreement between the cryoEM and the biochemical models showcases the effectiveness of 2-AP based mapping method in deciphering the architecture of the helicase-DNA polymerase complex at the leading strand fork.

3.1 Equipment

- FluoroMax-4 (Horiba, Ltd., Kyoto, Japan) or equivalent spectrofluorometer with temperature control
- Heat block to keep the reaction mixture in microcentrifuge tubes at the desired temperature

3.2 Buffers and reagents

- Buffer B: 50mM Tris–HCl, pH 7.6, 40mM NaCl, 10mM $MgCl_2$, 5mM DTT.
- Annealed fork DNA substrates.
- dTMP-PCP or dTDP.AlF_3.
- Purified T7 helicase and DNA polymerase proteins.

3.3 Procedure

1. Prepare the samples by mixing 100nM of fork DNA substrate, 200nM T7 helicase (where present), 10μM dTMP-PCP, and 200nM T7 DNA polymerase (where present) in buffer B. Pre-incubate the samples at 25°C for 10min before starting the fluorescence measurements.

2. Transfer the sample in a quartz cuvette placed in a cuvette holder of spectrofluorometer set at 25 °C.
3. Carry out the equilibrium fluorescence measurements by exciting the sample at 315 nm (at 2 mm slit width) and measuring the emission at 370 nm (at 6 mm slit width).
4. Make corrections for the observed fluorescence for the buffer background and the protein bound to the unlabeled replication fork substrate. Check for the inner filter effect and correct it if needed (see the following section on data analysis).

3.4 Analysis of the 2-AP fluorescence data

Unstacking of the 2-AP base by the bound DNA polymerase or helicase results in an increase in fluorescence. The observed fluorescence intensity is corrected for background from the buffer, fluorescence changes due to protein binding to the DNA independent of the fluorescence change brought by 2-AP unstacking, and volume and inner filter volume effects. The fluorescence changes caused by protein binding to the DNA can be determined by performing equilibrium fluorescence measurements with mock DNA substrates lacking the 2-AP probe. Fluorescence data are corrected for inner filter and volume effects using the following equation:

$$F_c = F\,x\,V_f/V_o \times 10^{0.5(A_x+A_m)} \tag{3}$$

where F_c is 2-AP fluorescence intensity corrected for volume and inner filter effects, F is observed 2-AP fluorescence intensity, V_f and V_o are total and original volume of the reaction mixture, respectively, A_x is the absorbance of the mixture at the excitation wavelength, and A_m is the absorbance at the emission wavelength.

Correction for protein binding is performed as follows:

$$F_{pc} = F_c - F_{mc} \tag{4}$$

Where, F_{pc} is the fluorescence intensity of 2AP DNA corrected for protein fluorescence change, inner filter effect, and volume, and F_{mc} is the fluorescence intensity obtained with mock DNA without the 2AP probe. The fluorescence change of 2AP DNA in response to protein binding is expressed as "fold change" and can be quantified (Eq. 5). Here $F_{2AP\ DNA}$ is the fluorescence intensity of 2AP DNA in the absence of protein.

$$\text{Fold Change} = F_{pc}/F_{2AP\ DNA} \tag{5}$$

3.5 Notes

1. The samples should be maintained at a preset temperature during the measurement to minimize the effect of temperature fluctuations.
2. The extent of quenching of 2-AP fluorescence by neighboring bases is dependent on sequence context in which 2-AP is incorporated (Jean & Hall, 2001; Jones & Neely, 2015; Larsen et al., 2004). This information can be used to strategically design the DNA substrates.
3. The 2-AP fluorescence intensity increase can be due to neighboring base-pair unstacking or specific amino acids interacting with the 2-AP base. Proper control experiments should be conducted to eliminate these possibilities.
4. The proteins and any nucleotides present in the reaction may contribute to the inner filter effect, and this must be checked for and corrected when necessary.

4. Simultaneous measurement of polymerase and exonuclease activities using minicircle fork DNA

Replicative DNA polymerases misincorporate with a frequency of 10^{-4} to 10^{-6} (Kunkel, 2004; Kunkel & Bebenek, 2000). Therefore, replicative DNA polymerases have a proofreading exonuclease activity that excises the 3'-end mismatched nucleotide in the nascent DNA to increase the accuracy of DNA synthesis. The exonuclease activity is either part of the same polypeptide as the DNA polymerase, like in T7 DNA polymerase, or a separate subunit, like in *E. coli* DNA polymerase III (Scheuermann & Echols, 1984). The polymerase and exonuclease active sites are typically 30–40 Å apart (Beese, Derbyshire, & Steitz, 1993; Doublie et al., 1998). Therefore, 3–4 base-pairs at the primer-template junction must melt to partition the primer-end into the exonuclease site, which happens during misincorporation when DNA synthesis is inhibited. However, the probability of spontaneous melting/partitioning of a correctly matched primer-end is expected to be low, estimated to excise less than 0.1% of correctly incorporated nucleotides (Donlin et al., 1991). However, studies show that replicative DNA polymerases, including T7 DNA polymerase, excise close to 7% of correctly incorporated nucleotides into dNMPs during leading strand synthesis. This implies that primer-end shuttling between the exonuclease and polymerase sites happens more frequently than expected during DNA synthesis (Fersht, Knill-Jones, & Tsui, 1982; Singh et al., 2020).

The recent study showed that high excision events are partly due to translocation hurdles encountered by the helicase-DNA polymerase during DNA replication (Singh et al., 2020). Therefore, measuring the polymerase to exonuclease ratio (Pol/Exo) provides a convenient way to monitor replication hurdles during DNA synthesis.

We describe a facile way to measure the kinetics of nucleotide incorporation and excision during leading strand DNA synthesis (Singh et al., 2020). There are many ways to measure the kinetics of single nucleotide incorporation during DNA synthesis, including gel-based methods and single-molecule tweezers and nanopore methods (Lieberman et al., 2010; Manosas, Spiering, Ding, Bensimon, et al., 2012; Pandey & Patel, 2014; Sun et al., 2015). However, these methods cannot measure the excision of single nucleotides during processive DNA synthesis. We use a 70bp minicircle fork DNA (Table 1) that supports efficient rolling circle DNA synthesis of kb-sized DNA products (Pandey et al., 2009). T7 helicase and DNA polymerase are assembled on the minicircle fork with a 40-nt long 5' ssDNA tail (Fig. 4A). Reactions are started by adding a mixture of dVTPs and α-^{32}P-dGTP, Mg^{2+}, and T7 gp2.5 and quenched with formic acid at pre-decided time-points. The radiolabeled dGMP incorporated into the nascent DNA, dGMP excised by the DNA polymerase's exonucleolytic activity, dGDP produced from helicase's dNTP hydrolysis activity, and the unused dGTP are simultaneously resolved on the PEI-cellulose TLC sheet (Fig. 4B). The spot intensities of dGTP, dGDP, incorporated dGMP, and excised dGMP are quantified using ImageQuant TL software to determine

Table 1 Oligonucleotide sequences used for the radiometric TLC-based assay.

Oligonucleotide	Sequence
70 nucleotide minicircle DNA	5' CACCATATCCTCGACCATCCCCAATATGGTC CATCAACCCTTCACCTCACTTCACTCCAC TATACC ACTC 3'
Primer for the 70 nucleotide minicircle DNA	5' TTTTTTTTTTTTTTTTTTTTTTTTTTTTTTTTT TTTTTTTGAGTGGTATAGTGGAGTGAA GTGAGGTGAAGGGTTGATGGACCATATTGG GGATGGTCGAGGATATGGTG 3'
Linear 70 nucleotide ssDNA	/5'Phos/CACCATATCCTCATCAACCCTTCACGA CCATCCCCAATATGGTCCCTCACTTCACTCCAC TATACCACTC 3'
Splint DNA	5' ATGGTGGAGTGG 3'

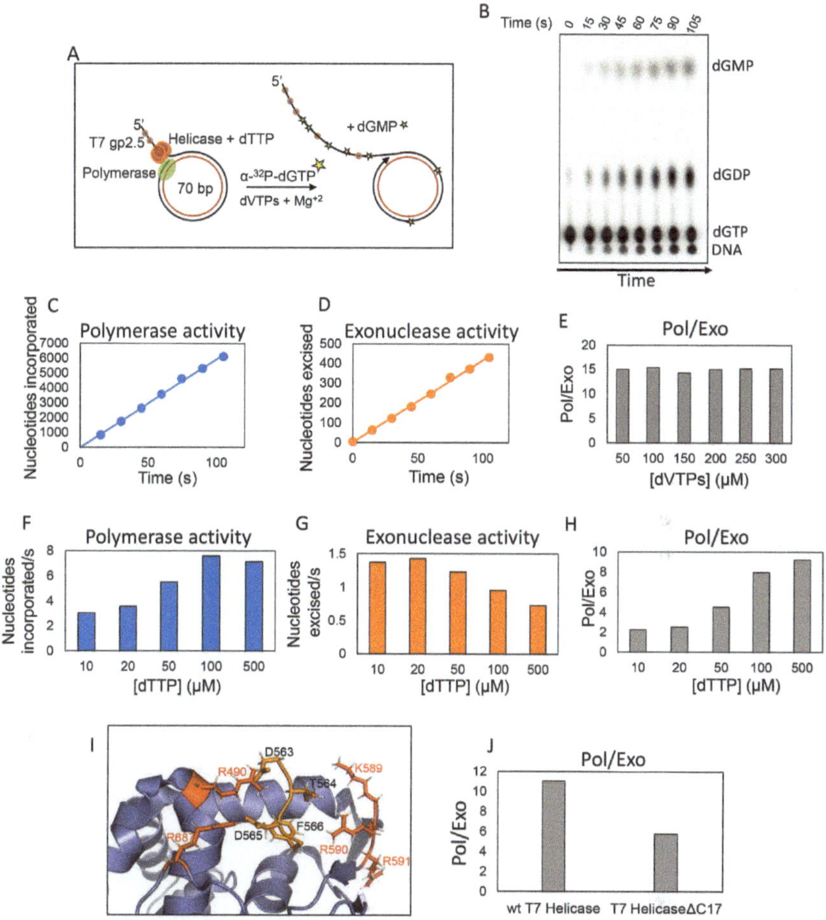

Fig. 4 Thin layer chromatography (TLC)-based assay to measure the coupling between polymerase and exonuclease activities of T7 DNA polymerase during leading strand DNA synthesis (Singh et al., 2020). (A) Experimental design to simultaneously measure polymerase and exonuclease activities during leading strand synthesis. T7 helicase and DNA polymerase were assembled on a primed-minicircle DNA substrate to measure rolling circle leading strand synthesis. Asterisks represent incorporated, excised, and unused radiolabeled nucleotides. (B) Representative TLC image shows the spots from dGMP incorporated into the nascent DNA, dGMP excised by the DNA polymerase's exonucleolytic activity, dGDP, and the unused dGTP. (C, D) Time courses of polymerase activity (C) and exonuclease activity (D) expressed as moles of nucleotides incorporated and excised per mole of leading strand complex. Experiments were performed at a constant 500 μM dTTP concentration and 150 μM dVTPs. (E) Pol/Exo ratios from leading strand DNA synthesis reactions conducted at a constant 500 μM dTTP concentration and increasing concentrations of dVTPs. (F, G) Moles of nucleotides incorporated (F) and excised (G) per mole of leading strand complex per second during leading strand

(Continued)

the amounts of nucleotide added to the synthesized DNA and excised during each reaction period. The time courses provide the rates of DNA synthesis and exonuclease reactions to obtain the Pol/Exo ratio (Fig. 4C and D).

The leading strand synthesis reaction by T7 helicase-polymerase on the minicircle shows a Pol/Exo ratio of ~15 under optimum nucleotide concentration conditions (Fig. 4E). Thus, T7 DNA polymerase excises one out of 15 correctly incorporated nucleotides from the nascent DNA. To understand the mechanistic basis for the high excision rate, one can measure the Pol/Exo ratio at increasing dNTPs. Partitioning of the primer-end from the polymerase site into the exonuclease site occurs from the pre-translocated state (Lieberman, Dahl, & Wang, 2014; Singh et al., 2020), and high concentrations of dNTPs stabilize the post-translocated state. If excision reactions are simply due to sub-saturating levels of dNTPs, then high dNTPs should reduce the exonuclease activity. On the other hand, if excision reactions are due to translocation problems, then Pol/Exo should be insensitive to dNTP concentrations, which we found (Fig. 4E). Therefore, Pol/Exo ratio is an indicator of translocation problems during DNA synthesis.

Translocation problems can be created by replication blocks or occasional uncoupling of the helicase and DNA polymerase activities during leading strand synthesis. Stalling of the replisome may result in transient reannealing of the junction base-pair and primer-end fraying, stimulating the exonuclease activity (Singh et al., 2020). This was tested by deliberating slowing down the helicase with reduced dTTP levels (preferred fuel for T7 helicase). Indeed, reduced dTTP levels decreased the polymerase activity and increased the exonuclease activity, reducing the Pol/Exo ratio (Fig. 4F–H). Uncoupling of helicase and polymerase activities can be caused by occasional disruptions in the physical interactions between the helicase and polymerase. The acidic C-tail of T7 helicase interacts with the basic front patch of the T7 DNA polymerase to physically couple the two proteins in

Fig. 4—Cont'd synthesis is shown for reactions performed with 10 μM dVTPs and different dTTP concentrations. (H) Pol/Exo ratios for leading strand synthesis reactions are shown for reactions performed with different dTTP concentrations, as determined from (F, G). (I) Structure shows the basis for physical coupling between T7 helicase-DNA polymerase through interactions between the acidic C-tail of T7 helicase and the basic front patch of T7 gp5 (PDB entry 6P7E) (Foster et al., 2019). (J) Pol/Exo ratio during leading strand synthesis reactions is shown for reactions performed with WT T7 helicase and its C-tail deletion mutant. Reactions were conducted with 500 μM dTTP and 150 μM dVTPs.

the leading strand replisome (Fig. 4I) (Foster et al., 2019; Hamdan et al., 2007). The ΔC17 helicase mutant showed a lower Pol/Exo ratio of 6 relative to 12 in WT helicase (Fig. 4J). Thus, the Pol/Exo ratio measurement provides a convenient way to monitor how well the helicase and DNA polymerase are physically and functionally coupled during leading strand synthesis. Such studies can help understand replication defects in the disease-causing mutants of mitochondrial helicase Twinkle and DNA polymerase γ (Chan & Copeland, 2009; Peter & Falkenberg, 2020), and provide a better understanding of the ultramutational phenotypes of the exonuclease site mutants of human DNA polymerase ε linked with cancer progression (Kane & Shcherbakova, 2014; Li et al., 2018).

4.1 Equipment

- Cooling block to set the temperature of microcentrifuge tubes.
- TLC chambers.
- Air dryer.
- Typhoon FLA 9500 imaging system (GE Healthcare, Chicago, IL).

4.2 Reagents

- Buffer C: 50 mM Tris–HCl, pH 7.5, 40 mM NaCl, 10% glycerol and 2 mM DTT
- Primed minicircle DNA substrate
- Purified T7 replication proteins
- dNTPs
- Formic acid
- PEI Cellulose F TLC sheets (EMD-Millipore, Burlington, MA)
- Phosphor-imaging screens (GE Healthcare, Chicago, IL)
- Potassium phosphate buffer (pH 3.8)

4.3 Procedure

1. Incubate the minicircle DNA substrate (32 nM) with 400 nM T7 helicase, 500 μM dTTP, and 1.5 mM EDTA for 30 min. Add 400 nM T7 DNA polymerase to prepare "Mixture A". Incubate Mixture A for 1 h at 25 °C.
2. Prepare "Mixture B" consisting of 20 μM T7 gene 2.5 protein (gp2.5), 300 μM each of dVTPs, and 10 mM $MgCl_2$. Spike Mixture B with a small amount of radiolabeled nucleotide, α-^{32}P-dGTP. Variations of this experimental strategy can be employed to test the effect of helicase-DNA polymerase uncoupling by limiting the concentration

of dTTP (at 10 μM dVTPs and different concentrations of dTTP ranging from 10 to 500 μM) or by using C-tail deletion mutant of T7 helicase.

3. When starting the leading strand DNA synthesis reactions, transfer the tubes to a microcentrifuge tube cooler set at 18 °C and allow the mixtures to arrive at the temperature.
4. Initiate the leading strand DNA synthesis reaction by mixing equal volumes of Mixtures "A" and "B." Mix the reaction thoroughly with the help of a pipette.
5. Take out 5 μL of the reaction at the stated time-points and add 5 μL of 8 M formic acid to quench DNA synthesis.
6. To prepare reactions for background correction, 5 μL of formic acid was added to 2.5 μL Mixture A before adding 2.5 μL Mixture B.
7. Keep the quenched reactions on ice or store them at −80 °C for later use.
8. Lightly mark a line with a pencil 2 cm from one of the sides of the TLC sheet. With the help of a pipette, spot 1 μL of the quenched reaction on the line, leaving about 1.5 cm space between subsequent spots.
9. Let the reaction spots completely dry at room temperature. Fill the phosphate buffer in the TLC chamber. Place the TLC sheet inside the chamber so that the side with spots directly contacts the buffer. Keep the buffer low enough not to touch the spots directly and run through them when the buffer rises.
10. Let the buffer reach the TLC sheet's top edge (20 cm).
11. Dry the TLC sheets with an air blow dryer and wrap the sheets with a layer of Saran wrap.
12. Expose the blanked phosphor-imaging screens with the wrapped TLC sheets for 2–8 h.
13. Scan the exposed phosphor-imaging screens using Typhoon FLA 9500 imaging system.
14. Dispose of the TLC sheets, reactions, and the pipette tips used for handling radioactive material according to the radioactivity safety guidelines.

4.4 Analysis

1. Quantitate the spot intensities for dGMP incorporated in the synthesized DNA, unused dGTP, dGDP, and dGMP excised (I_{inc}, I_{dGTP}, I_{dGDP} and I_{ex}, respectively) for reactions quenched at different time-points using ImageQuant TL image analysis software (GE Healthcare, Chicago, IL). Correct for the background by subtracting background spot intensities.

2. Calculate the fractions of dGMP incorporated (F_s) and excised (F_e) as:

$$F_s = I_{\text{inc}}/(I_{\text{inc}} + I_{\text{dGTP}} + I_{\text{dGDP}} + I_{\text{ex}})$$
$$F_e = I_{ex}/(I_{inc} + I_{dGTP} + I_{dGDP} + I_{ex}) \quad (6)$$

3. Amounts of dNMP incorporated and excised can be quantified from the factions of labeled dGMP incorporated and excised during leading strand DNA synthesis and are expressed as moles of nucleotides incorporated or excised per mole of the replisome complex.
4. Plot the data obtained from step 3 against time and fit it to a line to obtain slopes (nucleotides incorporated or excised per mole of the replisome complex per second). We assume the amount of replisome complex to be the same as the amount of minicircle substrate DNA used in the assay.
5. Determine the Pol/Exo ratio using the obtained slopes.

4.5 Notes

1. DNA minicircle can be synthesized by ligating two ends of a linear ssDNA using T4 DNA ligase. A single-stranded splint DNA stabilizes the circular conformation bringing the two ends of the ssDNA together to facilitate the ligation reaction. Additionally, a DNA secondary structure strategically positioned in the middle of the linear DNA also helps in increasing the yields of circularized DNA. The 70-nt minicircle is synthesized as described (An et al., 2017). The initial reaction contained 2 μM 5' phosphorylated linear 70 nt ssDNA, 25 μM splint DNA (see Table 1 for DNA sequences), 10 U of T4 DNA ligase and 0.05X T4 DNA ligase buffer (Thermo Scientific, PA, USA) in a 20 μL reaction. Linear DNA is added to the reaction intermittently in small batches to final concentration of 10 μM as described (An et al., 2017). The reaction is incubated at 20 °C for 12 h. Circularized DNA is separated from remaining linear DNA and higher-order linear polymers on a 15% urea TBE gel and is electroeluted from the gel to obtain pure DNA minicircle. Using a lower concentration of T4 DNA ligase buffer and the gradual addition of linear DNA in the reaction reduces the formation of higher-order linear products.
2. Use caution while using radioactive nucleotides and follow the institutional guidelines for the usage and disposal of radioactive material.

3. Preincubation time and temperature for preassembly of the leading strand complex should be optimized for replication proteins from other systems.
4. It is essential to keep the temperature constant during the reaction.
5. We considered replacing the radioactive dNTP with a fluorescent dNTP analog. However, replicative DNA polymerases do not incorporate fluorescent dNTP analogs with the same kinetics as normal dNTPs. Hence, results with dNTP analogs are unreliable for testing physiologically relevant models.

Acknowledgments

This work was supported by NIH grant R35 GM118086 to S.S.P.

References

Abdelhaleem, M. (2010). Helicases: An overview. *Methods in Molecular Biology*, *587*, 1–12. https://doi.org/10.1007/978-1-60327-355-8_1.

Ahnert, P., & Patel, S. S. (1997). Asymmetric interactions of hexameric bacteriophage T7 DNA helicase with the 5'- and 3'-tails of the forked DNA substrate. *The Journal of Biological Chemistry*, *272*(51), 32267–32273. https://doi.org/10.1074/jbc.272.51.32267.

An, R., Li, Q., Fan, Y., Li, J., Pan, X., Komiyama, M., et al. (2017). Highly efficient preparation of single-stranded DNA rings by T4 ligase at abnormally low Mg(II) concentration. *Nucleic Acids Research*, *45*(15), e139. https://doi.org/10.1093/nar/gkx553.

Beese, L. S., Derbyshire, V., & Steitz, T. A. (1993). Structure of DNA polymerase I Klenow fragment bound to duplex DNA. *Science*, *260*(5106), 352–355. https://doi.org/10.1126/science.8469987.

Benkovic, S. J., & Spiering, M. M. (2017). Understanding DNA replication by the bacteriophage T4 replisome. *The Journal of Biological Chemistry*, *292*(45), 18434–18442. https://doi.org/10.1074/jbc.R117.811208.

Benkovic, S. J., Valentine, A. M., & Salinas, F. (2001). Replisome-mediated DNA replication. *Annual Review of Biochemistry*, *70*, 181–208. https://doi.org/10.1146/annurev.biochem.70.1.181.

Bourgeois, C. F., Mortreux, F., & Auboeuf, D. (2016). The multiple functions of RNA helicases as drivers and regulators of gene expression. *Nature Reviews. Molecular Cell Biology*, *17*(7), 426–438. https://doi.org/10.1038/nrm.2016.50.

Burgers, P. M. J., & Kunkel, T. A. (2017). Eukaryotic DNA replication fork. *Annual Review of Biochemistry*, *86*, 417–438. https://doi.org/10.1146/annurev-biochem-061516-044709.

Chan, S. S., & Copeland, W. C. (2009). DNA polymerase gamma and mitochondrial disease: Understanding the consequence of POLG mutations. *Biochimica et Biophysica Acta*, *1787*(5), 312–319. https://doi.org/10.1016/j.bbabio.2008.10.007.

Chib, S., Byrd, A. K., & Raney, K. D. (2016). Yeast helicase Pif1 unwinds RNA:DNA hybrids with higher processivity than DNA:DNA duplexes. *The Journal of Biological Chemistry*, *291*(11), 5889–5901. https://doi.org/10.1074/jbc.M115.688648.

Delagoutte, E., & von Hippel, P. H. (2002). Helicase mechanisms and the coupling of helicases within macromolecular machines. Part I: Structures and properties of isolated helicases. *Quarterly Reviews of Biophysics*, *35*(4), 431–478. https://doi.org/10.1017/s0033583502003852.

Delagoutte, E., & von Hippel, P. H. (2003). Helicase mechanisms and the coupling of helicases within macromolecular machines. Part II: Integration of helicases into cellular processes. *Quarterly Reviews of Biophysics*, *36*(1), 1–69. https://doi.org/10.1017/s0033583502003864.

Donlin, M. J., Patel, S. S., & Johnson, K. A. (1991). Kinetic partitioning between the exonuclease and polymerase sites in DNA error correction. *Biochemistry*, *30*(2), 538–546. https://doi.org/10.1021/bi00216a031.

Donmez, I., & Patel, S. S. (2006). Mechanisms of a ring shaped helicase. *Nucleic Acids Research*, *34*(15), 4216–4224. https://doi.org/10.1093/nar/gkl508.

Doublie, S., Tabor, S., Long, A. M., Richardson, C. C., & Ellenberger, T. (1998). Crystal structure of a bacteriophage T7 DNA replication complex at 2.2 A resolution. *Nature*, *391*(6664), 251–258. https://doi.org/10.1038/34593.

Duderstadt, K. E., Geertsema, H. J., Stratmann, S. A., Punter, C. M., Kulczyk, A. W., Richardson, C. C., et al. (2016). Simultaneous real-time imaging of leading and lagging strand synthesis reveals the coordination dynamics of single replisomes. *Molecular Cell*, *64*(6), 1035–1047. https://doi.org/10.1016/j.molcel.2016.10.028.

Fersht, A. R., Knill-Jones, J. W., & Tsui, W. C. (1982). Kinetic basis of spontaneous mutation. Misinsertion frequencies, proofreading specificities and cost of proofreading by DNA polymerases of Escherichia coli. *Journal of Molecular Biology*, *156*(1), 37–51. https://doi.org/10.1016/0022-2836(82)90457-0.

Foster, B. M., Rosenberg, D., Salvo, H., Stephens, K. L., Bintz, B. J., Hammel, M., et al. (2019). Combined solution and crystal methods reveal the electrostatic tethers that provide a flexible platform for replication activities in the bacteriophage T7 replisome. *Biochemistry*, *58*(45), 4466–4479. https://doi.org/10.1021/acs.biochem.9b00525.

Gao, Y., Cui, Y., Fox, T., Lin, S., Wang, H., de Val, N., et al. (2019). Structures and operating principles of the replisome. *Science*, *363*(6429). https://doi.org/10.1126/science.aav7003.

Georgescu, R. E., Langston, L., Yao, N. Y., Yurieva, O., Zhang, D., Finkelstein, J., et al. (2014). Mechanism of asymmetric polymerase assembly at the eukaryotic replication fork. *Nature Structural & Molecular Biology*, *21*(8), 664–670. https://doi.org/10.1038/nsmb.2851.

Hamdan, S. M., Johnson, D. E., Tanner, N. A., Lee, J. B., Qimron, U., Tabor, S., et al. (2007). Dynamic DNA helicase-DNA polymerase interactions assure processive replication fork movement. *Molecular Cell*, *27*(4), 539–549. https://doi.org/10.1016/j.molcel.2007.06.020.

Hamdan, S. M., & Richardson, C. C. (2009). Motors, switches, and contacts in the replisome. *Annual Review of Biochemistry*, *78*, 205–243. https://doi.org/10.1146/annurev.biochem.78.072407.103248.

Jean, J. M., & Hall, K. B. (2001). 2-Aminopurine fluorescence quenching and lifetimes: Role of base stacking. *Proceedings of the National Academy of Sciences of the United States of America*, *98*(1), 37–41. https://doi.org/10.1073/pnas.011442198.

Johnson, K. A. (2009). Fitting enzyme kinetic data with KinTek global kinetic explorer. *Methods in Enzymology*, *467*, 601–626. https://doi.org/10.1016/S0076-6879(09)67023-3.

Johnson, D. S., Bai, L., Smith, B. Y., Patel, S. S., & Wang, M. D. (2007). Single-molecule studies reveal dynamics of DNA unwinding by the ring-shaped T7 helicase. *Cell*, *129*(7), 1299–1309. https://doi.org/10.1016/j.cell.2007.04.038.

Johnson, K. A., Simpson, Z. B., & Blom, T. (2009). Global kinetic explorer: A new computer program for dynamic simulation and fitting of kinetic data. *Analytical Biochemistry*, *387*(1), 20–29. https://doi.org/10.1016/j.ab.2008.12.024.

Jones, A. C., & Neely, R. K. (2015). 2-Aminopurine as a fluorescent probe of DNA conformation and the DNA-enzyme interface. *Quarterly Reviews of Biophysics*, *48*(2), 244–279. https://doi.org/10.1017/S0033583514000158.

Jose, D., Datta, K., Johnson, N. P., & von Hippel, P. H. (2009). Spectroscopic studies of position-specific DNA "breathing" fluctuations at replication forks and primer-template junctions. *Proceedings of the National Academy of Sciences of the United States of America, 106*(11), 4231–4236. https://doi.org/10.1073/pnas.0900803106.

Kane, D. P., & Shcherbakova, P. V. (2014). A common cancer-associated DNA polymerase epsilon mutation causes an exceptionally strong mutator phenotype, indicating fidelity defects distinct from loss of proofreading. *Cancer Research, 74*(7), 1895–1901. https://doi.org/10.1158/0008-5472.CAN-13-2892.

Kim, S., Dallmann, H. G., McHenry, C. S., & Marians, K. J. (1996). Coupling of a replicative polymerase and helicase: A tau-DnaB interaction mediates rapid replication fork movement. *Cell, 84*(4), 643–650. https://doi.org/10.1016/s0092-8674(00)81039-9.

Kose, H. B., Xie, S., Cameron, G., Strycharska, M. S., & Yardimci, H. (2020). Duplex DNA engagement and RPA oppositely regulate the DNA-unwinding rate of CMG helicase. *Nature Communications, 11*(1), 3713. https://doi.org/10.1038/s41467-020-17443-7.

Kunkel, T. A. (2004). DNA replication fidelity. *The Journal of Biological Chemistry, 279*(17), 16895–16898. https://doi.org/10.1074/jbc.R400006200.

Kunkel, T. A., & Bebenek, K. (2000). DNA replication fidelity. *Annual Review of Biochemistry, 69*, 497–529. https://doi.org/10.1146/annurev.biochem.69.1.497.

Larsen, O. F. A., van Stokkum, I. H. M., de Weerd, F. L., Vengris, M., Aravindakumar, C. T., van Grondelle, R., et al. (2004). Ultrafast transient-absorption and steady-state fluorescence measurements on 2-aminopurine substituted dinucleotides and 2-aminopurine substituted DNA duplexes. *Physical Chemistry Chemical Physics, 6*(1), 154–160. https://doi.org/10.1039/b308992d.

Law, S. M., Eritja, R., Goodman, M. F., & Breslauer, K. J. (1996). Spectroscopic and calorimetric characterizations of DNA duplexes containing 2-aminopurine. *Biochemistry, 35*(38), 12329–12337. https://doi.org/10.1021/bi9614545.

Levin, M. K., Hingorani, M. M., Holmes, R. M., Patel, S. S., & Carson, J. H. (2009). Model-based global analysis of heterogeneous experimental data using gfit. *Methods in Molecular Biology, 500*, 335–359. https://doi.org/10.1007/978-1-59745-525-1_12.

Li, H. D., Cuevas, I., Zhang, M., Lu, C., Alam, M. M., Fu, Y. X., et al. (2018). Polymerase-mediated ultramutagenesis in mice produces diverse cancers with high mutational load. *The Journal of Clinical Investigation, 128*(9), 4179–4191. https://doi.org/10.1172/JCI122095.

Li, H., & O'Donnell, M. E. (2019). DNA replication from two different worlds. *Science, 363*(6429), 814–815. https://doi.org/10.1126/science.aaw6265.

Lieberman, K. R., Cherf, G. M., Doody, M. J., Olasagasti, F., Kolodji, Y., & Akeson, M. (2010). Processive replication of single DNA molecules in a nanopore catalyzed by phi29 DNA polymerase. *Journal of the American Chemical Society, 132*(50), 17961–17972. https://doi.org/10.1021/ja1087612.

Lieberman, K. R., Dahl, J. M., & Wang, H. (2014). Kinetic mechanism at the branchpoint between the DNA synthesis and editing pathways in individual DNA polymerase complexes. *Journal of the American Chemical Society, 136*(19), 7117–7131. https://doi.org/10.1021/ja5026408.

Lionnet, T., Spiering, M. M., Benkovic, S. J., Bensimon, D., & Croquette, V. (2007). Real-time observation of bacteriophage T4 gp41 helicase reveals an unwinding mechanism. *Proceedings of the National Academy of Sciences of the United States of America, 104*(50), 19790–19795. https://doi.org/10.1073/pnas.0709793104.

Manosas, M., Spiering, M. M., Ding, F., Bensimon, D., Allemand, J. F., Benkovic, S. J., et al. (2012). Mechanism of strand displacement synthesis by DNA replicative polymerases. *Nucleic Acids Research, 40*(13), 6174–6186. https://doi.org/10.1093/nar/gks253.

Manosas, M., Spiering, M. M., Ding, F., Croquette, V., & Benkovic, S. J. (2012). Collaborative coupling between polymerase and helicase for leading-strand synthesis. *Nucleic Acids Research, 40*(13), 6187–6198. https://doi.org/10.1093/nar/gks254.

Manosas, M., Xi, X. G., Bensimon, D., & Croquette, V. (2010). Active and passive mechanisms of helicases. *Nucleic Acids Research*, *38*(16), 5518–5526. https://doi.org/10.1093/nar/gkq273.

Nandakumar, D., Pandey, M., & Patel, S. S. (2015). Cooperative base pair melting by helicase and polymerase positioned one nucleotide from each other. *eLife*, *4*. https://doi.org/10.7554/eLife.06562.

Nandakumar, D., & Patel, S. S. (2016). Methods to study the coupling between replicative helicase and leading-strand DNA polymerase at the replication fork. *Methods*, *108*, 65–78. https://doi.org/10.1016/j.ymeth.2016.05.003.

O'Donnell, M. E., & Li, H. (2018). The ring-shaped hexameric helicases that function at DNA replication forks. *Nature Structural & Molecular Biology*, *25*(2), 122–130. https://doi.org/10.1038/s41594-018-0024-x.

Ogawa, T., & Okazaki, T. (1980). Discontinuous DNA replication. *Annual Review of Biochemistry*, *49*, 421–457. https://doi.org/10.1146/annurev.bi.49.070180.002225.

Okazaki, R., Okazaki, T., Sakabe, K., Sugimoto, K., & Sugino, A. (1968). Mechanism of DNA chain growth. I. Possible discontinuity and unusual secondary structure of newly synthesized chains. *Proceedings of the National Academy of Sciences of the United States of America*, *59*(2), 598–605. https://doi.org/10.1073/pnas.59.2.598.

Ozes, A. R., Feoktistova, K., Avanzino, B. C., Baldwin, E. P., & Fraser, C. S. (2014). Real-time fluorescence assays to monitor duplex unwinding and ATPase activities of helicases. *Nature Protocols*, *9*(7), 1645–1661. https://doi.org/10.1038/nprot.2014.112.

Pandey, M., Levin, M. K., & Patel, S. S. (2010). Experimental and computational analysis of DNA unwinding and polymerization kinetics. *Methods in Molecular Biology*, *587*, 57–83. https://doi.org/10.1007/978-1-60327-355-8_5.

Pandey, M., & Patel, S. S. (2014). Helicase and polymerase move together close to the fork junction and copy DNA in one-nucleotide steps. *Cell Reports*, *6*(6), 1129–1138. https://doi.org/10.1016/j.celrep.2014.02.025.

Pandey, M., Syed, S., Donmez, I., Patel, G., Ha, T., & Patel, S. S. (2009). Coordinating DNA replication by means of priming loop and differential synthesis rate. *Nature*, *462*(7275), 940–943. https://doi.org/10.1038/nature08611.

Patel, S. S., Pandey, M., & Nandakumar, D. (2011). Dynamic coupling between the motors of DNA replication: hexameric helicase, DNA polymerase, and primase. *Current Opinion in Chemical Biology*, *15*(5), 595–605. https://doi.org/10.1016/j.cbpa.2011.08.003.

Patel, S. S., & Picha, K. M. (2000). Structure and function of hexameric helicases. *Annual Review of Biochemistry*, *69*, 651–697. https://doi.org/10.1146/annurev.biochem.69.1.651.

Peter, B., & Falkenberg, M. (2020). TWINKLE and other human mitochondrial DNA helicases: Structure, function and disease. *Genes (Basel)*, *11*(4). https://doi.org/10.3390/genes11040408.

Picha, K. M., & Patel, S. S. (1998). Bacteriophage T7 DNA helicase binds dTTP, forms hexamers, and binds DNA in the absence of Mg2+. The presence of dTTP is sufficient for hexamer formation and DNA binding. *The Journal of Biological Chemistry*, *273*(42), 27315–27319. https://doi.org/10.1074/jbc.273.42.27315.

Reha-Krantz, L. J. (2009). The use of 2-aminopurine fluorescence to study DNA polymerase function. *Methods in Molecular Biology*, *521*, 381–396. https://doi.org/10.1007/978-1-60327-815-7_21.

Scheuermann, R. H., & Echols, H. (1984). A separate editing exonuclease for DNA replication: The epsilon subunit of Escherichia coli DNA polymerase III holoenzyme. *Proceedings of the National Academy of Sciences of the United States of America*, *81*(24), 7747–7751. https://doi.org/10.1073/pnas.81.24.7747.

Schlierf, M., Wang, G., Chen, X. S., & Ha, T. (2019). Hexameric helicase G40P unwinds DNA in single base pair steps. *eLife*, *8*. https://doi.org/10.7554/eLife.42001.

Singh, A., Pandey, M., Nandakumar, D., Raney, K. D., Yin, Y. W., & Patel, S. S. (2020). Excessive excision of correct nucleotides during DNA synthesis explained by replication hurdles. *The EMBO Journal*, e103367. https://doi.org/10.15252/embj.2019103367.

Stano, N. M., Jeong, Y. J., Donmez, I., Tummalapalli, P., Levin, M. K., & Patel, S. S. (2005). DNA synthesis provides the driving force to accelerate DNA unwinding by a helicase. *Nature*, *435*(7040), 370–373. https://doi.org/10.1038/nature03615.

Sun, B., Johnson, D. S., Patel, G., Smith, B. Y., Pandey, M., Patel, S. S., et al. (2011). ATP-induced helicase slippage reveals highly coordinated subunits. *Nature*, *478*(7367), 132–135. https://doi.org/10.1038/nature10409.

Sun, B., Pandey, M., Inman, J. T., Yang, Y., Kashlev, M., Patel, S. S., et al. (2015). T7 replisome directly overcomes DNA damage. *Nature Communications*, *6*, 10260. https://doi.org/10.1038/ncomms10260.

Sun, B., Singh, A., Sultana, S., Inman, J. T., Patel, S. S., & Wang, M. D. (2018). Helicase promotes replication re-initiation from an RNA transcript. *Nature Communications*, *9*(1), 2306. https://doi.org/10.1038/s41467-018-04702-x.

Ward, D. C., Reich, E., & Stryer, L. (1969). Fluorescence studies of nucleotides and polynucleotides. I. Formycin, 2-aminopurine riboside, 2,6-diaminopurine riboside, and their derivatives. *The Journal of Biological Chemistry*, *244*(5), 1228–1237.

Yang, W., Seidman, M. M., Rupp, W. D., & Gao, Y. (2019). Replisome structure suggests mechanism for continuous fork progression and post-replication repair. *DNA Repair (Amst)*, *81*, 102658. https://doi.org/10.1016/j.dnarep.2019.102658.

Yuan, Q., & McHenry, C. S. (2009). Strand displacement by DNA polymerase III occurs through a tau-psi-chi link to single-stranded DNA-binding protein coating the lagging strand template. *The Journal of Biological Chemistry*, *284*(46), 31672–31679. https://doi.org/10.1074/jbc.M109.050740.

CHAPTER FIVE

Assembling bacteriophage T7 leading-strand replisome for structural investigation

Chen-Yu Lo and Yang Gao*
Department of BioSciences, Rice University, Houston, Texas, United States
*Corresponding author: e-mail address: yg60@rice.edu

Contents

Abstract

Replicative helicase and polymerase form the leading-strand replisome that unwinds parental DNA and performs continuous leading-strand DNA synthesis. Uncoupling of the helicase-polymerase complex results in replication stress, replication errors, and genome instability. Although numerous replisomes from different biological systems have been reconstituted and characterized, structural investigations of the leading-strand replisome complex are hindered by its large size and dynamics. We have determined the first replisome structure on a fork substrate with bacteriophage T7 replisome as a model system. Here, we summarized our protocols to prepare and characterize the coupled T7 replisome complex. Similar methods can potentially be applied for structural investigations of more complicated replisomes.

Methods in Enzymology, Volume 672
ISSN 0076-6879
https://doi.org/10.1016/bs.mie.2022.03.009

1. Introduction

The accurate and efficient replication of genetic DNA is fundamental to all living systems (O'Donnell, Langston, & Stillman, 2013). DNA polymerases are replicases that perform template-dependent DNA synthesis on single-stranded DNA (ssDNA) (Rothwell & Waksman, 2005). Most DNA polymerases have limited strand-displacement synthesis activity, which limit their replications on the parental dsDNA (Singh et al., 2020). DNA helicases are motor proteins that couple the ATP hydrolysis to parental dsDNA unwinding (Gao & Yang, 2020; O'Donnell & Li, 2018; Singleton, Dillingham, & Wigley, 2007). In archaeal and eukaryotic cells, the helicase migrates on the leading strand in front of polymerase. In contrast, bacterial replicative helicases have opposite polarities and translocate on the lagging-strand DNA. Although with different architectures, the helicase and the leading-strand polymerase work synergistically in all systems (Graham, Marians, & Kowalczykowski, 2017; Langston et al., 2014; Manosas, Spiering, Ding, Croquette, & Benkovic, 2012; Nandakumar, Pandey, & Patel, 2015). Together, the two constitute the leading-strand replisome that powers fork progression. Moreover, there are many roadblocks on genomic DNA due to DNA damages, secondary structures, and tightly bound proteins. Decoupling of the leading-strand replisome in front of roadblocks will trigger replication stress responses, inducing disease-related replication errors and genome rearrangements (Cortez, 2019; Gao et al., 2019).

The mechanistic basis of helicase and polymerase coupling in bacterial and phage systems has been extensively studied (Graham et al., 2017; Manosas et al., 2012; Manosas, Spiering, Ding, Croquette, & Benkovic, 2012; Nandakumar et al., 2015). Both the helicase and polymerase can translocate along ssDNA. When they work together, they stimulate the activity and processivity of each other. Moreover, kinetic and single-molecule studies suggested that the DNA unwinding transit from passive to active when helicase and polymerase are coupled (Manosas, Spiering, Ding, Bensimon, et al., 2012; Manosas, Xi, Bensimon, & Croquette, 2010; Morin et al., 2015; Nandakumar et al., 2015). DNA footprinting experiments suggested that the helicase and polymerase are located closely at the fork (Pandey & Patel, 2014). The cooperative binding of the polymerase and helicase can induce local duplex melting (Nandakumar et al., 2015). Direct or indirect physical interactions between the helicase and polymerase hold the two together during processive DNA unwinding (Hamdan et al., 2007;

Kim, Dallmann, McHenry, & Marians, 1996; Lo & Gao, 2021). Although biochemical and biophysical studies provided rich information about helicase-polymerase coupling, there is only limited success in structural investigations of the coupled helicase-polymerase complex at the replication fork (Gao et al., 2019; Santos, 2021; Yuan, Georgescu, Schauer, O'Donnell, & Li, 2020), which is possibly due to the structural dynamics of the helicase and polymerase during their action, the flexible interaction between helicase and polymerase, and difficulties in designing and testing proper fork DNA substrate to capture helicase and polymerase simultaneously.

Due to its simplicity, the T7 replisome has been used extensively as a model system to understand DNA replication mechanisms (Hamdan & Richardson, 2009). The full-length T7 gp4 harbors primase activities on its N-terminal domain and helicase activity on its C-terminal domain, which will assemble into a homohexameric ring (Singleton, Sawaya, Ellenberger, & Wigley, 2000). T7 gp5 polymerase and host thioredoxin (trx) form a holo-replicase to catalyze processive DNA replication (Doublie, Tabor, Long, Richardson, & Ellenberger, 1998). The gp4 on the lagging strand and the gp5-trx on the leading strand work synergistically (Lo & Gao, 2021; Nandakumar et al., 2015). The two enzymes are sufficient to catalyze processive DNA replication with products up to 20kb long (Lee, Chastain 2nd, Kusakabe, Griffith, & Richardson, 1998). We have recently captured the first structure of the T7 leading-strand replisome with cryo-electron microscope (cryo-EM) (Gao et al., 2019). The structure illustrated fundamental principles for helicase-polymerase coupling in DNA replication. This review describes the experimental details of assembling a coupled T7 leading-strand replisome complex for structural investigations. Similar methods can potentially be applied to structural studies of other replication systems.

2. Major considerations

2.1 Fork DNA substrate design

A simplified replication fork contains parental dsDNA, leading-strand dsDNA with possible ssDNA gap, and lagging-strand ssDNA (Fig. 1; Table 1). Previous DNA protection assays suggested that the T7 replisome interacts with over 30 base pairs of dsDNA region near the fork (Pandey & Patel, 2014). T7 gp5-trx can interact with ~25 base pairs of dsDNA on the leading strand, as suggested by both kinetic and structural studies (Brieba et al., 2004; Patel, Wong, & Johnson, 1991). The ssDNA gap between the primer end and the fork opening can be a significant variable for

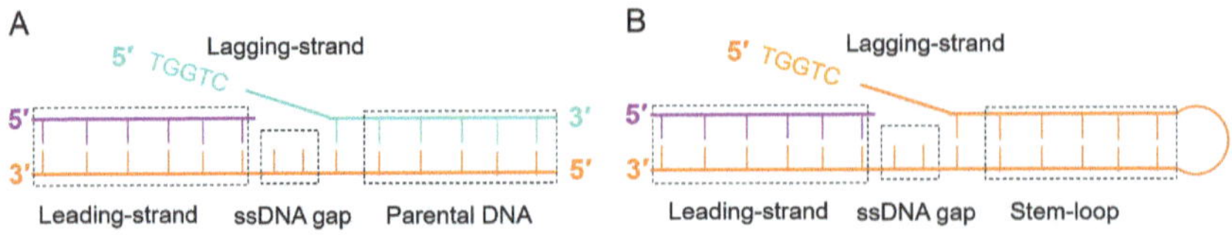

Fig. 1 Fork DNA substrate design. (A) A scheme of the fork DNA substrate. (B) A scheme of the fork DNA substrate with a hairpin structure on the parental dsDNA.

Table 1 DNA substrates.

DNA Oligoes	Sequence
Primer1 (Fig. 1A, B; Fig. 5)	5′-CTA GTA ACA GTG TGA ACT TGA CGA CAT AGC G $_{dd}$A – 3′
Leading1 (Fig. 1A; Fig. 4)	5′-TAC AAG CAA TCG ATC ACA GCA CTA GCG TTA TCG CTA TGT CGT CAA GTT CAC ACT GTT ACT AG-3′
Lagging1 (Fig. 1A; Fig. 4)	5′-TGG TCT TTT TTT TTT TTT TTT TTT TTT TTC GCT AGT GCT GTG ATC GAT TGC TTG TA-3′
Leading2 (Fig. 1B)	5′-TGG TCT TTT TTT TTT TTT TTT TTT TTT TTC GGA GTC GTT TCG ACT CCG ATT TCG CTA TGT CGT CAA GTT CAC ACT GTT ACT AG-3′
Primer2 (Fig. 4)	5′-/Cy5/CTA GTA ACA GTG TGA ACT TGA CGA CAT AGC G$_{dd}$A-3′
Lagging2 (Fig. 5)	5′-TTT TTT TTT TTT TTT TTT TTT TTT TTT TTT G/2AP/T CGC TAG TGC TGT GAT CGA TTG CTT GTA-3′
Leading3 (Fig. 5, coupled)	5′-TAC AAG CAA TCG ATC ACA GCA CTA GCG ATC ATC GCT ATG TCG TCA AGT TCA CAC TGT TAC TAG-3′
Primer3 (Fig. 5, uncoupled)	5′-TAC AAG CAA TCG ATC ACA GCA CTA GCG ATC TTA TCG CTA TGT CGT CAA GTT CAC ACT GTT ACT AG-3′

DNA design. Biochemical studies suggested that one nucleotide difference in the ssDNA gap can alter helicase-polymerase coupling (Nandakumar et al., 2015). Interestingly, the cryo-EM structure of gp4-gp5 complex assembled on a fork DNA substrate with 10 nucleotides ssDNA gap failed to capture DNA binding (Kulczyk, Moeller, Meyer, Sliz, & Richardson, 2017). Moreover, gp5 was located on the primase side of gp4 in the complex, far from the fork junction. Previous crystal structures of gp5-trx-DNA complex suggested that two nucleotides ssDNA are bound to gp5, one serving as the template and the other being held by protein sidechains (Doublie et al., 1998). Cooperative binding of gp4 and gp5 to the fork substrate can induce local melting of parental dsDNA, providing the ssDNA gaps required for gp5 interaction. From our tests, 25–35 bp dsDNA and 0–2 nt ssDNA gap on the leading strand can support proper gp5 binding and interaction with gp4. Bacterial or phage helicase is on the lagging strand. The helicase domain of bacterial DnaB or T7 gp4 occupies 12 nucleotides ssDNA, with each subunit interacting with 2 nucleotides (Gao & Yang, 2020; Itsathitphaisarn, Wing, Eliason, Wang, & Steitz, 2012). In addition, the N-terminal domain of gp4 encodes a primase that also interacts with ssDNA and utilizes "3′-CTG" site for priming (Lee, Zhu, Hamdan, & Richardson, 2010). Biochemical studies suggested that ~30 nucleotides of ssDNA will provide optimal gp4 binding (Pandey & Patel, 2014). Thus, ~30 nt ssDNA with "3′-CTGGT" at the 5′-end of the lagging strand was used. Furthermore, 8–20 bp parental dsDNA can support the proper assembling of the gp4-gp5 complex. A hairpin stem-loop on the parental dsDNA can minimize the number of DNA substrates and help the annealing of the short parental dsDNA. DNA substrates should be designed and experimentally tested for proper replisome complex formation.

2.2 Strategies to pause replisome progression

Both helicase and polymerase are motor proteins that can migrate on DNA. These motors need to be paused to capture a homogenous complex for structural studies. The DNA polymerase can be stopped by a dideoxy primer end, non-hydrolyzable incoming dNTP, inhibitory metal ions, or active site mutations. The helicase can be captured by non-hydrolyzable NTP analogs or active site mutations. In addition, protein roadblocks on parental dsDNA have also been used to halt helicase translocation (Eickhoff et al., 2019; Georgescu et al., 2017). However, a bulky protein on the parental DNA will be located between helicase and polymerase in bacterial and phage replisomes and thus may sterically affect the helicase-polymerase interaction. Moreover, the helicase hydrolyzes the phosphodiester bond between β- and γ-phosphate groups, whereas the polymerase cleaves the α-β linkage. Under *in vitro* conditions, T7 gp4 prefers dTTP over ATP but can still hydrolyze other types of NTPs and dNTPs (Crampton, Mukherjee, & Richardson, 2006; Satapathy, Kulczyk, Ghosh, van Oijen, & Richardson, 2011).

Thus, it is difficult to find proper non-hydrolyzable analogs for helicase-polymerase complex formation. Previous structural studies of DNA polymerases suggested that the dideoxy primer end can stabilize the polymerase ternary complex formation (Doublie et al., 1998). Moreover, the dideoxynucleotide can be easily incorporated into the primer end by a polymerase at the designed location. A Gp4 helicase active site mutation, E343Q, inhibits hydrolysis while not affecting dTTP or DNA binding (Crampton et al., 2006). Thus, the fork DNA with a dideoxynucleotide primer end on the leading strand and an E343Q gp4 helicase were selected for structural studies of the T7 replisome. In addition, we designed the leading strand nascent template to be dA so that only dTTP is needed for complex assembling.

2.3 Assays for examining the helicase-polymerase complex formation

The helicase or polymerase can bind to the fork DNA substrate individually. In addition, T7 gp4 and gp5 interact with each other in the absence of DNA (Hamdan et al., 2007; Kulczyk et al., 2017; Wallen et al., 2017). However, gp4 and gp5 couple their actions only when both proteins are located right at the replication fork (Nandakumar et al., 2015; Pandey & Patel, 2014). Thus, co-migration of proteins in gel-filtration, Electrophoretic Mobility Shift Assay (EMSA), or sedimentation is not sufficient to guarantee proper complex formation. The cooperative binding of gp4 and gp5 to the fork DNA can be reflected in binding constant changes in the absence or presence of their partners (Lo & Gao, 2022). Moreover, some assays, such as DNA footprinting or fluorescence assays, can simultaneously provide binding and positional information (Nandakumar et al., 2015; Pandey & Patel, 2014). 2-aminopurine (2AP) is a fluorescence nucleotide analog that can report DNA unwinding through a change in fluorescence when a base pair is destabilized. Previous experiments suggested that the cooperative binding of gp4 and gp5 can induce a 2AP signal at the fork junction, while individual gp4 or gp5 cannot. We employed the EMSA supershift assay and the 2AP assay to optimize DNA substrates and examine replisome complex formation.

2.4 Structural methods for visualizing replisome complex

X-ray crystallography and cryo-EM can both yield high-resolution structures of large protein complexes. However, the replisome's large size and highly dynamic nature pose significant challenges for crystallization.

Although a gp4-gp5 complex structure has been determined by crystallography (Wallen et al., 2017), gp4 in the complex exists as a heptamer, which is not competent for DNA binding and translocation, and the DNA substrate was not included in the structure. Cryo-EM can determine structures of molecular complexes in solution without lattice contacts. Furthermore, cryo-EM requires much less sample than crystallography. The large size of the replisome complex and the oligomeric nature of helicase indeed aid the computational alignment and classification of the complex. In addition, multiple conformations can be separated during data processing, thus partially overcoming the flexibility issue in replisome structural determination. One major disadvantage of cryo-EM is that the hydrophobic interactions at the air-water interface during cryo-EM sample freezing may denature proteins and induce complex disassembly. Advanced sample preparation techniques are under rapid development and may further facilitate structural investigations of replisome (D'Imprima et al., 2019; Noble et al., 2018).

3. Experimental protocols

3.1 DNA substrate purification and annealing

Materials and equipment

- Handhold UV lamp (Biorad®)
- Digital Waving Rotators (Thermo Scientific)
- Spectrophotometer (Nanodrop2000c, Thermo Scientific)
- Prep: FreeZone 4.5 L Freeze Dry System (LABCONCO®)
- Oligos (Integrated DNA Technologies)
- 10% Urea polyacrylamide gel (24 g Urea, 12.5 mL of 40% acrylamide [29:1], 166 μL of 30% Ammino Ammonium persulfate, 20 μL TEMED, 5 mL of 10 × TBE buffer, and 20 mL deionized water)
- 2 × Denaturing buffer (80% formamide, 0.025% Bromophenol Blue)
- 1 × TBE buffer (100 mM Tris base, 100 mM Boric acid, 2 mM EDTA)
- Desalting column (Biorad®)
- 1 × TE buffer (10 mM Tris, 1 mM EDTA)
- 10 × Annealing buffer (100 mM Tris, pH 7.5, 500 mM NaCl)
- SYBR Gold nucleic acid stain (Sigma-Aldrich®)

Step-by-step method details

1. Primers, templates, lagging strand ssDNA (Fig. 1), or the corresponding DNA substrates with fluorescence Cy5 label at primer end or 2AP within

the parental DNA were ordered from Integrated DNA Technologies (Table 1.). Gp5-trx can incorporate dideoxy primer end during sample preparation
2. Mix an equal volume of oligo and denaturing buffer for denaturing the oligo at 95 °C for 5 min and cool down the reaction on the ice for 2 min. Load the denatured oligo onto the 10% urea-polyacrylamide gel for electrophoresis in 1 × TBE buffer at 250 voltage for 1 h. Cut off the band containing DNA under the handhold UV lamp (258 nm).
3. The gel with DNA will be smashed and soaked in TE buffer (10 mM Tris, 1 mM EDTA) overnight.
4. The solution will be frozen at −80 °C and attached to the Prep: FreeZone 4.5 L Freeze Dry System (LABCONCO®) for freeze-drying.
5. The DNA pellet will later be re-suspended by 1 mL of deionized water to go through a desalting column (BioRad®) to remove the urea salt.
6. After the template, primer, and the lagging strand ssDNA have been purified separately, the oligos will be mixed at a 1:1:1 ratio for annealing. DNA substrates were annealed by incubating at 95 °C for 3 min and gradually cooling to 20 °C (drop 1 °C per minute) in 1 × annealing buffer (10 mM Tris, pH 7.5, 50 mM NaCl).
7. The quality of the annealed substrate will be confirmed with 6–10% native polyacrylamide gel and SYBR-Gold staining.

3.2 Protein purification

Expression and purification of the gp5-trx complex followed the steps outlined in Fig. 2A. SDS-PAGE gel of a typical run of gp5-trx purification is shown in Fig. 2B.

Materials and equipment

- Incubator/shaker (MaxQ8000, Thermo Fisher)
- Spectrophotometer (Nanodrop2000c, Thermo Scientific)
- Centrifuge (BC Avanti J-E, Beckman Coulter)
- Sonicator (Model FB505, Fisher Brand®)
- Centrifuge (BC Avanti J-E, Beckman Coulter)
- AKTA purifier (Cytiva)
- Spectrophotometer (Nanodrop 2000c, Thermo Scientific)
- Heparin HP column (5 mL, GE Healthcare)
- MonoQ-10/100 GL column (GE Healthcare)
- T7 gp5 (D5A, E7A) in pET15b vector
- TrxA in pRSFDuet vector

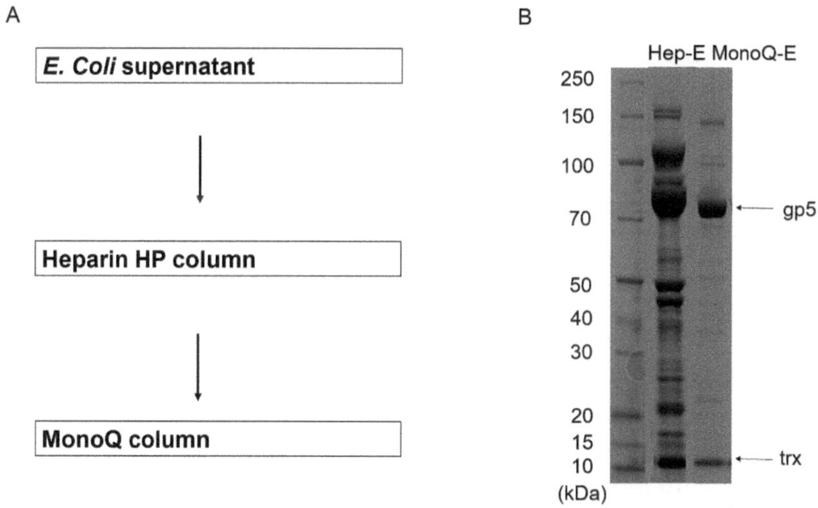

Fig. 2 Purification of gp5-trx. (A) A diagram of purification steps. (B) Coomassie-stained SDS-PAGE gel. The lanes from left to right are MW marker (M), elution from Heparin column (Hep-E), and elution from MonoQ (MonoQ-E).

- *E. coli* BL21 (DE3) competent cells (Novagen)
- LB media
- 100 mg/mL Ampicillin stock, filtered with a 0.22-μm syringe filter (ThermoFisher)
- 100 mg/mL Kanamycin stock, filtered with a 0.22-μm syringe filter (ThermoFisher)
- 0.75 M IPTG stock, filtered with a 0.22-μm syringe filter
- 50 mg/mL DNase I (Sigma)
- Complete™, EDTA-free Protease Inhibitor Cocktail Tablets (Roche)
- 100 mM PMSF stock dissolved in isopropanol
- Hep-A buffer: 25 mM Tris (pH 7.5), 50 mM KCl, 0.1 mM EDTA, and 3 mM Dithiothreitol (DTT)
- Hep-B buffer: 25 mM Tris (pH 7.5), 1 M KCl, 0.1 mM EDTA, and 3 mM DTT
- MonoQ-A buffer: 25 mM Tris (pH 7.5), 50 mM KCl, 0.1 mM EDTA, and 3 mM DTT
- MonoQ-B buffer: 25 mM Tris (pH 7.5), 1 M KCl, 0.1 mM EDTA, and 3 mM DTT
- Stock buffer: 20 mM Tris (pH 7.5), 200 mM KCl, 80% glycerol, and 3 mM DTT

Step-by-step method details

1. Co-transform gp5-pET15b and TrxA-pRSFDuet into *E. coli* BL21 (DE3) competent cells. Spread the cells on the LB agarose plate with 50 μg/mL ampicillin and 25 μg/mL kanamycin at 37 °C for 12 h.
2. A single colony is picked and inoculated in 100 mL LB media with 50 μg/mL ampicillin and 25 μg/mL kanamycin at 37 °C for 8–10 h.
3. Transfer 5 mL of cells to 1 L LB media in 2-L flasks (six in total) each with 50 μg/mL ampicillin and 25 μg/mL kanamycin, and grow at 37 °C to an OD_{600} of 0.6–0.8.
4. Add IPTG to a final concentration of 0.75 mM to co-express the T7 gp5 (D5A E7A) and trxA at 16 °C for 20 h.
5. Spin down cells as pellets by centrifuging at 6000 rpm for 20 min at 4 °C. Freeze cell pellets at −80 °C.
6. Resuspend the cell pellet from 1 L LB media in 40 mL Hep-A buffer. Add one protease inhibitor tablet, 1 mM PMSF, and 50 μg/mL DNase I to mix with the cell suspension.
7. Break cells by Sonicator with the setting (3 s on, 6 s off at 60% amplitude) for 3 min at 4 °C. Cells are spun down at 4 °C by centrifugation at 18,000 rpm for 1 h. The supernatant was filtered using a 0.22-μm syringe filter.
8. Use 50 mL Hep-A buffer to equilibrate a 5 mL Heparin HP column. First, load the filtered-supernatant solution to the Heparin column at 1 mL/min. Later, wash the column with 50 mL Hep-A buffer using a peristaltic pump at the rate of 3 mL/min.
9. Set up the AKTA Purifier for eluting the complex of gp5-trx at the rate of 1.5 mL/min, with the gradient elution of0%–100% Hep-B buffer in 90 mL
10. Combine the protein fractions according to OD_{280} (∼20 mg proteins can be acquired at this procedure). Then, dilute the collected solution to 10-fold with the addition of MonoQ-A buffer.
11. Load the diluted protein onto the MonoQ column that has been equilibrated by 50 mL of MonoQ-A buffer on an AKTA purifier. The column will be washed with 50 mL MonoQ-A buffer. Then, elute the proteins with a 0%–100% MonoQ-B buffer gradient solution in 100 mL at a 1 mL/min flow rate.
12. Mix the protein fractions (usually ∼10–15 mg proteins can be obtained after this step). Concentrate the protein sample to ∼2 mg/mL in a 50 mL Amicon Centrifugal Filter Units (Millipore) and aliquot the protein sample with stock buffer to a final concentration of 50% glycerol to 250 μL per tube. Flash-freeze the samples in liquid nitrogen (LN2) and store them at −80 °C.

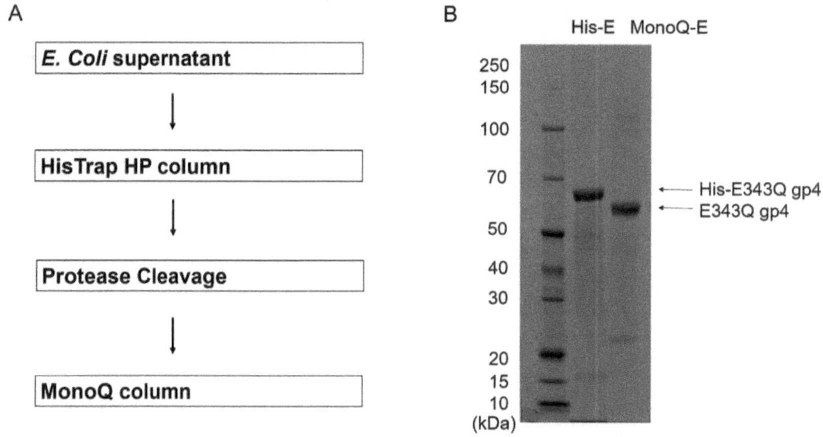

Fig. 3 Purification of E343Q gp4. (A) A diagram of purification steps. (B) Coomassie-stained SDS-PAGE gel. The lanes from left to right are Molecular weight (MW), elution from Histrap column (His-E), and elution from MonoQ (MonoQ-E).

Purification of the gp4 followed the steps outlined in Fig. 3A. SDS-PAGE gel of a typical run of the complex of gp4 purification is shown in Fig. 3B.

Materials and equipment

- Incubator/shaker (MaxQ8000, Thermo Fisher)
- Spectrophotometer (Nanodrop2000c, Thermo Scientific)
- Centrifuge (BC Avanti J-E, Beckman Coulter)
- Sonicator (Model FB505, Fisher Brand®)
- Centrifuge (BC Avanti J-E, Beckman Coulter)
- AKTA Purifier (Cytiva)
- Spectrophotometer (Nanodrop 2000c, Thermo Scientific)
- MonoQ-10/100 GL column (GE Healthcare)
- Clone E343Qgp4 in a modified pET28a vector, with N-terminal histidine tag and preScission protease cleavage site
- C3013I cells derived from T7 Express lysY/Iq Competent *E. coli* (Novagen)
- LB media
- 100 mg/mL Kanamycin stock, filtered with a 0.22-μm syringe filter (ThermoFisher)
- 0.75 M IPTG stock, filtered with a 0.22-μm syringe filter
- 50 mg/mL DNase I
- Complete™, EDTA-free Protease Inhibitor Cocktail Tablets (Roche)
- 100 mM PMSF stock dissolved in isopropanol
- Ni-A buffer: 25 mM Tris (pH7.5), 1 M NaCl, 20 mM Imidazole, 0.1 mM EDTA, and 5 mM β-mercaptoethanol (β-ME)

- Ni—B buffer: 25 mM Tris (pH 7.5), 1 M NaCl, 50 mM Imidazole, 0.1 mM EDTA, and 5 mM β-ME
- Ni—C buffer: 25 mM Tris (pH 7.5), 1 M NaCl, 300 mM Imidazole, 0.1 mM EDTA, and 3 mM DTT
- MonoQ-A buffer: 25 mM Tris (pH 7.5), 50 mM KCl, 0.1 mM EDTA, and 3 mM DTT
- MonoQ-B buffer: 25 mM Tris (pH 7.5), 1 M KCl, 0.1 mM EDTA, and 3 mM DTT
- Stock buffer: 20 mM Tris (pH 7.5), 200 mM KCl, 80% glycerol, and 3 mM DTT

Step-by-step method details

1. Transform E343Qgp4-pET28 into C3013I cells. Spread the cells on the LB agarose plate with 50 μg/mL kanamycin at 37 °C for 12 h.
2. A single colony is picked and inoculated in 100 mL LB media with 50 μg/mL kanamycin at 37 °C for 8–10 h.
3. Transfer 5 mL of cells to 1 L LB media in 2-L flasks (six in total) each with 50 μg/mL kanamycin, and grow at 37 °C to an OD_{600} of 0.6–0.8.
4. Add IPTG to a final concentration of 0.75 mM to express the E343Qgp4 at 16 °C for 20 h.
5. Spin down cells as pellets by centrifuging at 6000 rpm for 20 min at 4 °C. Freeze cell pellets at −80 °C.
6. Resuspend the cell pellet from 1 L LB media in 40 mL Ni-A buffer. Add one protease inhibitor tablet, 1 mM PMSF, and 50 μg/mL DNase I to mix with the cell suspension.
7. Break cells by Fisherbrand™ Model Sonicator with the setting (3 s on, 6 s off at 60% amplitude) for 3 min at 4 °C. Cells are spun down at 4 °C by centrifugation at 18,000 rpm for 1 h. Filter supernatant using a 0.22-μm syringe filter.
8. Use 50 mL Ni-A buffer to equilibrate a 5 mL Histrap FF column. First, load the filtered-supernatant solution to the Heparin column at 1 mL/min. Then wash the column with 250 mL Ni—B buffer using a peristaltic pump at the rate of 2 mL/min.
9. Set up the AKTA Purifier for eluting the E343Q gp4 at the rate of 1.5 mL/min, with the gradient elution of 0%–100% Ni—C buffer in 90 mL.
10. Combine the protein fractions according to OD_{280} (~20 mg proteins can be acquired at this procedure). Buffer-exchange the collected fractions into MonoQ-A Buffer by using the Amicon Centrifugal Filter Units (Millipore). Add Prescission Protease to the mass ratio of 1:50

(protease to protein) to cut off the His-tag at 4 °C overnight. Dilute the collected solution 10-fold with the addition of MonoQ-A buffer.

11. The MonoQ column shall be equilibrated by 50 mL of MonoQ-A buffer on an AKTA Purifier. First, wash the column with 50 mL MonoQ-A buffer. Then, elute the proteins with a 0%–100% MonoQ-B buffer gradient solution in 100 mL at a 1 mL/min flow rate.
12. Mix the protein fractions (usually ~10–15 mg proteins can be obtained after this step). Concentrate the protein sample to ~2 mg/mL in a 50 mL Amicon Centrifugal Filter Units (Millipore) and aliquot the protein sample with stock buffer to a final concentration of 50% glycerol to 250 μL per tube. Flash-freeze the samples in liquid nitrogen (LN2) and store them at −80 °C.

3.3 Electrophoretic mobility supershift assay

A representative gel of gp4-gp5 complex assembling on EMSA supershift assay is shown in Fig. 4A. The binding of gp4 to the fork DNA is enhanced by ~3 fold in the presence of gp5 (Fig. 4B).

Materials and equipment

- Electrophoresis power supply (PowerEase Touch 350 W Power Supply, Thermo Fisher)
- Vertical electrophoresis apparatus (Thermo Scientific)
- Side-arm flask & access to a mild vacuum, degas gel polymerization solution before polymerization

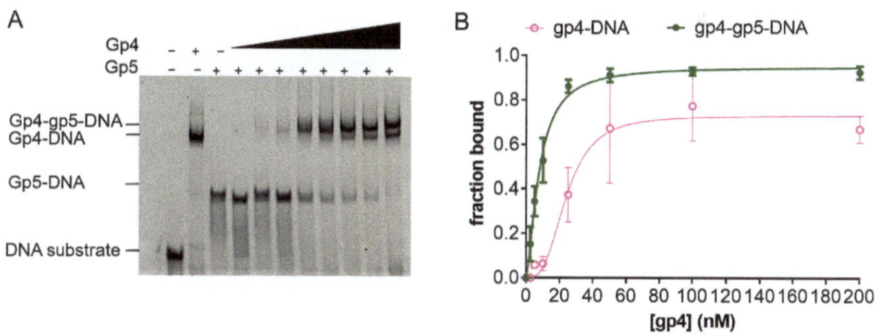

Fig. 4 Cooperative gp4-gp5 binding to the fork template. (A) A representative gel of mobility shift assay for gp4-gp5 binding. (B). Quantification of the supershift gp4 binding to DNA or gp5-DNA. *Panels were adapted and modified from Lo, C.-Y., and Gao, Y. (2022). DNA polymerase-parental DNA interaction is essential for helicase-polymerase coupling during bacteriophage T7 DNA replication.* International Journal of Molecular Sciences 23, *1342.*

- Gel Scanner (Sapphire Biomolecular imager, Azure Biosystem®)
- Protein stock of the complex of T7 gp5 (D5A, E7A)-trx
- Protein stock of the E343Q gp4
- 40% *w/v* acrylamide-bisacrylamide solution (37.5:1 acrylamide: bisacrylamide [Sigma])
- 10 × TBE (1 M Tris, 1 M Boric acid, and 0.02 M EDTA)
- 80% glycerol
- TEMED (N′,N′,N′-Tetramethylethylenediamine) [Sigma]
- 10% Ammonium persulfate [Sigma]
- 10 × EMSA binding buffer: 200 mM Hepes (pH 7.5 at 20 °C), 10 mM EDTA, 1 M KCl, 1 mM βME, 50% *v*/v glycerol, 0.10 mg/mL BSA
- DNA substrate with cyanine 5 (Cy5) labeled (Table 1.)
- 10 × Annealing buffer: 100 mM Tris, pH 7.5, 500 mM NaCl
- 6 × orange G loading dye 10 mM Tris-HCl (pH 7.5), 0.15% orange G, 30% glycerol

Step-by-step method details

1. The DNA substrate with Cy5 label and dideoxy nucleotide on the leading strand primer was prepared similarly to Section 3.1 (Table 1).
2. Make gel polymerization solution of 1 × TBE, 6% polyacrylamide, 2.5% glycerol, and degas the solution for 20 min. Next, place the gel solution on ice and add APS to the final concentration of 0.0375% and TEMED to 0.03%. Later, pour the solution into a gel cassette for gel polymerization.
3. 5 nM DNA substrate with or without gp5 (60 nM) were incubated for 15 min at 25 °C under the buffer containing 1 × binding buffer (pH 7.5 at 20 °C).
4. An increasing amount of hexameric E343Qgp4 (0–400 nM) is added to each premixed aliquot (DNA or DNA-gp5) for incubation at 25 °C for 15 min.
5. Mix the reaction solution with loading buffer for loading the reaction samples into a 6% polyacrylamide gel.
6. Set up the gel electrophoresis with 1 × TBE buffer at 150 voltage, 10 mA, for 70 min at 4 °C or on ice.
7. Place the gel on the scanner and set up the program to detect Cy5 fluorescence to calculate the Kd value.

3.4 2-aminopurine fluorescence assays

Representative scans of 2AP fluorescence emission is shown in Fig. 5A. The 2AP at the fork junction becomes unwound only in the presence of coupled

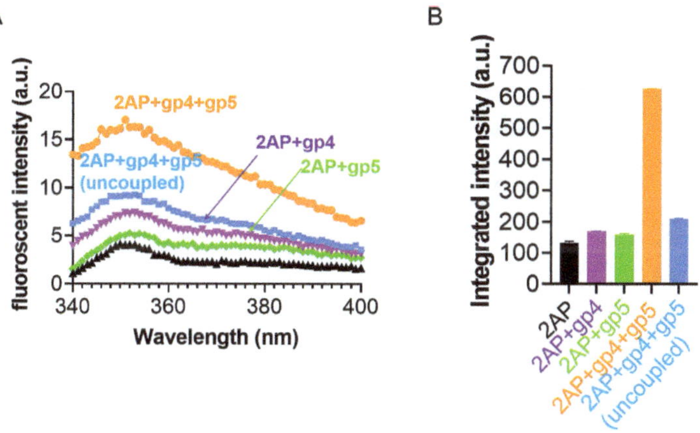

Fig. 5 2AP assay for studying gp4-gp5 coupling. (A) Fluorescent emission signals at 340–400 nm. (B) Integration and comparison of fluorescent intensities. *Panels were adapted and modified from Lo, C.-Y., and Gao, Y. (2022). DNA polymerase-parental DNA interaction is essential for helicase-polymerase coupling during bacteriophage T7 DNA replication.* International Journal of Molecular Sciences 23, *1342.*

gp4 and gp5 binding (Fig. 5B). A DNA substrate with too long of a ssDNA gap leads to gp4-gp5 uncoupling (Fig. 5B).

Materials and equipment

- Cary Eclipse Fluorescence Spectrophotometer (Agilent®)
- Spectrophotometer (Nanodrop 2000c, Thermo Scientific)
- Centrifuge (BC Avanti J-E, Beckman Coulter)
- Suprasil® quartz, spectral range 200–2500 nm, 130 μL, Hellma® fluorescence cuvettes

Reagents and solutions

- Protein stock of the complex of T7 gp5 (D5A, E7A)-trx
- Protein stock of the E343Q gp4
- 10 × Binding buffer: 500 mM Tris (pH 7.5 at 20 °C), 0.1 mM EDTA, 1.5 M KCl, 1 mM βME
- DNA substrate (Table 1).
- 10 × Annealing buffer: 100 mM Tris, pH 7.5, 500 mM NaCl

Step-by-step method details

1. A single 2AP, which will be base-paired to a thymine, is included on the parental dsDNA proximate to the fork junction. The fork DNA substrate with 2AP and dideoxy nucleotide on the leading strand was prepared similarly to Section 3.1 (Table 1).

2. T7 gp5-trx and E343Q gp4 are buffer-exchanged to remove glycerol and concentrated by Amicon Centrifugal Filter Units (Millipore) to the concentration of 8 μM.
3. After the Fluorescence Spectrophotometer is initiated, the excitation wavelength is set to 315 nm, and the emission is monitored from 340 to 400 nm. The data model is set as fluorescence with the intensity unit using (a.u.).
4. 200 nM of DNA substrate without or with 2AP labeling are added in the cuvette to estimate the background signal.
5. 200 nM of gp4, gp5, or gp4-gp5 were added to the 2AP DNA in the presence of 0.5 mM dTTP and 2 mM $MgCl_2$, and the signals of each sample is measured as in step3.
6. The fluorescent intensity is collected and integrated under the wavelength 340 nm to 400 nm. The signals are normalized with respect to DNA alone control to compare the different effects of helicase, polymerase, and helicase-polymerase on DNA unwinding. The background fluorescence intensities (a.u.) are usually between ~100–150 a.u. A fluorescence intensity between 200 and 400 a.u. is considered as an indication for base separation.

3.5 Complex assembling, gel filtration, and cryo-EM freezing

Materials and equipment

- AKTA Purifier (Cytiva)
- Spectrophotometer (Nanodrop 2000c, Thermo Scientific)
- PCR machine
- Centrifuge (BC Avanti J-E, Beckman Coulter)
- Superose 6 10/300 gl (GE Healthcare)
- Vitrobot plunge freezer (Thermo Fisher)
- EasiGlow glow discharger (TedPella)
- Protein stock of the complex of T7 gp5 (D5A, E7A)-trx
- Protein stock of the E343Q gp4
- 10× Binding buffer: 500 mM Tris (pH 7.5 at 20 °C), 0.1 mM EDTA, 1.5 M KCl, 1 mM βME
- Size-exclusion chromatography (SEC) buffer: 25 mM Tris (pH 7.5 at 20 °C), 0.1 mM EDTA, 100 mM KCl, 1 mM DTT
- DNA substrate
- 10× Annealing buffer: 100 mM Tris, pH 7.5, 500 mM NaCl
- Quantifoil R1.2/1.3300 mesh grids (Quantifoil)

Step-by-step method details

1. T7 gp5-trx and E343Q gp4 are buffer-exchanged to remove glycerol and concentrated by Amicon Centrifugal Filter Units (Millipore) to the concentration of 10 μM.
2. T7 gp5-trx and DNA substrate are mixed at 1: 1.1 ratio in the 1 × buffer containing 50 mM Tris (pH 7.5 at 20 °C), 0.01 mM EDTA, 150 mM KCl, 0.1 mM βME.
3. Add E343Q gp4 and 0.2 mM dTTP into the reaction. The final concentration of protein and DNA: T7gp5-trx (1 μM), E343Qgp4 (1 μM), DNA substrate (1.1 μM).
4. Use 50 mL SEC buffer to equilibrate a 26 mL Superose 6 column at the 0.5 mL/min rate.
5. Inject 0.4 mL of the reaction mixture into the 0.5 mL loop on AKTA Purifier. Set up the AKTA Purifier for eluting the gp4-gp5-DNA complex at the rate of 0.5 mL/min, and collect the eluted protein at a 96-well plate using 25 mL of the SEC buffer.
6. The gel filtration profile is shown in Fig. 6A. Typically, the gp4-gp5-DNA complex elutes at 13 mL (Fig. 6). Take the major peak from gel filtration. Add dTTP and $MgCl_2$ to a final concentration of 0.5 mM and 2 mM, respectively. Concentrate protein to 0.5 mg/mL. The E343Qgp4 was with detectable dTTPase activity and long-time incubation will lead dTTP consumption. Addition of extra dTTP will stabilize the complex.
7. Glow discharge quantifoil grids with EasiGlow at 10 mA for 60 s.
8. Initialize Vitrobot and equilibrate the chamber at 100% humidity for 10 min.

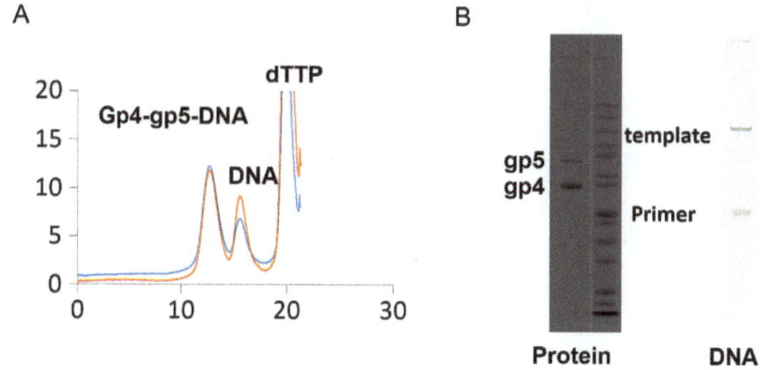

Fig. 6 (A) Gel filtration of the gp4-gp5-DNA complex. (B) SDS PAGE and urea PAGE for protein and DNA in the gel filtration peak.

9. Add 3 μL of 0.5 mg/mL protein sample to the glow-discharged grids, blot at force 4 for 3–4 s. Freeze the grids first in liquid ethane and then transfer them to liquid nitrogen for storage.

4. Cryo-EM data analysis

We have been using cryo-EM as a final assay for screening DNA substrates and verifying replisome complex formation. For screening purposes, 100–200 micrographs from a screening cryo-EM microscope can provide sufficient information for evaluating the replisome sample. The replisome particles can be easily visible on raw micrographs (Fig. 7A). However, it is hard to distinguish gp4 alone from gp4-gp5 complex visually. 2D-classification can be powerful in separating different forms of complexes. As shown in Fig. 7B, we usually capture gp4 alone, gp5, and gp4-gp5 complexes with different conformations from a single preparation. The percentage of particles in the gp4-gp5 complex form can be a suitable parameter for evaluating complex formation. However, even with the optimized substrate and sampling

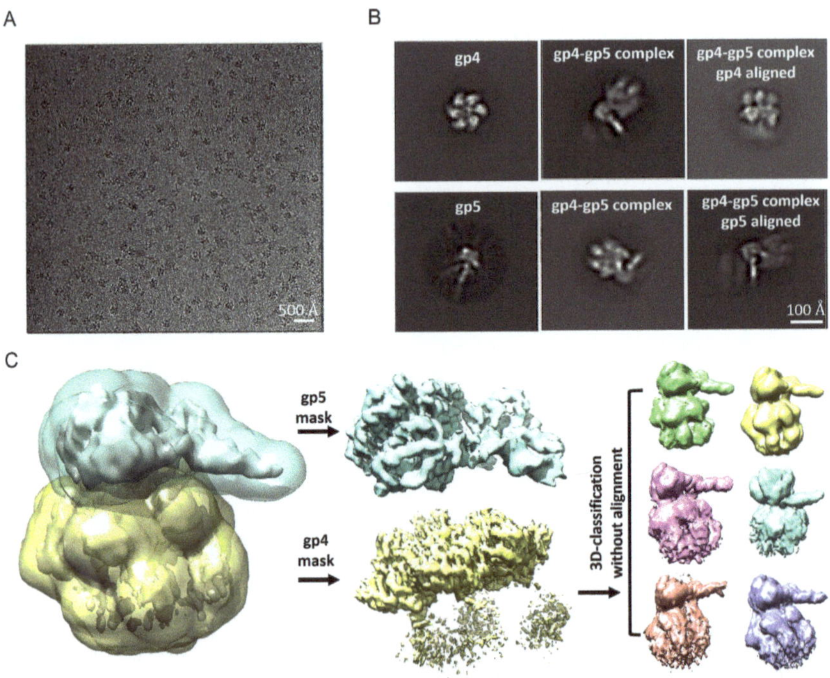

Fig. 7 Cryo-EM data processing. (A) A representative micrograph. (B). Representative 2D-classification images of gp4-gp5 complex. (C). 3D classification and local refinements.

assemble conditions, we obtain only 30–40% particles in the complex form, possibly due to sample denaturation and disassembly during freezing.

Gp4-gp5 interaction is mediated by the acidic C-terminal tails of gp4 and the positively charged surface patches on gp5 (Hamdan et al., 2007; Kulczyk et al., 2017). However, the charge-charge interaction is relatively weak; moreover, gp4 exists as a hexamer with six tails. Thus, the gp4-gp5 interaction appears highly dynamic. As shown in Fig. 6B, only half of the complex is well-aligned, while the other half is flexible in many 2D classes, indicating significant relative motions between the two parts. The flexible interaction limits the final resolution of the gp4-gp5 complex. On the other hand, the gp4 and gp5 portions of the complex can be separately refined to high resolution with proper softmasks and signal subtraction (Fig. 7C). Moreover, aligning all particles first according to the gp4 part of the complex can help separate different conformations through 3D classifications (Fig. 7C). Rigid body docking of the well-refined gp4 and gp5 structures to the lower resolution gp4-gp5 complex structures can help build the final model of the replisome. The many conformations of gp4-gp5 complexes at the replication fork suggested significant dynamics of the replisome, which may be essential for replisome operation.

5. Summary

In this review, we have summarized the methods for assembling bacteriophage T7 replisome for structural investigations. The T7 replisome structures revealed important mechanistic details for helicase-polymerase coupling during DNA replication. The methods and strategies can be easily adapted for bacterial and mitochondrial DNA replication, which share similar features as the phage replisome. However, because of the different polarity of helicase, the eukaryotic replisomes are structurally distinct from the T7 replisome, and their structural visualization may require extensive DNA substrate design and experimental optimization. Nevertheless, with the recent advances in reconstituting replisome systems *in vitro* and cryo-EM sample preparation and data analysis, we believe structural visualization of DNA replication at atomic resolution can be achieved in the near future.

Acknowledgments

This work is supported by National Institutes of Health (NIH) grant 1R35GM142722. Y.G. is a Cancer Prevention & Research Institute of Texas (CPRIT) Scholar (Award RR190046). We highly appreciate Dr. Wei Yang from NIH for her advise and support.

References

Brieba, L. G., Eichman, B. F., Kokoska, R. J., Doublie, S., Kunkel, T. A., & Ellenberger, T. (2004). Structural basis for the dual coding potential of 8-oxoguanosine by a high-fidelity DNA polymerase. *The EMBO Journal, 23*, 3452–3461.

Cortez, D. (2019). Replication-coupled DNA repair. *Molecular Cell, 74*, 866–876.

Crampton, D. J., Mukherjee, S., & Richardson, C. C. (2006). DNA-induced switch from independent to sequential dTTP hydrolysis in the bacteriophage T7 DNA helicase. *Molecular Cell, 21*, 165–174.

D'Imprima, E., Floris, D., Joppe, M., Sanchez, R., Grininger, M., & Kuhlbrandt, W. (2019). Protein denaturation at the air-water interface and how to prevent it. *Elife, 8*.

Doublie, S., Tabor, S., Long, A. M., Richardson, C. C., & Ellenberger, T. (1998). Crystal structure of a bacteriophage T7 DNA replication complex at 2.2 A resolution. *Nature, 391*, 251–258.

Eickhoff, P., Kose, H. B., Martino, F., Petojevic, T., Abid Ali, F., Locke, J., et al. (2019). Molecular basis for ATP-hydrolysis-driven DNA translocation by the CMG helicase of the eukaryotic replisome. *Cell Reports, 28*, 2673–2688 e2678.

Gao, Y., Cui, Y., Fox, T., Lin, S., Wang, H., de Val, N., et al. (2019). Structures and operating principles of the replisome. *Science, 363*.

Gao, Y., & Yang, W. (2020). Different mechanisms for translocation by monomeric and hexameric helicases. *Current Opinion in Structural Biology, 61*, 25–32.

Georgescu, R., Yuan, Z., Bai, L., de Luna Almeida Santos, R., Sun, J., Zhang, D., et al. (2017). Structure of eukaryotic CMG helicase at a replication fork and implications to replisome architecture and origin initiation. *Proceedings of the National Academy of Sciences of the United States of America, 114*, E697–E706.

Graham, J. E., Marians, K. J., & Kowalczykowski, S. C. (2017). Independent and stochastic action of DNA polymerases in the replisome. *Cell, 169*, 1201–1213 e1217.

Hamdan, S. M., Johnson, D. E., Tanner, N. A., Lee, J. B., Qimron, U., Tabor, S., et al. (2007). Dynamic DNA helicase-DNA polymerase interactions assure processive replication fork movement. *Molecular Cell, 27*, 539–549.

Hamdan, S. M., & Richardson, C. C. (2009). Motors, switches, and contacts in the replisome. *Annual Review of Biochemistry, 78*, 205–243.

Itsathitphaisarn, O., Wing, R. A., Eliason, W. K., Wang, J., & Steitz, T. A. (2012). The hexameric helicase DnaB adopts a nonplanar conformation during translocation. *Cell, 151*, 267–277.

Kim, S., Dallmann, H. G., McHenry, C. S., & Marians, K. J. (1996). Coupling of a replicative polymerase and helicase: A tau-DnaB interaction mediates rapid replication fork movement. *Cell, 84*, 643–650.

Kulczyk, A. W., Moeller, A., Meyer, P., Sliz, P., & Richardson, C. C. (2017). Cryo-EM structure of the replisome reveals multiple interactions coordinating DNA synthesis. *Proceedings of the National Academy of Sciences of the United States of America, 114*, E1848–E1856.

Langston, L. D., Zhang, D., Yurieva, O., Georgescu, R. E., Finkelstein, J., Yao, N. Y., et al. (2014). CMG helicase and DNA polymerase epsilon form a functional 15-subunit holoenzyme for eukaryotic leading-strand DNA replication. *Proceedings of the National Academy of Sciences of the United States of America, 111*, 15390–15395.

Lee, J., Chastain, P. D., 2nd, Kusakabe, T., Griffith, J. D., & Richardson, C. C. (1998). Coordinated leading and lagging strand DNA synthesis on a minicircular template. *Molecular Cell, 1*, 1001–1010.

Lee, S. J., Zhu, B., Hamdan, S. M., & Richardson, C. C. (2010). Mechanism of sequence-specific template binding by the DNA primase of bacteriophage T7. *Nucleic Acids Research, 38*, 4372–4383.

Lo, C. Y., & Gao, Y. (2021). DNA helicase-polymerase coupling in bacteriophage DNA replication. *Viruses*, *13*.

Lo, C.-Y., & Gao, Y. (2022). DNA polymerase-parental DNA interaction is essential for helicase-polymerase coupling during bacteriophage T7 DNA replication. *International Journal of Molecular Sciences*, *23*, 1342.

Manosas, M., Spiering, M. M., Ding, F., Bensimon, D., Allemand, J. F., Benkovic, S. J., et al. (2012). Mechanism of strand displacement synthesis by DNA replicative polymerases. *Nucleic Acids Research*, *40*, 6174–6186.

Manosas, M., Spiering, M. M., Ding, F., Croquette, V., & Benkovic, S. J. (2012). Collaborative coupling between polymerase and helicase for leading-strand synthesis. *Nucleic Acids Research*, *40*, 6187–6198.

Manosas, M., Xi, X. G., Bensimon, D., & Croquette, V. (2010). Active and passive mechanisms of helicases. *Nucleic Acids Research*, *38*, 5518–5526.

Morin, J. A., Cao, F. J., Lazaro, J. M., Arias-Gonzalez, J. R., Valpuesta, J. M., Carrascosa, J. L., et al. (2015). Mechano-chemical kinetics of DNA replication: Identification of the translocation step of a replicative DNA polymerase. *Nucleic Acids Research*, *43*, 3643–3652.

Nandakumar, D., Pandey, M., & Patel, S. S. (2015). Cooperative base pair melting by helicase and polymerase positioned one nucleotide from each other. *eLife*, *4*.

Noble, A. J., Wei, H., Dandey, V. P., Zhang, Z., Tan, Y. Z., Potter, C. S., et al. (2018). Reducing effects of particle adsorption to the air-water interface in cryo-EM. *Nature Methods*, *15*, 793–795.

O'Donnell, M., Langston, L., & Stillman, B. (2013). Principles and concepts of DNA replication in bacteria, archaea, and eukarya. *Cold Spring Harbor Perspectives in Biology*, *5*.

O'Donnell, M. E., & Li, H. (2018). The ring-shaped hexameric helicases that function at DNA replication forks. *Nature Structural & Molecular Biology*, *25*, 122–130.

Pandey, M., & Patel, S. S. (2014). Helicase and polymerase move together close to the fork junction and copy DNA in one-nucleotide steps. *Cell Reports*, *6*, 1129–1138.

Patel, S. S., Wong, I., & Johnson, K. A. (1991). Pre-steady-state kinetic analysis of processive DNA replication including complete characterization of an exonuclease-deficient mutant. *Biochemistry*, *30*, 511–525.

Rothwell, P. J., & Waksman, G. (2005). Structure and mechanism of DNA polymerases. *Advances in Protein Chemistry*, *71*, 401–440.

Santos, R.d. L. A. (2021). *Structural studies of DNA replication proteins and the human immunoproteasome* (p. 141). Ann Arbor, State University of New York at Stony Brook.

Satapathy, A. K., Kulczyk, A. W., Ghosh, S., van Oijen, A. M., & Richardson, C. C. (2011). Coupling dTTP hydrolysis with DNA unwinding by the DNA helicase of bacteriophage T7. *The Journal of Biological Chemistry*, *286*, 34468–34478.

Singh, A., Pandey, M., Nandakumar, D., Raney, K. D., Yin, Y. W., & Patel, S. S. (2020). Excessive excision of correct nucleotides during DNA synthesis explained by replication hurdles. *The EMBO Journal*, *39*, e103367.

Singleton, M. R., Dillingham, M. S., & Wigley, D. B. (2007). Structure and mechanism of helicases and nucleic acid translocases. *Annual Review of Biochemistry*, *76*, 23–50.

Singleton, M. R., Sawaya, M. R., Ellenberger, T., & Wigley, D. B. (2000). Crystal structure of T7 gene 4 ring helicase indicates a mechanism for sequential hydrolysis of nucleotides. *Cell*, *101*, 589–600.

Wallen, J. R., Zhang, H., Weis, C., Cui, W., Foster, B. M., Ho, C. M. W., et al. (2017). Hybrid methods reveal multiple flexibly linked DNA polymerases within the bacteriophage T7 replisome. *Structure*, *25*, 157–166.

Yuan, Z., Georgescu, R., Schauer, G. D., O'Donnell, M. E., & Li, H. (2020). Structure of the polymerase epsilon holoenzyme and atomic model of the leading strand replisome. *Nature Communications*, *11*, 3156.

CHAPTER SIX

In vivo fluorescent TUNEL detection of single stranded DNA gaps and breaks induced by *dnaB* helicase mutants in *Escherichia coli*

Megan S. Behrmann and Michael A. Trakselis*
Department of Chemistry and Biochemistry, Baylor University, Waco, TX, United States
*Corresponding author: e-mail address: michael_trakselis@baylor.edu

Contents

Abstract

The genome of prokaryotes can be damaged by a variety of endogenous and exogenous factors, including reactive oxygen species, UV exposure, and antibiotics. To better understand these repair processes and the impact they may have on DNA replication, normal genome maintenance processes can be perturbed by removing or editing associated genes and monitoring DNA repair outcomes. In particular, the replisome activities of DNA unwinding by the helicase and DNA synthesis by the polymerase must be tightly coupled to prevent any appreciable single strand DNA (ssDNA) from accumulating and amplifying genomic stress. If decoupled, vulnerable ssDNA would persist, likely leading to double strand breaks (DSBs) or requiring replication restart mechanisms

Methods in Enzymology, Volume 672
ISSN 0076-6879
https://doi.org/10.1016/bs.mie.2022.02.021

downstream of a stall. In either case, free 3′-OH strands would exist, resulting from ssDNA gaps in the leading strand or complete DSBs. Terminal deoxyribonucleotide transferase (TdT)-mediated dUTP nick end labeling (TUNEL) can enzymatically label ssDNA ends with bromo-deoxy uridine triphosphate (BrdU) to detect free 3′-OH DNA ends in the *E. coli* genome. Labeled DNA ends can be detected and quantified using fluorescence microscopy or flow cytometry. This methodology is useful in applications where *in situ* investigation of DNA damage and repair are of interest, including effects from enzyme mutations or deletions and exposure to various environmental conditions.

1. Introduction

Terminal deoxynucleotidyl transferase (TdT) dUTP Nick-End Labeling (TUNEL) assay was originally designed to detect eukaryotic apoptotic cells and tissues, which undergo extensive DNA degradation and fragmentation during the late stages of apoptosis (Gavrieli, Sherman, & Ben-Sasson, 1992). Since then, the technique has been adapted and honed for more fine-tuned identification of DNA breaks, nicks, and gaps, and utilized in a variety of organisms and systems (Dwyer, Camacho, Kohanski, Callura, & Collins, 2012; Li, Melamed, & Darzynkiewicz, 1996; Rohwer & Azam, 2000; Sarac & Hollenstein, 2019). The linchpin of the TUNEL method is the TdT enzyme, which adds nucleotides to the 3'-OH ends of DNA breaks, including blunt ends, and is independent of a template (Michelson & Orkin, 1982; Motea & Berdis, 2010). This enzyme is an X-family DNA polymerase and biologically catalyzes the addition of random nucleotides to 3′-overhangs during V(*D*)J recombination (Sarac & Hollenstein, 2019). The unusual biological function and biochemical properties intrinsic to the TdT polymerase have motivated development of various oligonucleotide-based tools, notably synthetic tagging when combined with non-canonical nucleotides.

Several variants of TUNEL methodology exist utilizing direct or indirect fluorochrome-labeling of 3′-OH termini of DNA strand breaks *in situ* with the use of exogenous TdT and different synthetic nucleotides. Here, we discuss strand gap or break labeling with 5-bromo-2′-deoxyuridine triphosphate (BrdU), which is subsequently detected by antibody labeling with an anti-BrdU antibody (Fig. 1). TUNEL with BrdU detects free DNA ends with greater sensitivities compared to other DNA break labeling techniques (Shee et al., 2013) or different modified nucleotide additions (Darzynkiewicz, Galkowski, & Zhao, 2008). Simultaneous staining of DNA

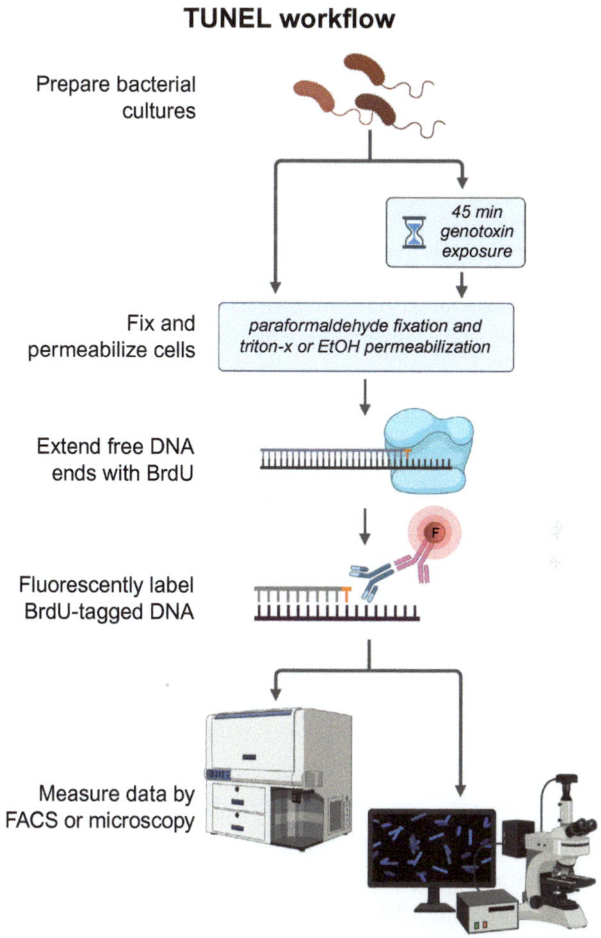

Fig. 1 TUNEL assay flowchart. A diagram showing method progression from overnight sample strains to final analysis. Cultures of the test and control strains are grown to log phase, then introduced to a genotoxic or blank solution of DMSO for 45 min. Cells are fixed with paraformaldehyde, permeabilized, and subsequently labeled with TUNEL incorporated BrdU. The labeled DNA is then tagged with a fluorescent antibody and analyzed by a combination of fluorescence microscopy and flow cytometry. Figure created with the help of BioRender.

with propidium iodide, Sytox Green, or 4′,6-diamidino-2-phenylindole (DAPI) and quantification of signal by flow cytometry enables correlation of cell growth or phase with the frequency of nicks, gaps, and breaks.

Cell fixation and permeabilization is critical to successfully label DNA strand breaks. Cells are first fixed with a crosslinking agent, *i.e.* paraformaldehyde, which prevents the extraction or loss of low molecular weight

DNA fragments during the frequent centrifugation and washing steps (Darzynkiewicz et al., 2008). Fixed cells are then permeabilized by suspension in ethanol or by using detergents to allow TdT and BrdU access to the chromosome. The 3′-termini of any DNA nick, break, or single-stranded DNA (ssDNA) gap serves as a primer and becomes labeled with BrdU by TdT (Li et al., 1996; Li & Darzynkiewicz, 1995). The incorporated BrdU is then immunochemically detected by a BrdU antibody conjugated with a fluorescent dye for easy detection (Dolbeare & Selden, 1994).

In *E. coli*, replication initiates from a single origin (called *oriC*) and migrates bi-directionally around the circular chromosome with the two replication forks meeting opposite the origin at the replication terminus. During replication, the replisome frequently encounters replication blocks such as DNA crosslinks or lesions, bound or crosslinked proteins, alkylation, or complex structures like G-quadruplexes (Perera, Behrmann, Hoang, Griffin, & Trakselis, 2019; Romero et al., 2020). When lesions or blocks are encountered by the replisome, resolution and repair can take many forms, and most involve the generation of free DNA ends as an intermediate. In conditions with high levels of genomic stress, these broken DNA lesions or repair intermediates can persist and even be left unrepaired entirely (Erental, Sharon, & Engelberg-Kulka, 2012). TUNEL can be utilized to detect the presence of these ignored or unrepaired DNA damage intermediates.

Daughter ssDNA gaps can be generated on either the leading or lagging strand by damage-tolerant lesion bypass. While translesion synthesis (TLS) is a well-documented DNA damage tolerant repair pathway (Cranford, Kaszubowski, & Trakselis, 2020; Kath et al., 2016), some lesions can be bypassed by lesion skipping and repriming mechanisms. This involves the replisome 'skipping over' the replication block on either the leading or lagging strand and reinitiating with helicase-dependent downstream repriming, similar to regular Okazaki fragment synthesis (Laureti, Demol, Fuchs, & Pages, 2015; McInerney & O'Donnell, 2004; Mezzina, Menck, Courtin, & Sarasin, 1988). This type of damage tolerant lesion bypass leaves a gap behind the replisome in one of the "daughter strands," which is then repaired post-replicatively by TLS, RecA-dependent pathways (Henrikus et al., 2018), RecFOR (Morimatsu & Kowalczykowski, 2003), or a specific subset of the PriA/B/C pathways (Sandler, 2000). Downstream repriming is used to couple lagging strand synthesis to unwinding and may also be involved in maintaining coordination between leading strand synthesis and unwinding during polymerase switching or helicase-polymerase decoupling events (Beattie et al., 2017; Kim, Dallmann, McHenry, & Marians, 1996; Lewis et al., 2017).

Replicative helicases unwind DNA, interact with the replisome dynamically, and are critically involved in replisome function, coordination, and genome conservation. In bacteria, the replicative hexameric helicase (DnaB) serves as a stable anchor for dynamic and transient interactions of other fork-associated proteins. Replisome-coupled helicase-polymerase activity is strongly favored for efficient replication. If a helicase decouples from the DNA polymerase, a molecular mechanism engages to slow the unwinding rate by over threefold in order to protect the DNA duplex and to encourage recoupling (Graham, Marians, & Kowalczykowski, 2017; Strycharska et al., 2013). Recently, we investigated the genomic impact of four *in vivo* DnaB mutations that disrupted this molecular break, promoting in a more constricted hexameric structure that correlates with faster unwinding, promoting replisome decoupling (Behrmann et al., 2021). For those *dnaB* mutants, there was evidence of increased frequencies of ssDNA gaps and DNA breaks detected by this TUNEL method. The TUNEL method is useful for specifically investigating the impact of helicase mutations on genome maintenance, coupled unwinding and synthesis activity, and other diverse DNA damage events but can also be utilized to probe the impact of many other enzymes involved in replisome coupling, restart, or repair in both prokaryotes and eukaryotes.

2. Materials and equipment

2.1 Materials

- Overnight cultures of bacterial strains to be tested (Table 1) (Behrmann et al., 2021; Costantino & Court, 2003)
- LB media: 10 g tryptone, 10 g NaCl, 5 g yeast extract per L, pH 7.0
- Genotoxins for inducing damage, such as:
  Mitomycin C (MMC): FisherScientific, BP25312
  Nitrofurazone (NFZ): TCI America, N020025G
- Sterile 1 × PBS: 0.137 M NaCl, 2.7 mM KCl, 10 mM Na_2HPO_4, 1.8 mM KH_2PO_4
- Sterile ddH_2O water
- Triton-X: ThermoScientific, A16046AP
- Tween-20: (Fisher BioReagents, BP151-100)

Table 1 Strains.

MG1655	***E. coli* K12**	**One of the two parent strain derivatives of *E. coli* K12**
W3110	*E. coli* K12	One of the two parent strain derivatives of *E. coli* K12
HME6	W3110 $galK_{tyr145UAG}$ $\Delta lacU169$ [λ *cI857* Δ(*cro-bioA*)]	This strain contains the *galK* assay system for oligo recombination.
HME63	HME6 *mutS<>amp*	Defective for MMR thus gives high frequency oligo recombination.
MSB5	HME63 *dnaB:R328/9A*	Contains a *dnaB* point mutation

- Paraformaldehyde (PFA): Acros, 41678-5000
- Cold 77% ethanol at −20 °C
- 5-Bromo-2'-Deoxyuridine (BrdU): FisherScientific, B23151
- Terminal Deoxynucleotidyl Transferase (TdT) enzyme and buffer: NEB, M0315S
- Cobalt chloride (CoCl2) (*optional*)
- 1 × TBST: 50 mM Tris–Cl, pH 7.5, 150 mM NaCl, 1% Tween-20
- Bovine Serum Albumin (BSA): FisherScientific, BP1600-100
- 0.45 μm syringe filters or other filtration sterilization method
- Mouse-α-BrdU: BD Biosciences, 347,580 at 1:100 in 2% BSA in TBST
- α-mouse IgG-Alexa647: Thermo Fisher, A-21235 at 1:500 in 2% BSA in TBST
- Sodium citrate ($Na_3C_6H_5O_7$)
- 1,4-Diazabicyclo[2.2.2]octane (DABCO): ThermoFisher, 112471000
- DABCO mounting medium: (25 mg/mL DABCO in 90% glycerol/10% 1xPBS, pH 8.6)
- DAPI fluorescent DNA stain for microscopy: Invitrogen, D1306
- Glass slides, coverslips, and clear nail polish or other sealant
- Sytox Green DNA stain for FACS: Invitrogen, S7020
- Sterilized 250 mL glass Erlenmeyer flasks (one for each strain)
- 50 mL sterile conical tubes, such as falcon tubes, or equivalent sterile container

2.2 Equipment

- Temperature controlled shaker (*able to maintain 225 RPM and 37 °C*)
- 37 °C water bath or equivalent benchtop incubator
- Olympus brightfield microscope IX-81 with a 60 × oil immersion lens (*or equivalent*)
- BD Biosciences FACSverse (*or equivalent*)

3. Culture preparation

3.1 Growth and treatment

This assay is used to investigate the presence of genomic single-stranded DNA ends or double -strand breaks, and so, it is important to include both positive and negative controls to eliminate any background signals of native repair in the strains. When testing strains with genomic mutations, deletions or insertions, it is important to also test the parental strain, such as W3110, MG1655, or others containing any secondary mutations. Based on our experimentation, the presence of additional genomic material, such as plasmids, does not appear to have a significant effect on either the amount or location of BrdU foci signal (as seen in Fig. 2), and so, it is not absolutely necessary to remove vectors before performing the TUNEL assay. However, if vectors are utilized, we do recommend that they be present in both the control and test strains.

3.1.1 Plasmid curing

1. To cure an *E. coli* strain of plasmid(s), first check to see if the plasmid has an internal curing mechanic, such as a heat sensitive origin or inducible expression of targeted restriction enzymes. Many modern plasmid systems have these "self-curing" mechanisms to increase the efficiency of removal. If no in-built curing mechanism exists, plasmids can be cured by repeated growth in the absence of the selective agent(s) (*i.e. without selective antibiotics*).
2. Begin by streaking out a glycerol stock onto LB-agar that lacks the plasmid selective agent. From this plate, pick a single colony to grow overnight in LB media absent of plasmid-specific selective agent(s).
3. Once the culture is turbid, streak out once again onto agar lacking the selective agent(s) and place in a 37 °C incubator overnight.
4. Pick a minimum of eight individual colonies and streak dually on to two separate agar plates marked into eight sections, one with and one without the selective agent.
5. Select a plasmid-free colony based on sensitivity to the vector's resistance marker. Grow overnight in liquid culture before adding 15% glycerol and flash freezing in liquid nitrogen.
6. If no sensitivity is detected, repeat the passaging process (*steps 2–4*) until sensitive colonies are identified.

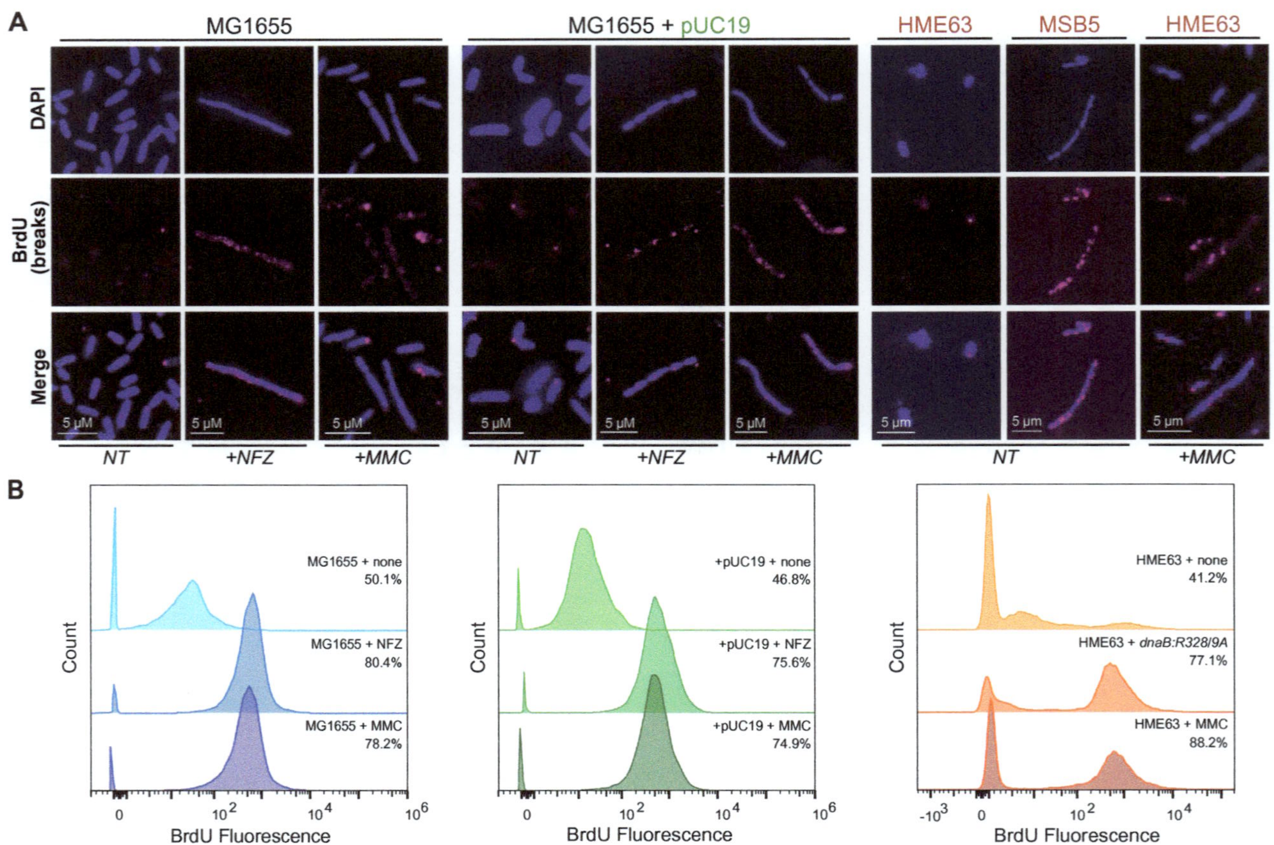

Fig. 2 BrdU foci increase with exogenous damage but not extragenomic DNA. (A) Microscopy images of MG1655 with and without the high-copy plasmid pUC19 (*middle* and *left*, respectively), HME63 (HME6 *mutS*<>*amp*), and the conformationally constricted helicase mutant strain MSB5 (HME63 *dnaB:R328/9A*). Cells were probed for DNA breaks using TUNEL; blue (DAPI) represents DNA staining, and pink (BrdU) represents tagged DNA breaks. NT—nontreated; +MMC—treated with 0.01 μg/mL MMC. Images shown are representative of the population observed. The fraction of BrdU labeled cells was (B) measured and quantified by flow cytometry (FACS), utilizing the same exponential growth cultures as the microscopy images. Cells were gated for +/-BrdU signal; BrdU positive populations are recorded on the right of each plot.

3.1.2 Growing up test and control strains

1. Start with overnight cultures of the test and control strains grown in appropriate media.
2. Dilute each strain 1:100 in LB media (with selective antibiotics if necessary) into 30 mL of media in a 250 mL sterile Erlenmeyer flask. This will grow enough sample for two test conditions.
3. Incubate at 37 °C and shake at 220 RPM until the optical density (OD_{600}) reads between 0.3 and 0.4, early log phase.
4. As each test culture reaches the appropriate OD_{600}, split the contents of the flask in to two ~10 mL aliquots in 50 mL tubes and discard any remaining culture according to appropriate biohazard and containment protocols. For the control strain, separate into three 10 mL aliquots, reserving one to be an "unstained" control for flow cytometry. *Do not expose this third control aliquot to genotoxins or TdT extension.*
5. Treat one tube with an appropriate test genotoxic agent and the other with mock control conditions before returning to shaking incubation. Examples include 0.01 μg/mL MMC or 5 μM NFZ for 45 min, although both the concentrations and exposure times can be varied.

   Note: While an appropriate control strain serves as an effective baseline, hydroxyurea can also be used for an additional or alternative negative control. For this method, once the strain reaches early log phase, add 10 mM HU and return to incubation with shaking for 4 h.
6. Harvest strains by spinning down at 4000 × *g* for 12–15 min until a pellet forms. Discard the media and begin washing by resuspending the cell pellet in 1 mL of filter sterilized PBS and transfer to a 1.5 mL microcentrifuge tube. Repeat this pelleting and resuspension procedure a total of three times. After the initial centrifugation, all subsequent spins listed in this method are performed in a microcentrifuge at 20,000 × *g* for 1 min. From this point forward, all solutions and samples are kept cold and on ice unless otherwise stated.

3.2 Fixation and permeabilization

After fixation, there are two different methods for cell permeabilization that can be used.

3.2.1 Paraformaldehyde fixation

Paraformaldehyde (PFA) fixation is necessary to crosslink and prevent the loss of small DNA fragments.

1. Pellet cells by microcentrifugation at 20,000 × *g* for 1 min, wash in PBS, and then resuspend in 1 mL PFA fixing solution *(4% paraformaldehyde in 1× PBS).*
2. Let cells incubate in the fixing solution for 20 min at room temperature.
3. Spin down fixed cells, wash in cold PBS, then proceed immediately to the chosen permeabilization method.

   Note: Paraformaldehyde dissolves in water best when provided with mild heat, alkalinity, agitation, and time. This solution should be made ahead of time and chilled. Paraformaldehyde degrades within a few weeks when kept in the fridge but can last in aliquots in the freezer at −20 °C for several months. Avoid freeze-thaw cycles.

3.2.2 Detergent permeabilization

PFA fixation followed by Triton-X permeabilization allows for downstream utilization of the cells and has been shown to preserve cell surface ultrastructures (Chao & Zhang, 2011); however, it requires sample analysis within a short time frame from fixation as it can be reversible.

1. Resuspend fixed pelleted cells in 500 μL permeabilization solution (*0.1% Triton X-100 and 0.1% sodium citrate in* ddH_2O) and let sit for 2 min on ice.
2. Pellet permeabilized cells, wash with 1 mL cold PBS, and then keep on ice or at 4 °C until ready to proceed with DNA end labeling

   Note: An alternative detergent like Tween-20 can be substituted for Triton-X.

3.2.3 Ethanol permeabilization/fixation

Ethanol is a quick and effective permeabilization method that results in permanent fixation but can cause significant cell lysis in some genera (Zhu, Rajendram, & Huang, 2021). Alcohol preservation methods have been shown to reduce damage and impact to DNA because of rapid entrance into the cell, condensing DNA into a stable form that readily resumes native structure when rehydrated (Srinivasan, Sedmak, & Jewell, 2002). Because it causes permeant fixation, it allows for long term storage with minimal to no change in sample overtime, and for these reasons, it is the preferred method. While the presence of ethanol can quench fluorophores, this can be nullified by thorough washing before introducing dyes and antibodies.

1. Pellet cells by centrifugation, wash once in PBS, and then resuspend in ice cold 70% ethanol, immediately pipetting up and down several times to avoid aggregation and clumping.

2. This fixation method is permanent, and cells have an extremely good shelf life when stored in ethanol in the freezer or fridge. We have kept cells in the fridge for up to 4 months and the freezer for over a year with no detectable change in sample integrity.

At this point, the fixed cells can be stored at 4 °C until ready to proceed to the DNA end labeling. It is not recommended to store PFA and Triton-X treated cells for more than 7–10 days in total (Lanier & Warner, 1981; Sedek et al., 2020), and so, it is recommended to label and image samples within a few days when using this method. For longer storage, use PFA fixation and ethanol permeabilization.

4. TUNEL labeling

4.1 TdT extension assay

Terminal deoxynucleotidyl transferase (TdT) enzyme adds a single nucleotide to free 3'-OH DNA ends, including synthetic nucleotides such as BrdU (Brouwer et al., 2001).

1. To make TdT reaction buffer, combine 100–300 μM BrdU (or other synthetic nucleotide) with 5 U of TdT in 1 × commercial TdT buffer with optional 2.5 mM cobalt chloride.

   Note: the optional addition of cobalt chloride enhances the tailing reaction.
2. Spin down fixed and permeabilized cells, wash two-three times in PBS, and then resuspend in 100 μL of TdT reaction buffer.
3. Incubate cells in a water bath or water-filled heat block at 37 °C for 60 min to allow 3′ extension to occur.
4. After elongation, pellet cells, wash with 1 mL PBS, and then resuspend in 1 mL PBS.

   Note: At this point cells can again be stored at 4 °C, but it is recommended to proceed directly to antibody probing.

   Note: from this point forward samples are kept in PBS or TBST buffers and thus are not suitable for long term storage. Samples should be imaged within 10–14 days of performing TUNEL extension, or sooner if treated with PFA and Triton-X.

4.2 Antibody probing

TdT tagged DNA nicks, breaks, or gaps with BrdU can be detected with appropriate fluorescent secondary anti-BrdU antibodies. Permeabilized cells allow antibodies to freely flow in and interact with internal components; however, over-fixation in PFA can create a cross-linking network that prevents effective diffusion within the cell. If antibody probing is weak on the positive control, we recommend experimenting with shorter fixation times.

1. Spin down cells and resuspend in an antibody blocking solution of 4% BSA in TBST. Let cells block by rocking at room temperature for 60 min.
2. Pellet cells and resuspend in 100 μL of 1:100 anti-BrdU primary antibody in 2% BSA in TBST. Incubate while rocking at room temperature for 60 min.
3. Wash three times by pelleting and resuspending in TBST. Then, resuspend in 100 μL of 1:500 fluorescent secondary antibody in TBST and incubate for 60 min at room temperature.

   Note: At this point, remember to protect everything from light for minimal signal loss.
4. Pellet the cells and wash 3 × with TBST to remove any unbound antibody. Resuspend cells in PBS and keep on ice or in the fridge

   Note: BSA can take a while to dissolve in water, and even after extended mixing there are often a few remaining small chunks and clumps. Prepare the BSA solution ahead of time and filter sterilize using a 0.45 μM syringe filter or equivalent. Aliquot and store in the −20 °C freezer indefinitely.

   Note: Any of the blocking and antibody incubation steps can also be performed at 4 °C overnight.

5. Analyzing your TUNEL samples

Fluorescent microscopic imaging and flow cytometry allow for both quantitative and qualitative analysis of a bacterial TUNEL assay. While neither method is dependent on the other, the data pairs nicely and provides two independent datasets for comparisons. To minimize variation from environmental conditions and instrument variation during flow cytometry, all samples from one data set must be run and analyzed during one session. Therefore, if you do plan to do both methods, perform microscopy first to screen the samples and re-prepare any that were not harvested or treated correctly before moving on to flow cytometry.

5.1 Microscopy

1. Spin down fixed cells, wash once, and then resuspend in 1 mL PBS. Add 10 μL of this solution to 90 μL of sterile PBS.

2. Working with clean labeled microscope slides, spot 2 μL each of the concentrated and 1:10 diluted solutions onto different areas of the slide. Cover and leave on the benchtop for ~10 min until dry.
3. Top each spot with 2–3 μL of DABCO with DAPI mounting solution and let rest for ~15 min. Cover each spot with a coverslip and seal with clear nail polish.
4. Let the sealant dry, and then store the slides protected from light at 4 °C overnight.
5. Image on an epifluorescent light microscope using a 60 × or 100 × oil immersion lens using DAPI and Cy5 preset filters. Bacteria exposed to MMC or NFZ should have a subpopulation of cells with evident foci, as seen in Fig. 2.
6. Collect and store images from multiple fields of view. Overlay DAPI and Cy5 channels according to microscope software, adjusting the signal intensity as needed to reduce background.
7. The total amount of Cy5 fluorescence or the number of Cy5 foci per cell can be quantified using commercial microscopy software or using Image J (Rasband, 1997–2016, 17 October 2015).

As seen in Fig. 2A (*left* panels), exogenous damage of MG1655 by either NFZ or MMC shows cellular elongation and prevalent BrdU foci throughout the cell. Transfection of a high copy number plasmid does not appear to change the distribution of foci through the cell (Fig. 2A, *middle* panels) Interestingly, genomic mutation of *dnaB* to create DnaB(R328/329) contained within the MSB5 strain (Table 1) shows prevalent BrdU TUNEL foci and cellular lengthening, even in the absence of any exogenous damage (Fig. 2A, *right* panels). This *dnaB* mutation along with several others has been thoroughly characterized (Behrmann et al., 2021).

5.2 Flow cytometry

1. After removing a few microliters of sample for microscopy, spin down the remaining sample and wash three times in PBS to remove any lingering ethanol. Resuspend test strains in 1 mL sterile 1 × PBS with 1.5 μM Sytox Green (SG).
2. Split both the BrdU-tagged control and the reserved unstained sample into two separate 1.5 mL tubes. Resuspend one tube in 1 mL sterile 1 × PBS with 1.5 μM SG, and the other in an equal volume of PBS only. There are now four controls: unstained, BrdU+, SG+, and a regular BrdU+/SG+ control strain sample.
3. Incubate samples for 30 min at room temperature while protecting from light.

4. Depending on cell density and the instrument specifications, dilute samples with sheath fluid or sterile 1 × PBS *(usually ~3–5 mL)* immediately prior to analysis.
5. Set the forward scatter (FSC) and side scatter (SSC) gain by running the reserved sample of unstained cells. Ensure that the entire cell population is visible on the scatter plot.
6. Using the unstained control, SG+ only control, and BrdU+ only control, set the gain for the FITC *(Sytox Green)* and Cy5.5 *(AlexaFluor 647)* laser presets using histograms (Fig. 3). Signal intensity should be optimized to make best use of the static X-axis range, as shown for the FITC laser preset in Fig. 3A
7. Once parameters are set, collect data from 100,000 events for all samples.
8. Using an analysis software for .fcs files, such as FloJo, begin by gating all samples for single cells. To do this, plot a scatter plot of FSC-H by FSC-A and gate for the population of cells that fall linearly on the diagonal, as shown in Fig. 3B.
9. Next, use the SG+ and unstained controls to gate all single cell events for +/− SG staining to exclude any data from impermeabilized cells or cell contaminants (Fig. 3C). As with FACS signal gain optimization, correct gating should be visually intuitive and clearly differentiate between controls.
10. Selecting the subpopulation of SG-stained single cell events, set a universal gate for +/− BrdU signal using the BrdU+ and SG+ *(which is also BrdU-)* controls as reference (Fig. 3D). The +/− gated populations *(in this case, the fraction of cells with DNA gaps)* are listed in the correlating top corners of each graph and can be exported to excel directly from the FloJo workspace.
11. Check BrdU gating against the control strain (*MG1655*) and control strain exposed to genotoxin (*MG1655+MMC*); ensure that the location of the gate falls naturally between peaks for cells with and without BrdU staining. Examples histograms are shown in Fig. 2B, with MG1655 +/− genotoxins (*left*) and the parental test strain HME63 +/− a genomic *dnaB* helicase mutation (*right*).

As seen in Fig. 2B, the addition of exogenous DNA damage agents, NFZ or MMC, shows a increase in BrdU fluorescence indicated by a large shift to the right for MG1655 and HME63. Interestingly, when *dnaB* is mutated on the genome, there is also a significant increase in BrdU even in the absence of any exogenous damage. The similarity in BrdU staining and

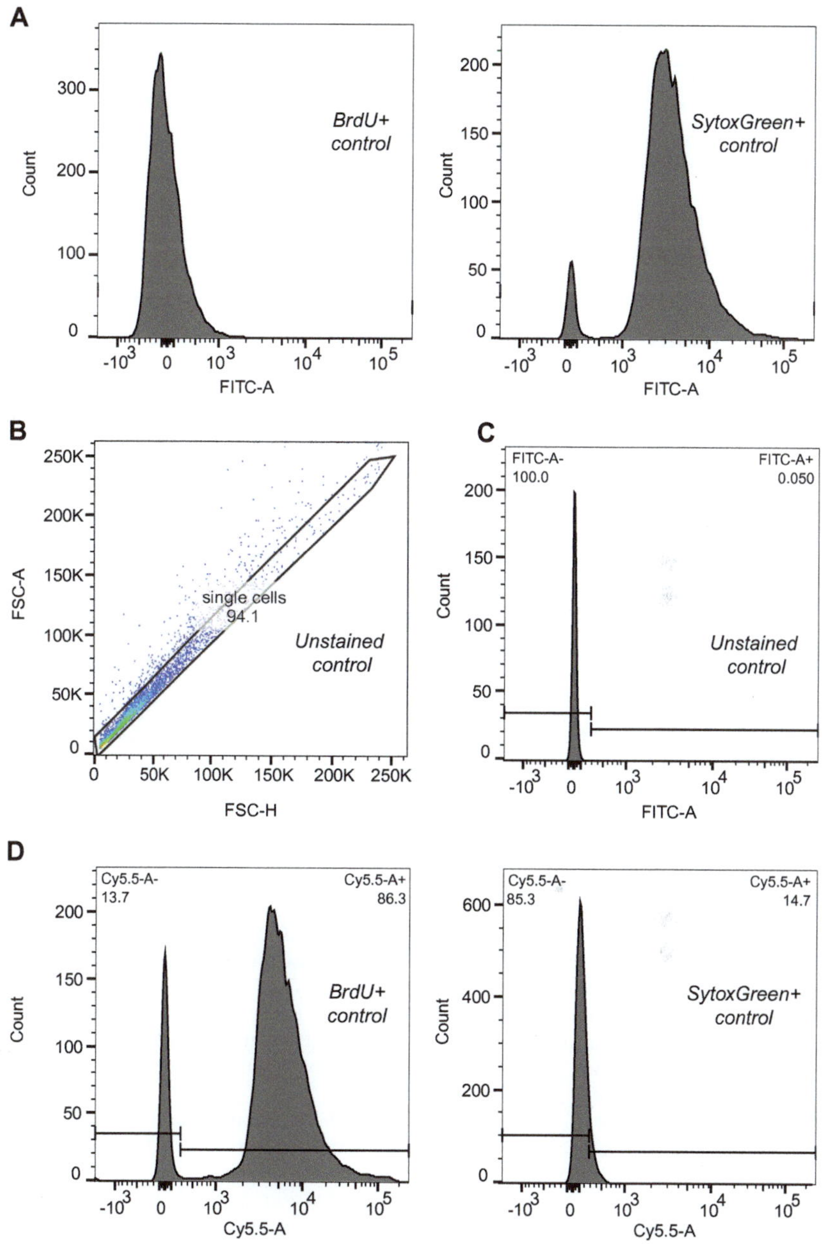

Fig. 3 Flow cytometry data collection and gating. Examples showing how to gate flow cytometry data in FloJo. (A) Prior to data collection, adjust the signal voltage so that the controls show up appropriately as high or low total signal based on their respective lasers, as shown for the green FITC laser. (B) The first gate applied to all samples is along the FSC-A/FSC-H diagonal for single cell events, followed by (C) gating for cells positive for SG DNA staining based on the unstained and SG+ controls. (D) To quantify the fraction of BrdU-containing cells, gate for +/− BrdU signal using the BrdU+ and unstained control, adjusting so that the SG+ control shows up as negative for BrdU fluorescence to avoid signal overlap.

intensities for either exogenously damaged cells or those of MSB5 suggest that this helicase mutant increases the frequency of decoupling, contributing to prevalent ssDNA gaps, nicks, or breaks.

6. Summary

DNA end labeling by TUNEL, an assay which has previously been utilized primarily for the detection of eukaryotic apoptosis, is a powerful technique for detecting and visualizing single stranded gaps and DNA ends in bacteria *in vivo*. Fluorescently tagging single strand DNA ends allows for detection of single strand nicks and gaps, which are important repair intermediates and fragile sites for replication or stress induced damage. While TUNEL does not specifically monitor ssDNA gaps, this method can be combined with an SOS induction assay to determine the likelihood of damage-induced foci as we did previously (Behrmann et al., 2021), or a secondary damage labeling technique like the DSB-specific binding of fluorescent Gam (Shee et al., 2013). This technique proves informative when generating a landscape of global replication and repair processes, especially when paired with targeted DNA damage assays and enzyme manipulation, such as DnaB helicase point mutations to test the consequences of altering the conformation to induce decoupling of unwinding synthesis.

References

Beattie, T. R., Kapadia, N., Nicolas, E., Uphoff, S., Wollman, A. J., Leake, M. C., et al. (2017). Frequent exchange of the DNA polymerase during bacterial chromosome replication. *eLife*, *6*, e21763.

Behrmann, M. S., Perera, H. M., Hoang, J. M., Venkat, T. A., Visser, B. J., Bates, D., et al. (2021). Targeted chromosomal Escherichia coli:dnaB exterior surface residues regulate DNA helicase behavior to maintain genomic stability and organismal fitness. *PLoS Genetics*, *17*, e1009886.

Brouwer, A. M., Frochot, C., Gatti, F. G., Leigh, D. A., Mottier, L., Paolucci, F., et al. (2001). Photoinduction of fast, reversible translational motion in a hydrogen-bonded molecular shuttle. *Science*, *291*, 2124–2128.

Chao, Y., & Zhang, T. (2011). Optimization of fixation methods for observation of bacterial cell morphology and surface ultrastructures by atomic force microscopy. *Applied Microbiology and Biotechnology*, *92*(2), 381–392. https://doi.org/10.1007/s00253-011-3551-5.

Costantino, N., & Court, D. L. (2003). Enhanced levels of lambda red-mediated recombinants in mismatch repair mutants. *Proceedings of the National Academy of Sciences of the United States of America*, *100*, 15748–15753.

Cranford, M. T., Kaszubowski, J. D., & Trakselis, M. A. (2020). A hand-off of DNA between archaeal polymerases allows high-fidelity replication to resume at a discrete intermediate three bases past 8-oxoguanine. *Nucleic Acids Research*, *48*, 10986–10997.

Darzynkiewicz, Z., Galkowski, D., & Zhao, H. (2008). Analysis of apoptosis by cytometry using TUNEL assay. *Methods, 44*, 250–254.

Dolbeare, F., & Selden, J. R. (1994). Immunochemical quantitation of bromodeoxyuridine: Application to cell-cycle kinetics. *Methods in Cell Biology, 41*, 297–316.

Dwyer, D. J., Camacho, D. M., Kohanski, M. A., Callura, J. M., & Collins, J. J. (2012). Antibiotic-induced bacterial cell death exhibits physiological and biochemical hallmarks of apoptosis. *Molecular Cell, 46*, 561–572.

Erental, A., Sharon, I., & Engelberg-Kulka, H. (2012). Two programmed cell death systems in *Escherichia coli*: An apoptotic-like death is inhibited by the *mazEF*-mediated death pathway. *PLOS Biology, 10*(3), e1001281.

Gavrieli, Y., Sherman, Y., & Ben-Sasson, S. A. (1992). Identification of programmed cell death *in situ* via specific labeling of nuclear DNA fragmentation. *The Journal of Cell Biology, 119*, 493–501.

Graham, J. E., Marians, K. J., & Kowalczykowski, S. C. (2017). Independent and stochastic action of DNA polymerases in the replisome. *Cell, 169*, 1201–1213.e1217.

Henrikus, S. S., Wood, E. A., McDonald, J. P., Cox, M. M., Woodgate, R., Goodman, M. F., et al. (2018). DNA polymerase IV primarily operates outside of DNA replication forks in *Escherichia coli*. *PLoS Genetics, 14*, e1007161.

Kath, J. E., Chang, S., Scotland, M. K., Wilbertz, J. H., Jergic, S., Dixon, N. E., et al. (2016). Exchange between *Escherichia coli* polymerases II and III on a processivity clamp. *Nucleic Acids Research, 44*, 1681–1690.

Kim, S., Dallmann, H. G., McHenry, C. S., & Marians, K. J. (1996). Coupling of a replicative polymerase and helicase: A tau-DnaB interaction mediates rapid replication fork movement. *Cell, 84*, 643–650.

Lanier, L. L., & Warner, N. L. (1981). Paraformaldehyde fixation of hematopoietic cells for quantitative flow cytometry (FACS) analysis. *Journal of Immunological Methods, 47*, 25–30.

Laureti, L., Demol, J., Fuchs, R. P., & Pages, V. (2015). Bacterial proliferation: Keep dividing and don't mind the gap. *PLoS Genetics, 11*, e1005757.

Lewis, J. S., Spenkelink, L. M., Jergic, S., Wood, E. A., Monachino, E., Horan, N. P., et al. (2017). Single-molecule visualization of fast polymerase turnover in the bacterial replisome. *eLife, 6*, e23932.

Li, X., & Darzynkiewicz, Z. (1995). Labelling DNA strand breaks with BrdUTP. Detection of apoptosis and cell proliferation. *Cell Proliferation, 28*, 571–579.

Li, X., Melamed, M. R., & Darzynkiewicz, Z. (1996). Detection of apoptosis and DNA replication by differential labeling of DNA strand breaks with fluorochromes of different color. *Experimental Cell Research, 222*, 28–37.

McInerney, P., & O'Donnell, M. (2004). Functional uncoupling of twin polymerases: Mechanism of polymerase dissociation from a lagging-strand block. *The Journal of Biological Chemistry, 279*, 21543–21551.

Mezzina, M., Menck, C. F., Courtin, P., & Sarasin, A. (1988). Replication of simian virus 40 DNA after UV irradiation: Evidence of growing fork blockage and single-stranded gaps in daughter strands. *Journal of Virology, 62*, 4249–4258.

Michelson, A. M., & Orkin, S. H. (1982). Characterization of the homopolymer tailing reaction catalyzed by terminal deoxynucleotidyl transferase. Implications for the cloning of cDNA. *The Journal of Biological Chemistry, 257*, 14773–14782.

Morimatsu, K., & Kowalczykowski, S. C. (2003). RecFOR proteins load RecA protein onto gapped DNA to accelerate DNA strand exchange: A universal step of recombinational repair. *Molecular Cell, 11*, 1337–1347.

Motea, E. A., & Berdis, A. J. (2010). Terminal deoxynucleotidyl transferase: The story of a misguided DNA polymerase. *Biochimica et Biophysica Acta, 1804*, 1151–1166.

Perera, H. M., Behrmann, M. S., Hoang, J. M., Griffin, W. C., & Trakselis, M. A. (2019). Contacts and context that regulate DNA helicase unwinding and replisome progression. *Enzyme, 45*, 183–223.

Rasband, W. S. (1997–2016, 17 October 2015). *ImageJ*. Bethesda, Maryland, USA: U. S. National Institutes of Health.

Rohwer, F., & Azam, F. (2000). Detection of DNA damage in prokaryotes by terminal deoxyribonucleotide transferase-mediated dUTP nick end labeling. *Applied and Environmental Microbiology*, *66*, 1001–1006.

Romero, Z. J., Chen, S. H., Armstrong, T., Wood, E. A., van Oijen, A., Robinson, A., et al. (2020). Resolving toxic DNA repair intermediates in every *E. coli* replication cycle: Critical roles for RecG, Uup and RadD. *Nucleic Acids Research*, *48*, 8445–8460.

Sandler, S. J. (2000). Multiple genetic pathways for restarting DNA replication forks in *Escherichia coli* K-12. *Genetics*, *155*, 487–497.

Sarac, I., & Hollenstein, M. (2019). Terminal deoxynucleotidyl transferase in the synthesis and modification of nucleic acids. *Chembiochem*, *20*, 860–871.

Sedek, L., Kulis, J., Slota, L., Twardoch, M., Pierzyna-SwitaLa, M., Perkowski, B., et al. (2020). The influence of fixation of biological samples on cell count and marker expression stability in flow cytometric analyses. *Central European Journal of Immunology*, *45*, 206–213.

Shee, C., Cox, B. D., Gu, F., Luengas, E. M., Joshi, M. C., Chiu, L. Y., et al. (2013). Engineered proteins detect spontaneous DNA breakage in human and bacterial cells. *eLife*, *2*, e01222.

Srinivasan, M., Sedmak, D., & Jewell, S. (2002). Effect of fixatives and tissue processing on the content and integrity of nucleic acids. *The American Journal of Pathology*, *161*, 1961–1971.

Strycharska, M. S., Arias-Palomo, E., Lyubimov, A. Y., Erzberger, J. P., O'Shea, V. L., Bustamante, C. J., et al. (2013). Nucleotide and partner-protein control of bacterial replicative helicase structure and function. *Molecular Cell*, *52*, 844–854.

Zhu, L., Rajendram, M., & Huang, K. C. (2021). Effects of fixation on bacterial cellular dimensions and integrity. *iScience*, *24*, 102348.

CHAPTER SEVEN

Biochemical methods to monitor loading and activation of hexameric helicases

Amy J. Fernandez and James M. Berger*
Department of Biophysics and Biophysical Chemistry, Johns Hopkins School of Medicine, Baltimore, MD, United States
*Corresponding author: e-mail address: jmberger@jhmi.edu

Contents

Abstract

Ring-shaped hexameric helicases are an essential class of enzymes that unwind duplex nucleic acids to support a variety of cellular processes. Because of their critical roles in cells, hexameric helicase dysfunction has been linked to DNA damage and genomic instability. Biochemical characterization of hexameric helicase activity and regulation *in vitro* is necessary for understanding enzyme function and aiding drug discovery efforts. In this chapter, we describe protocols for characterizing mechanisms of helicase loading, activation, and unwinding using the model replicative hexameric DnaB helicase and its cognate DnaC loading factor from *E. coli*.

Methods in Enzymology, Volume 672
ISSN 0076-6879
https://doi.org/10.1016/bs.mie.2022.03.061

1. Introduction

Helicases are a widespread class of enzymes that use the chemical energy from nucleoside triphosphate (NTP) hydrolysis to separate nucleic acid strands and displace proteins from DNA and RNA. These activities are central to cellular processes such as DNA replication, transcription, translation, and splicing. Helicases have been divided into six superfamilies, termed superfamily (SF) I to VI, based on their oligomeric state, translocation polarity, and motor domain fold (Gorbalenya & Koonin, 1993; Singleton, Dillingham, & Wigley, 2007). SF-III through SF-VI helicases act as ring-shaped hexamers that encircle DNA or RNA strands during translocation. Because some nucleic acid substrates lack free ends (e.g., internal chromosomal segments, circular chromosomes), many hexameric helicases depend on additional protein co-factors to assist with DNA or RNA strand loading and activation for unwinding (reviewed in Fernandez & Berger, 2021). Here, we describe a size-exclusion chromatography and gel-based assay to monitor the deposition of an archetypal, replicative hexameric helicase (*E. coli* DnaB) onto single-stranded (ss) DNA by its loading factor (DnaC). We also outline a simple fluorescent approach to measure DNA unwinding by DnaB using a plate reader. While these assays are described for a bacterial replication system, they can be easily adapted to characterize other helicase family members.

2. Hexameric helicase loading assay

This protocol describes a simple, benchtop gel-filtration based assay to monitor helicase loading onto ssDNA (Fig. 1A). By adding wildtype (WT) or mutant partner proteins along with various NTPs (e.g., ATP, ADP, ATPγS), this approach can be used to probe mechanistic aspects of the helicase loading reaction. Varying the salt type and concentration in the column buffer can be used as a control to ensure that the helicase is topologically closed around DNA and not just non-specifically associated with it. This assay has been adapted from O'Donnell and workers (Fang, Davey, & O'Donnell, 1999), as outlined in (Puri et al., 2021). Although the gel-filtration equipment used here relies on manual column packing, commercial pre-packed columns could in principle be used as well.

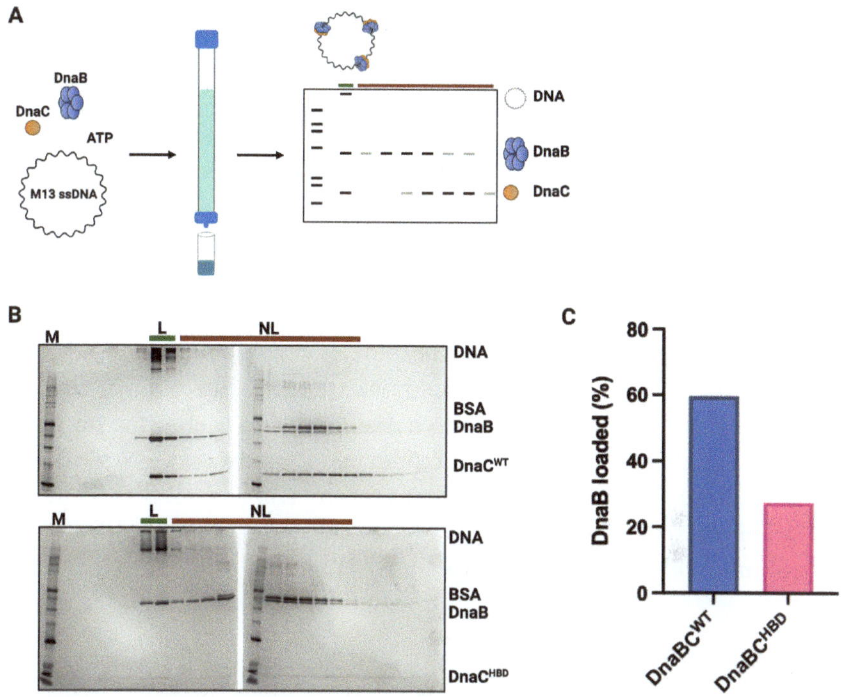

Fig. 1 (A) Overview of gel-filtration based helicase loading assay. (B) Sample silver-stained gel from helicase loading assay with DnaB and wild-type (WT) or isolated helicase binding domain (HBD) of DnaC. "M" in the molecular weight ladder, green bars with "L" indicate the void volume containing DNA and loaded DnaB, and red bars with "NL" indicate the free DnaB. Side labels (at right) indicate the migration points for different species in the reaction. (C) Quantification of the loading assay gels shown in (B).

2.1 Reagents, equipment, and software

Bio-gel A-1.5 m resin (*Bio-Rad*)

0.7 × 15 Flex-Columns and caps (*Kimble*)

12% poly-acrylamide Tris-glycine gels (*Invitrogen*)

1 × Tris-glycine buffer

SDS-PAGE apparatus (*Invitrogen*, XCell SureLock Mini-Cell)

Peristaltic pump, tubing, and column connectors

Sample collection tubes (1.5 mL snap-cap conical tubes, with mark at 250 μL).

Gel imaging system

ImageJ analysis software

2.2 Column preparation

i. Mark column at the 5 mL point to know how much resin to add to column.

ii. Gently invert resin to mix and then add ~5 mL slurry to Flex Column. Allow the column to start draining.

iii. As the resin settles, gently add more resin and allow to drain until a stable, 5 mL column bed is attained. Stop the column flow.

iv. Gently layer 2 mL ultrapure water on top of each column, disturbing the bed as minimally as possible. Hook the column up to a peristaltic pump that is connected to an ultrapure water reservoir.

v. Open the column release valve and turn on the pump to a low flow rate ($\leq$ 0.5 mL/min). As the column packs, periodically stop the flow, add small volumes of resin to the column, and then restart the system until the column is stable at 5 mL under pressure from the peristaltic pump.

vi. When the column is packed and washed with water, equilibrate the column with column buffer (see below).

2.3 Loading assay

i. Buffer preparation:

1. Column buffer: 40 mM HEPES-KOH pH 7.5, 100 mM NaCl, 10 mM $MgCl_2$, 5% glycerol, with or without nucleotide of choice.

2. Assay buffer: 40 mM HEPES-KOH pH 7.5, 10 mM $MgCl_2$, 5 mM DTT, 0.1 mg/mL BSA, 5 mM ATP (other nucleotides could be substituted if desired to monitor their effects). No salt is included in assay buffer as proteins are stored in high salt (500 mM KCl), resulting in a 20 mM final concentration of KCl in the reaction.

3. 5 × SDS-PAGE loading dye: 250 mM Tris-HCl pH 6.8, 30% glycerol, 10% SDS, 0.02% bromophenol blue, 5% BME.

4. **Note:** A standard loading assay is performed without nucleotide in the column buffer. To monitor helicase *un*loading, 1 mM ATP can be included in the column buffer. Additionally, the salt concentration in the column buffer can be increased (e.g., to 250 mM NaCl) to determine whether the helicase topologically encircles DNA.

ii. Set up 100 μL loading reactions in assay buffer that contain 10 μM helicase monomer, 10 μM loader, 1 μg M13 ssDNA, and 5 mM nucleotide. Incubate reactions at 37 °C for 15 min.

iii. Set up sample collection tubes (e.g., 1.5 mL conical snapcap tubes) under the column so that you can manually collect samples while applying the loading reaction to the column.

iv. Disconnect the peristaltic pump and gently aspirate any buffer from the top of column without disturbing the resin bed. Layer the reaction directly on top of the column bed without disturbing it or introducing bubbles.

v. Allow the reaction to enter the column while beginning to collect fractions. After the reaction volume has fully entered, carefully layer 150 μL column buffer atop the resin bed. After this fraction has fully entered the column, repeat this step with 1 mL column buffer. Move the fraction collector to the next tube for every 250 μL of eluent collected.

vi. Hook up the peristaltic pump and run column buffer through the column until you have collected 20–25 250 μL fractions.

vii. Analyze samples of each fraction by SDS-PAGE. Combine 20 μL of each fraction with 5 μL loading dye. Load 5 μL of each sample onto 12% Tris-glycine gels and run gels in 1 × Tris-Glycine buffer at 225 V for 40 min. For best resolution between DnaB and BSA, run gel until the dye front is just starting to come off the bottom of the gel.

2.4 Silver staining of SDS-PAGE gels

i. Solutions:

1. Fix solution: 50% methanol, 12% acetic acid
2. Silver stain solution, prepared fresh: 0.025% formaldehyde and 0.2 mg/mL $AgNO_3$ in ultrapure water
3. Developing solution, prepared fresh: 0.5 M Na_2CO_3, 0.018% formaldehyde, 16 mM sodium thiosulfate in ultrapure water
4. **Note:** Many components of silver staining solutions are unstable. A stock of 20% *w/v* silver nitrate can be prepared ahead of time and is best stored in a cold, dark place (such as wrapped in foil and stored in the fridge). Sodium thiosulfate can be prepared as a 10% stock solution and stored at room temperature for no longer than 1 week.

ii. After running, dismantle gel apparatus and fix gels for >1 h in fix solution with gentle shaking (can be left overnight).

iii. Wash gel for 15 min in 50% ethanol.

iv. Wash gel for 15 min in 30% ethanol.

v. Soak gel for 60 s in 800 μM sodium thiosulfate.

vi. Wash gel 3 times in ultrapure water, 20 s per wash.

vii. Develop gel in developing solution for ~3 min, until bands are easily visible but there is not high background.

viii. Wash gel quickly with ultrapure water and then stop developing with fix solution.

ix. Scan gels to generate image files.

x. **Note:** Overdeveloping gels leads to high background from the silver staining approach that interferes with gel quantification. It is critical to use ultrapure water and very clean gel boxes for all steps during the silver staining procedure. Avoid touching the gel surface (grasp only the edges if necessary) to avoid staining artifacts.

2.5 Data analysis

i. Using ImageJ or a comparable data analysis package, quantify and sum the intensities of the bands of DnaB that elute with DNA in the void volume of the column as compared to all fractions (void and otherwise) that contain DnaB (Fig. 1B).

ii. Helicase loading efficiency is calculated as the ratio of DnaB that elutes in the void volume compared to the total amount of DnaB in all fractions (Eq. 1, Fig. 1C).

$$DnaB\ Loaded\ (\%) = 100 \times \frac{Loaded\ DnaB}{Total\ DnaB} \tag{1}$$

3. Fluorescence-based fork unwinding assay

Helicase activity is commonly quantified by using radiolabeled nucleic acid substrates, manually quenching time points during the course of the unwinding reaction, and running the samples on native acrylamide gels. The fluorescence-based assay described here avoids the use of radioactivity, is faster, and allows for real-time data collection (Fig. 2A). Although DnaB unwinds double-stranded DNA in the 5′ to 3′ direction, the oligonucleotide substrates could be easily modified to have reversed polarity, RNA-DNA heteroduplexes, various modifications, or alternative structures. Final reactions contain 1.2 μM DnaB monomer, 1.2 μM DnaC monomer, 100 nM fork substrate, 200 nM capture strand that is complementary to a Cy3-labeled lead strand to prevent fork reannealing, and 1 mM ATP. To be able to record the initial unwinding phase, a separate mixture of the

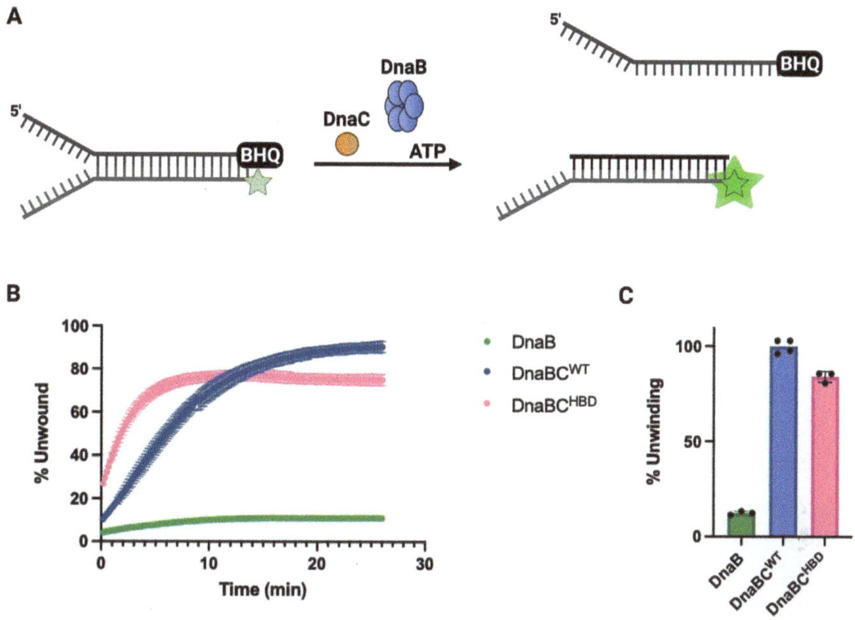

Fig. 2 (A) Overview of real-time fluorescence assay to monitor helicase-dependent DNA unwinding. (B) Sample data from unwinding assay. Note that *E. coli* DnaB is only marginally active for DNA unwinding on its own but can be stimulated by the addition of the WT DnaC loader, with the DnaC^{HBD} construct as a comparison. (C) Quantification of the maximum percent unwound relative to DnaBC^{WT}.

capture strand and ATP is made and added to the protein mixture at the plate reader immediately before data collection. This approach is based on a method developed for the SF-1 UvrD helicase by the Patel group (Nandakumar, Pandey, & Patel, 2015).

3.1 Reagents and equipment

Microplate reader with fluorescence and temperature control capabilities (e.g., *BMG Labtech* CLARIOstar)

37 °C heat block

384-well black flat-bottomed plate

Cy3-labeled leading strand: 5′-Cy3-TACGTAACGAGCCTGC$(T)_{25}$–3′

Black hole quencher-labeled lagging strand: 5′-$(T)_{25}$-GCAGGCTCG TTACGTA-BHQ2-3′

Capture strand: 5′-GCAGGCTCGTTACGTA-3′

Acetylated Bovine serum albumin (BSA) (*Sigma-Aldrich*)

3.2 Fork annealing

i. Reconstitute HPLC-purified DNA oligonucleotides to 100 μM in annealing buffer (20 mM Tris-HCl pH 7.5, 10 mM $MgCl_2$, 20 mM KCl).

ii. Set up annealing reaction with a final concentration of 20 μM leading strand and 24 μM lagging or capture strand in annealing buffer.

iii. Anneal oligos in a thermocycler using a protocol that incubates at 95 °C for 3 min and then decreases to 4 °C at a rate of 1 °C per min.

iv. Aliquot the annealed fork in 5 μL fractions, snap freeze in liquid nitrogen, and store at −80 °C.

v. **Note:** For this assay, two substrates are required—one with the Cy3-labeled lagging strand annealed to the BHQ-labeled leading strand (experimental substrate) and another with the Cy3-labeled lagging strand annealed to the capture strand (control substrate). The Cy3/BHQ fork is used to monitor unwinding, whereas the Cy3/capture substrate provides the maximum fluorescence signal that is needed to normalize the signal between experiments.

3.3 Unwinding assay

i. Prepare Reaction buffer: 20 mM HEPES-KOH pH 7.5, 5 mM magnesium acetate, 50 mM potassium glutamate, 5% glycerol, 0.2 mg/mL BSA.

ii. Turn on the plate reader, set to 37 °C, and put plate inside to pre-warm.

iii. Prepare a 75 μL control reaction containing 100 nM Cy3/capture substrate, 200 nM capture strand, and 1 mM ATP (or another nucleotide, if desired) in reaction buffer. Mix sample by gently resuspending and pre-warm for 5 min at 37 °C. Aliquot this sample into three 20 μL aliquots (technical replicates) in the pre-warmed plate and use for gain adjustment in the plate reader. This sample will also be used as a maximum fluorescence measurement to normalize between experiments.

iv. Prepare a 60 μL helicase/loader pre-mix containing 1.6 μM DnaB monomer, 1.6 μM DnaC monomer, and 133 nM fork substrate in reaction buffer on ice. Separately, prepare 20 μL of a pre-mix containing 4 mM ATP and 800 nM capture strand in reaction buffer. Incubate each solution at 37 °C for 5 min to pre-warm.

v. At the plate reader, initiate the reaction by adding the ATP/capture strand solution to the helicase/loader sample, pipette gently to mix, and immediately dispense as three 20 μL aliquots (technical replicates) into the 384 well plate.

vi. Monitor the control and reaction samples for change in Cy3 emission at 37 °C by collecting fluorescence data in 10 s increments for 26 min. Strand separation efficiency can be determined by measuring the difference in the percent of DNA unwound after a wildtype enzyme reaction has reached equilibrium (as evidenced by the plateauing of the curve) with respect to a control reaction (here, DnaBC vs. DnaB; Fig. 2B).

3.4 Data analysis

i. Average the raw fluorescence values obtained for each technical replicate at each collected time point for both the helicase unwinding reaction and the Cy3/capture control.

ii. Convert raw fluorescence readings to percent unwound values by normalizing each time point to the captured-fork raw fluorescence value. Plot percent unwinding against time (Eq. 2 and Fig. 2B).

$$\%Unwound = 100 \times \frac{Averaged\ fluorescence}{Captured\ fork\ fluorescence} \qquad (2)$$

4. Summary and discussion

The biochemical characterization of helicases *in vitro* is critical for understanding helicase mechanism, the role of partner proteins, and can aid in drug development. Here we describe two simple assays that can be used to study the bacterial replicative helicase, DnaB, and its loader protein, DnaC, in stimulating helicase loading and unwinding. By using some of the suggested modifications to reaction conditions, substrates and partner proteins, these assays can be easily adapted to study a variety of hexameric helicases. Additionally, the fluorescence-based DNA unwinding assay can be easily ported to a high-throughput format for conducting chemical inhibitor screens of helicase function.

Acknowledgments

This work was supported by the NIGMS ([R37-GM071747] to J.M.B. and [T32-GM008403] to A.J.F.) and the NSF ([DGE-1746891] to AJF). Figures were created with Biorender.

References

Fang, L., Davey, M. J., & O'Donnell, M. (1999). Replisome assembly at oriC, the replication origin of E. coli, reveals an explanation for initiation sites outside an origin. *Molecular Cell*, *4*(4), 541–553. https://doi.org/10.1016/s1097-2765(00)80205-1.

Fernandez, A. J., & Berger, J. M. (2021). Mechanisms of hexameric helicases. *Critical Reviews in Biochemistry and Molecular Biology*, *1–19*. https://doi.org/10.1080/10409238.2021.1954597.

Gorbalenya, A. E., & Koonin, E. V. (1993). Helicases: Amino acid sequence comparisons and structure-function relationships. *Current Opinion in Structural Biology*, *3*(3), 419–429. https://doi.org/10.1016/S0959-440X(05)80116-2.

Nandakumar, D., Pandey, M., & Patel, S. S. (2015). Cooperative base pair melting by helicase and polymerase positioned one nucleotide from each other. *eLife*, *4*. https://doi.org/10.7554/ELIFE.06562.

Puri, N., Fernandez, A. J., O'Shea Murray, V. L., McMillan, S., Keck, J. L., & Berger, J. M. (2021). The molecular coupling between substrate recognition and atp turnover in a aaa+ hexameric helicase loader. *eLife*, *10*. https://doi.org/10.7554/ELIFE.64232.

Singleton, M. R., Dillingham, M. S., & Wigley, D. B. (2007). Structure and mechanism of helicases and nucleic acid translocases. *Annual Review of Biochemistry*, *76*(1), 23–50. https://doi.org/10.1146/annurev.biochem.76.052305.115300.

CHAPTER EIGHT

DNA fiber analyses to study functional importance of helicases and associated factors during replication stress

Arindam Datta*,† and Robert M. Brosh Jr.*

Helicases and Genomic Integrity Section, Translational Gerontology Branch, National Institute on Aging, NIH, Baltimore, MD, United States

*Corresponding authors: e-mail address: arindam.datta@pennmedicine.upenn.edu; broshr@mail.nih.gov

Contents

Abstract

Helicases, DNA translocases, nucleases and DNA-binding proteins play integral roles in protecting replication forks in human cells. Perturbations to replication fork dynamics can be caused by genetic loss of key factor(s) or exposure to replication stress inducing agents that perturb the nucleotide pool, stabilize unusual DNA secondary structures, or inhibit protein function (typically catalytic activity performed by a DNA polymerase, nuclease or helicase). DNA fiber analysis is a highly resourceful and facile experimental approach to study the molecular dynamics of replication forks in living cells. In this chapter, we provide a detailed list of reagents, equipment and experimental strategies to

† Current Address: Department of Cancer Biology, Penn Center for Genome Integrity, Basser Center for BRCA, Perelman School of Medicine, University of Pennsylvania, Philadelphia, PA 19104-6160, USA

Methods in Enzymology, Volume 672
ISSN 0076-6879
https://doi.org/10.1016/bs.mie.2022.04.003

perform DNA fiber experiments. We have utilized these approaches to characterize the role of the Werner syndrome helicase (WRN) to protect replication forks in cells that are deficient in the tumor suppressor and genome stability factor BRCA2.

Abbreviations

CldU chlorodeoxyuridine
HR homologous recombination
HU hydroxyurea
IdU iododeoxyuridine
MMC mitomycin C
MMS methylmethane sulfonate
SL synthetic lethality
ssDNA single-stranded DNA
WRN Werner syndrome helicase
WRNi Werner inhibitor

1. Introduction

In the last couple decades, experimental evidence from a number of laboratories has demonstrated that the replication forks of mammalian cells can stall due to endogenous or exogenously induced stress, undergo remodeling, become degraded in a controlled or uncontrolled manner, and even be restarted with varying degrees of efficiency (for review, see (Berti, Cortez, & Lopes, 2020; Rickman & Smogorzewska, 2019; Scully, Elango, Panday, & Willis, 2021)). The fate of replication forks is a major determining factor for genomic instability, which is a leading cause of both cancer and aging. Characterizing the various intrinsic and extrinsic factors that influence fork stability and delineating the mechanisms involved is paramount to understand the very basis for preservation of chromosomal integrity and how fork instability relates to aberrant cellular phenotypes and human disease.

One of the more notable early discoveries in this area was the identification and characterization of a novel role for the breast cancer tumor suppressor BRCA2 in fork protection by the Jasin lab (Schlacher et al., 2011). Previously, BRCA2 was known for its role in double-strand break (DSB) repair to facilitate RPA removal from single-stranded DNA that had been resected at the sites of chromosomal breaks and subsequent loading of the major strand exchange protein Rad51 (Jensen, Carreira, & Kowalczykowski, 2010; Moynahan, Pierce, & Jasin, 2001). The removal of RPA and loading of Rad51 allows the resected single-stranded DNA to invade the site of homologous duplex in the sister chromatid of diploid cells by hydrogen bonding with

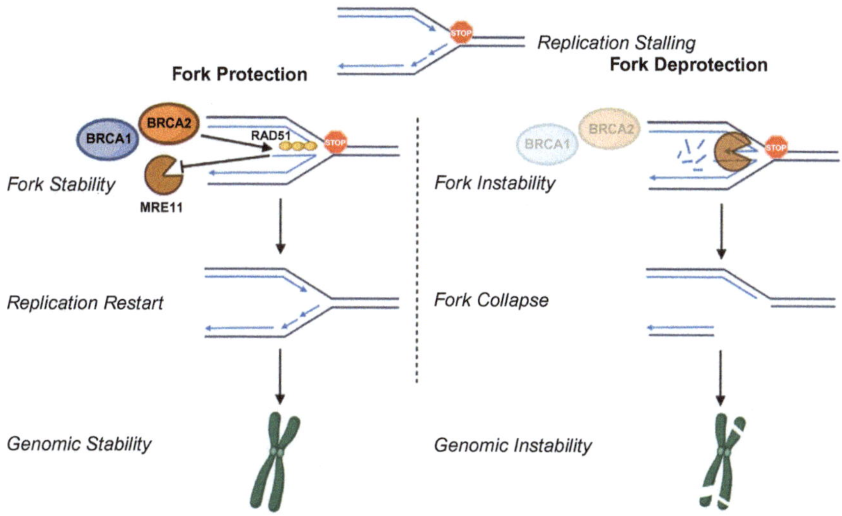

Fig. 1 Role of BRCA1 and BRCA2 in replication fork protection and genomic stability in human cells. Created with BioRender.com.

the complementary strand, ultimately permitting error-free DSB repair. However, BRCA2 has an additional function independent of DBS repair to prevent stalled replication forks from being degraded by the structure-specific endo-/*exo*-nuclease MRE11 (Schlacher et al., 2011) (Fig. 1). Elegant single-cell studies employing specialized nucleotide analogs that could be detected by specific antibodies and immunofluorescence techniques convincingly demonstrated that under conditions in which MRE11 was mutated or depleted in human cells, the nascently incorporated DNA at replication forks of living cells became susceptible to nucleolytic degradation by MRE11 nuclease. As it turns out, there are quite a few proteins that regulate fork stability in human cells including DNA-binding proteins, ATP-dependent translocases, nucleases, and helicases. For a detailed perspective of fork remodeling mechanisms, the readers are referred to several excellent review articles (Cortez, 2015; Quinet, Lemaçon, & Vindigni, 2017; Thakar & Moldovan, 2021).

Concomitant with the discoveries of proteins that influence the stability of replication forks, there has been a growing interest in synthetic lethality (SL) as a therapeutic strategy to combat cancer (Setton et al., 2021). SL is based on the principal that mutation, inactivation/depletion, or inhibition of two genes/gene products results in cell death, whereas loss of either gene by itself is compatible with viability. SL has been classically defined in relation to cell death, whereas loss of either gene by itself is compatible with viability. However, the loss of two genes can also be scored by a synergistic cellular phenotype that goes above simply an additive effect of loss of the two gene functions separately. Often SL can be conveniently measured by a cell proliferation assay, detection of a DNA damage marker (e.g., γ-H2AX, 53BP1), or some other readout for deleterious phenotype. With the advances of DNA fiber autoradiography in living cells, fork instability attributed to the co-deficiency of genetically interacting factors that operate in parallel pathways can be detected and quantitated. One class of genetic interactions underlying SL that we have been particularly interested in is that of the RECQ DNA helicases (Datta, Dhar, Awate, & Brosh, 2021).

In this article, we describe experimental methodology to assess genetic interactions important for replication fork stability in human cells. We employed the DNA fiber technique to investigate replication defects attributed to the SL interaction of BRCA2 with WRN (Datta, Biswas, et al., 2021), a RECQ helicase and ssDNA exonuclease encoded by the *WRN* gene. Bi-allelic mutations in *WRN* are linked to a premature aging disorder known as Werner syndrome (WS) (Yu et al., 1996) that is characterized by chromosomal instability, replication defects and sensitivity to DNA damaging agents (for review, see (Oshima, Sidorova, & Monnat, 2017)). Here, we provide the reader an itemized list of key resources and reagents, as well as detailed protocol instructions and tips for analyses and troubleshooting. We also discuss the limitations and advantages of DNA fiber studies to probe genetic interactions important for replication fork protection that upon perturbation potentially contribute to SL.

2. DNA fiber assay using fiber spreading technique

DNA fiber assay is a powerful technique to examine DNA replication genome-wide at single-molecule resolution. The technique has been widely used to study the immediate effects of DNA damaging agents or replication inhibitors on replication fork dynamics or how certain genetic mutations affect replication pattern in mammalian cells. Using permutations of the DNA fiber technique, researchers have gained mechanistic insights into the SL induced by PARP inhibitors in homologous recombination (HR)-deficient cancers (Panzarino et al., 2021; Ying, Hamdy, & Helleday, 2012) and a better appreciation for the mechanisms of acquired chemoresistance

(Liao, Ji, Helleday, & Ying, 2018; Ray Chaudhuri et al., 2016). We employed the DNA fiber spreading technique to unveil a critical function of WRN helicase in the recovery of unprotected forks following replication stalling in *BRCA2*-deficient cancer cells (Datta, Biswas, et al., 2021). Moreover, results obtained from our DNA fiber experiments suggest that the WRN inhibitor (WRNi)-induced SL in *BRCA2*-mutated cancer cells is attributed to increased fork stalling, exacerbated fork degradation, and compromised fork restart (Datta, Biswas, et al., 2021).

DNA fiber technique exploits the ability of actively dividing cells to incorporate halogenated thymidine analogs in newly synthesized DNA molecules. Briefly, replicating DNAs are sequentially labeled with chlorodeoxyuridine (CldU) and iododeoxyuridine (IdU) nucleoside analogs followed by stretching of labeled DNA fibers on charged glass slides or coverslips. CldU-IdU labeled replication events are probed with anti-CldU and anti-IdU antibodies and subsequently visualized by immunofluorescence microscopy. By modulating the CldU-IdU labeling scheme, researchers can study crucial replication events including replication fork speed, frequency of origin firing, nucleolytic degradation of stalled forks, efficiency of stalled fork restart, and single-stranded DNA (ssDNA) gaps generated during replication.

DNA fibers can be prepared by either molecular combing or a fiber spreading technique. In the combing technique (Bianco et al., 2012), labeled DNA fibers are combed onto charged coverslips at a regulated and constant speed using a DNA combing machine which ensures uniform stretching of the DNA molecules on the glass surface. Although DNA combing allows precise determination of DNA tract length because of evenly stretched straight fibers, the technique involves multiple steps, thereby making it time-consuming. Moreover, only a limited number of samples can be processed at a time. Preparation of DNA fibers using the fiber spreading technique utilizes gravity to stretch the labeled fibers on the charged microscope slides immediately after cell lysis (Parra & Windle, 1993) (Fig. 2). The fiber-spreading technique allows faster processing of multiple samples as compared to DNA combing method. In the following sections, we discuss the detailed methodologies of DNA fiber spreading technique which will allow researchers to quickly examine replication perturbations induced by genotoxic agents, small molecule enzyme inhibitors, or specific genetic mutations in human cells.

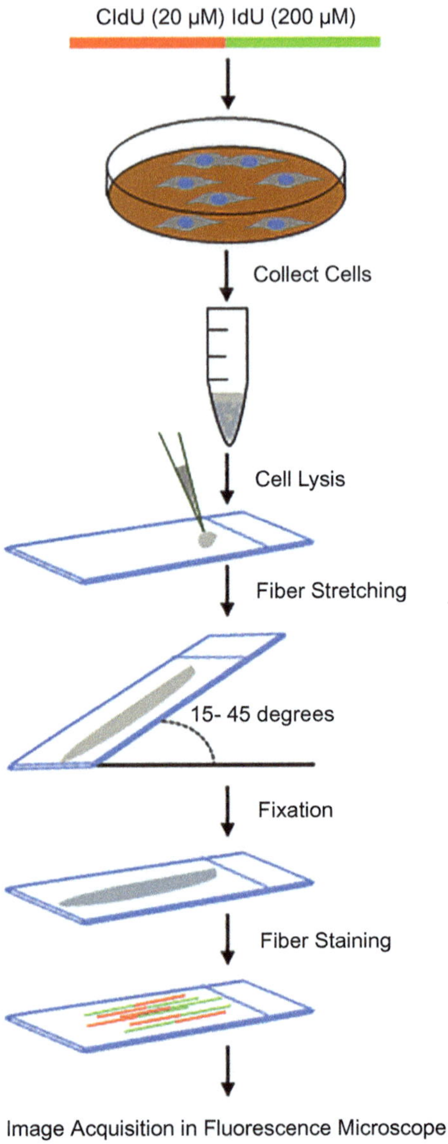

Fig. 2 Scheme of the DNA fiber spreading technique. DNA fibers are stretched along a glass slide by gravitational force. See text for details.

2.1 Reagents, equipment and buffers

- Dulbecco's Modified Eagle's Medium (DMEM, Cat. No. 11965092, Thermo Fisher Scientific) or RPMI-1640 (A1049101, ThermoFisher Scientific) tissue culture medium

- Penicillin-Streptomycin (Cat. No. 15070063, ThermoFisher Scientific)
- Trypsin (0.25%, Cat. No. 15050057, ThermoFisher Scientific)
- 1 × DPBS (Cat. No. 14190144, ThermoFisher Scientific)
- 5-Chloro-2′-deoxyuridine (Cat. No. C6891, Sigma-Aldrich)
- 5-Iodo-2′-deoxyuridine, Cat. No. I7125, Sigma-Aldrich)
- 2.5 M HCl (Stock solution—Cat. No. H1758, Sigma-Aldrich)
- ProLong Gold Antifade Mountant (Cat. No. P36930, ThermoFisher Scientific)
- Rat anti-CldU antibody (Cat. No. ab6326, Abcam)
- Mouse anti-IdU antibody (Cat. No. 347580, BD Pharmingen)
- AlexaFluor 647-conjugated anti-rat IgG secondary antibody (Cat No. A-21247, ThermoFisher Scientific)
- AlexaFluor 488-conjugated anti-mouse IgG secondary antibody (1:100, A-11001, ThermoFisher Scientific)
- 6-well tissue culture plate (Corning)
- Silane-coated microscope slides (5070, Newcomer Supply)
- Premiere Brand Coverslips (24 × 60 mm, Cat. No. 62307, Newcomer Supply)
- Coplin staining jar (S5641-12EA, Sigma-Aldrich)
- Zeiss Axio Observer.Z1/7 fluorescence microscope
- Cell lysis buffer (200 mM Tris–HCl pH 7.4, 50 mM EDTA, and 0.5% SDS)
- Fixing solution (Methanol: Acetic acid, 3:1)
- Neutralization buffer (400 mM Tris-HCl pH 7.4)
- Blocking buffer (5% BSA + 10% Goat serum in 1 × DPBS)

2.2 Methods

We describe here a step-by-step protocol of a DNA fiber experiment optimized to characterize fork progression in the BRCA2-deficient colorectal cancer cell line DLD1. The protocol can be used for other adherent cancer cell lines as well.

2.2.1 In vivo pulse-labeling of replication forks

Timing: 15–60 min

The first step in a DNA fiber assay is to pulse label the replication forks with IdU and CldU analogs in a sequential manner. In our experiments, we label the cells with CldU followed by IdU, and we designate CldU (red) and IdU (green) as 1st and 2nd label, respectively, in the subsequent sections. It is important to use the 2nd analog (IdU) at 10-fold greater concentration than that of the 1st analog to dilute out the residual amount of 1st label (CldU). Typically, asynchronously growing cells are sequentially incubated in complete cell culture medium containing 20 μM CldU and 200 μM IdU for 20 min each. However, the pulse labeling time could vary depending on the scheme of the specific experiment and cell types. For example, to evaluate the effects of a DNA damaging drug on replication fork progression, a short pulse of the 1st label (20 min) can be followed by a longer 2nd label pulse (40–60 min) in the presence of the drug.

1. Seed 1.5×10^5 cells in 2 mL of RPMI-1640 medium supplemented with 10% fetal bovine serum, 1% Penicillin-Streptomycin in a 6-well cell culture plate and grow overnight in a CO_2 (5%) incubator at 37 °C
2. Replace the medium with 2 mL of pre-warmed complete growth medium containing 20 μM CldU to each well and store for 15 min in incubator (Note 1).
3. Replace CldU with 2 mL of medium containing 200 μM IdU and desired concentration of the DNA damaging agent or replication inhibitor. Store for an additional 60 min in incubator. One control plate without drug should also be included (Note 2).

2.2.2 Preparation of DNA fibers

Timing: 1–2 h

1. Wash cells twice with ice-cold 1 × PBS, trypsinize, and collect in PBS in 15 mL tubes. Pellet down the cells by centrifugation at 500 × *g* for 5 min at 4 °C and resuspend in 1 × PBS to a final concentration of 10^6 cells/mL
2. Add 2 μL of cells suspension on the edge of the silane-coated slide next to the slide label and let it set for 1 min
3. Carefully add 10 μL of Cell lysis buffer to cells so that the tip does not touch the drop of cells on the slide. Wait for 2–3 min until the Cell lysis buffer on the slide is slightly dried out
4. Tilt the slides at around 10° to 25° angle and allow the liquid to slowly travel down the slide. Let the slides air-dry for 15–20 min (Note 3)
5. Fix the slides in Fixing solution in a coplin jar for 10 min. Let the slides air-dry
6. Proceed with the antibody staining. Alternatively, slides can be stored at −20 °C

2.2.3 Antibody staining of labeled fibers

Timing: 6–7 h.

1. Wash the slides with 1 × PBS at RT
2. Denature DNA with 2.5 M HCl for 1 h at RT

3. Neutralize the slides in 400mM Tris-HCl (pH 7.4) for 10min. Maintain the coplin jars on a rocker during neutralization
4. Wash the slides with 1 × PBS at RT (5min on a rocker). Wash with PBST (0.1% Tween 20) for 5min
5. Block the slides in blocking buffer at RT for 1h in a humid chamber. 100 μL of Blocking buffer is carefully placed in 3 drops along the length of the slide and evenly distributed over the entire slide using a strip of parafilm. The parafilm will create a thin layer of the Blocking buffer over the slide. Alternatively, a rectangular coverslip can be used (Note 4)
6. Add 100 μL of rat anti-CldU (1:200 dilution) primary antibody in a similar way described in step 5 and incubate for 1h at RT
7. Wash the slides with PBST for 5min at RT on a rocker. Repeat the PBST wash 3 times
8. Add 100 μL of AlexaFluor 647-conjugated anti-rat secondary antibody (1:100 dilution) and incubate the slides at RT for 45min in a humid chamber in the dark
9. Wash the slides as described in step 7
10. Add mouse anti-IdU primary antibody (1:40 dilution) and incubate at RT for 1h in a humid chamber in the dark
11. Wash the slides as described in step 7
12. Add 100 μL of AlexaFluor 488-conjugated anti-mouse secondary antibody (1:100 dilution) and incubate the slides at RT for 45min in a humid chamber in the dark
13. Wash the slides as described in step 7
14. Remove the excess liquid by tapping the slides on a tissue paper. Let the slides air dry in dark
15. Add 100 μL of ProLong Gold mountant and carefully place the coverslip on the top. Let the slides dry
16. Proceed with imaging. Slides can be stored at −20 °C

2.2.4 Image acquisition and data analysis

Visualize DNA fibers using a 63 × (oil) objective lens equipped with Cy5 and GFP filters for the detection of CldU and IdU labeled fibers, respectively (e.g., Zeiss Axio Observer.Z1/7). Score >200 untangled straight fibers for each experimental sample. Avoid imaging fields crowded with overlapping fibers. Fiber length and replication events (Fig. 3) can be measured or quantified using an imaging software (e.g., ZEN 3.0, Image J). A suitable statistical tool such as GraphPad Prism can be used for the quantitative

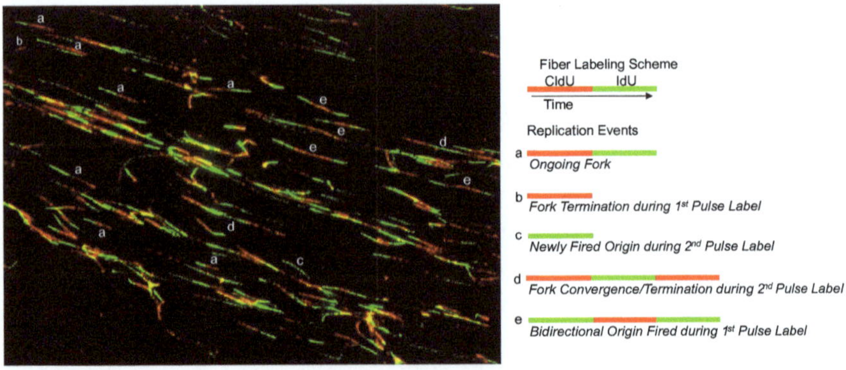

Fig. 3 DNA fibers from *BRCA2*-mutated ovarian cancer cells using the sequentially labeling protocol. Representative image (63 × oil objective lens) of DNA fibers prepared from *BRCA2*-mutated ovarian cancer cell line PEO1. Cells were sequentially pulse-labeled with CldU and IdU for 30 min each and DNA fibers were prepared using the fiber spreading technique. Replication events observed by the assay are shown.

comparison between experiment and controls samples by determining IdU/CldU tract length ratios or percentage of replication intermediates (e.g., stalled or restarted fork structures). We determined perturbation of replication fork progression in DLD1 cells upon WRN helicase inhibitor (WRNi) by measuring the mean tract length of IdU label normalized to that of CldU label. Compared to untreated cells, reduced IdU/CldU ratio in WRNi-treated cells implicates impaired fork progression induced by WRNi treatment of DLD1 cells (Fig. 4).

2.3 Biological questions addressed using the DNA fiber assay

The DNA fiber assay can be used to study different replication events in human cells upon loss of a critical protein or in response to a replication stress-inducing agent. In the following sections, we describe pulse-labeling strategies and data analysis to examine fork progression rate, nucleolytic degradation and replication restart following fork stalling. However, other biological questions such as replication origin firing and ssDNA gaps on daughter strand can be examined using DNA fiber spreading technique.

2.3.1 Replication fork progression and fork symmetry

Replication fork progression can be examined by sequential pulse-labeling of the ongoing forks by CldU (red, 1st label) and IdU (green, 2nd label) thymidine analogs. Fork progression between control and experimental samples can be compared by quantifying the mean tract length of IdU label

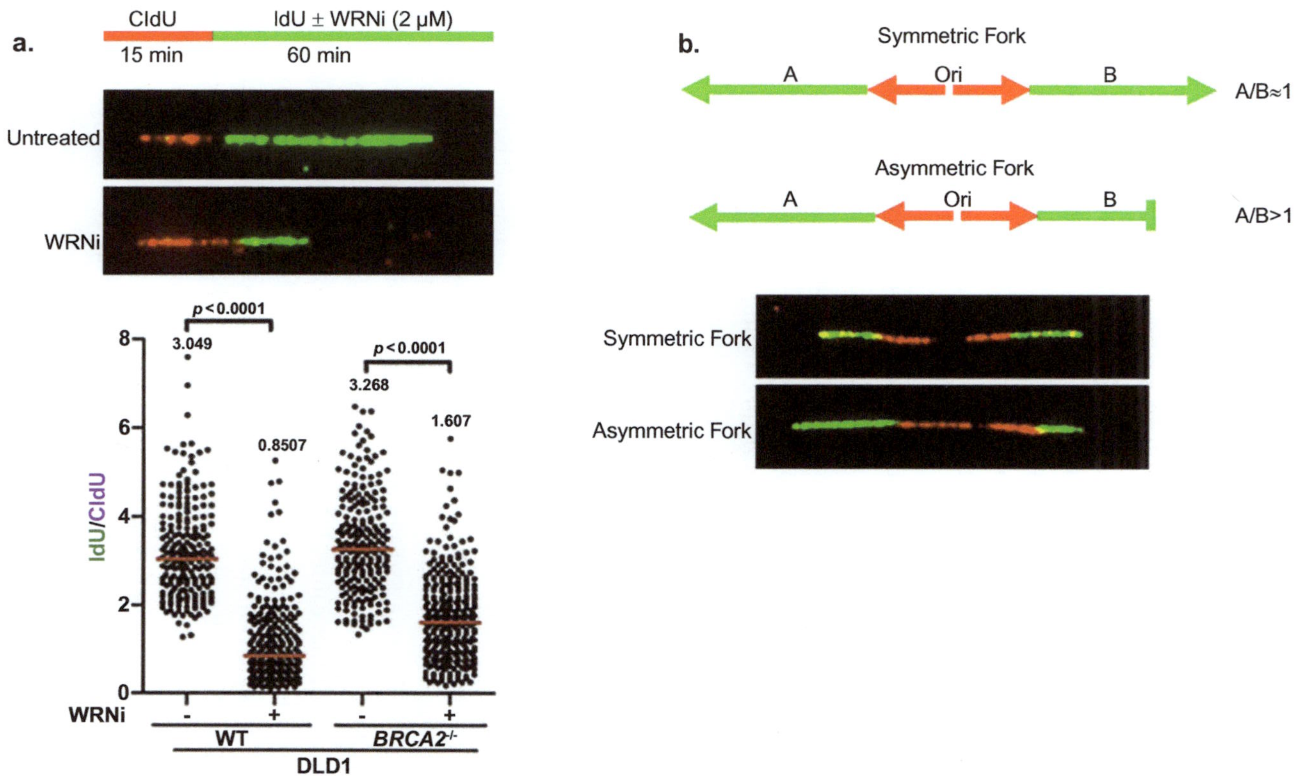

Fig. 4 DNA fiber studies used to detect replication abnormalities caused by exposure of the colorectal cancer cell line DLD1/*BRCA2*$^{-/-}$ to the WRN helicase inhibitor (WRNi). (A) IdU-labeled fibers are visibly shorter in cells exposed to the WRNi, indicative of reduced fork progression. Scatter dot plot showing reduced IdU/CldU ratios, indicating impaired fork progression upon WRN helicase inhibition. The graph was published in (Datta, Biswas, et al., 2021). This is an open-access article published under the Creative Commons Attribution 4.0 International (CC BY 4.0) https://creativecommons.org/licenses/by/4.0/. (B) Dissimilar tract lengths of IdU-labeled fibers originating from bidirectional forks, indicative of an asymmetric fork.

normalized to that of CldU of a contiguous CldU-IdU (red-green) labeled fiber. To study how exposure to a DNA damaging agent affects replication fork progression, forks are pulse-labeled with IdU concomitantly with the DNA damaging drug following a short pulse of CldU (Fig. 4A). Perturbation of fork progression induced by the drug can be determined by measuring the mean IdU tract length in treated and untreated samples. To accurately evaluate the effects of a DNA damaging agent on fork progression, IdU tract length can be normalized to CldU of individual contiguous CldU-IdU fibers to nullify the possible impacts of slow replicating genomic regions on the quantification. Compared to untreated cells, a decrease in IdU/CldU ratios in drug-treated cells suggest impaired fork progression induced by the drug.

Fork symmetry refers to the synchronized progression of two sister forks emanating from a single bi-direction origin (Fig. 4B). Forks are asymmetric when one of the two sister forks progresses bi-directionally at a different rate resulting in unequal sister fork length. Fork asymmetry is used as a parameter to determine the frequency of replication fork stalling or collapse events under condition of replication stress. Fork asymmetry is determined by calculating the sister fork (SF) tract (IdU) length ratio of individual bidirectional replication forks (IdU-CldU-IdU). A SF ratio of ~1 represents symmetric fork progression while SF ratio <1 indicates fork asymmetry.

2.3.2 Fork restart efficiency

The ability of stalled forks to resume DNA synthesis can be assessed by fork restart experiment using DNA fiber technique. In a classic fork restart assay, replication stalling agents such as hydroxyurea (HU) or aphidicolin are used to quickly stall all replication forks in cells. Fork restart following exposure to DNA damaging agents such as methylmethane sulfonate (MMS) or mitomycin C (MMC) that causes DNA lesions randomly in the replicating genome can also be studied using DNA fiber assay. However, relatively higher doses of these DNA damaging drugs should be used to generate DNA lesions sufficient such that a significant fraction of pulse-labeled forks encounters the damage. In a fork restart experiment, cells are first treated with the 1st analog (CldU) followed by treatment with replication stalling agent such as HU for a specific amount of time. After drug removal, cells are subsequently treated with the 2nd analog (IdU) to evaluate whether the stalled forks are able to restart as evaluated by the incorporation of the 2nd label (Fig. 5). With the labeling scheme described above, restarted fork are represented as DNA tracts with both CldU and IdU label (CldU-IdU,

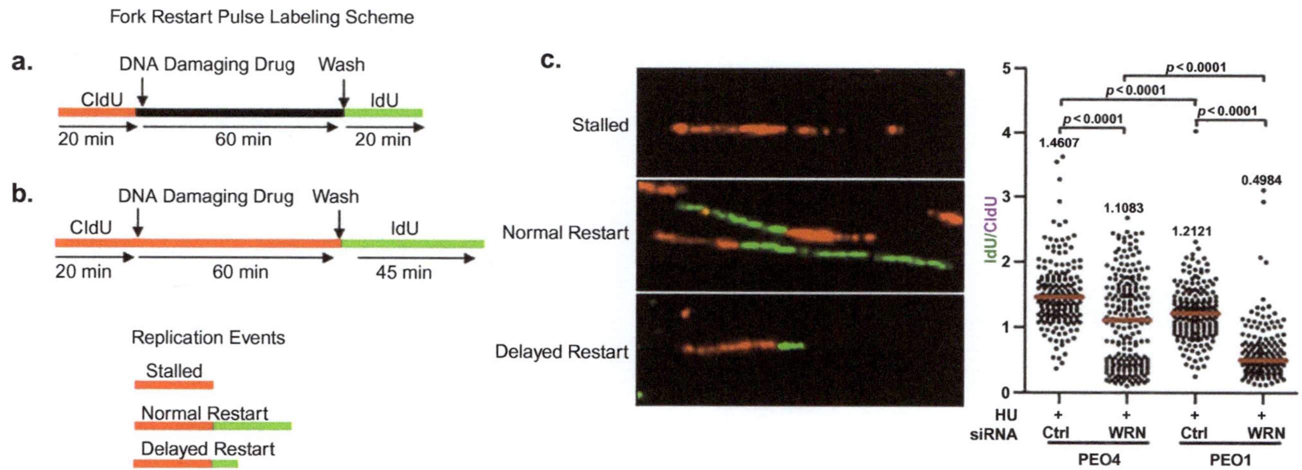

Fig. 5 Replication fork restart detection using the DNA fiber assay. (A and B) Scheme of fiber pulse-labeling for fork restart experiment. Replication forks can be pulse labeled as depicted in panel (A) or (B). Replication parameters observed by the fork restart assay are shown. (C) Representative DNA fibers showing a stalled (red) or normally restarted (red-green) fork following release from HU-induced replication stalling in *BRCA2*-mutated PEO1 cells. Delayed or impaired fork restart as evidenced by reduced incorporation of IdU analog (shorter green tract length) in *BRCA2*-mutated PEO1 cells upon WRN knockdown. Quantification of fork restart efficiency in *BRCA2*-wild type PEO4 and *BRCA2*-mutated PEO1 cells upon WRN genetic depletion. Fork restart efficiency following release from HU-induced replication stalling was determined by measuring the IdU/CldU tract length ratios in each experimental condition. A reduced IdU/CldU ratio indicates delayed fork restart or slow fork progression following replication restart. The graph was published in (Datta, Biswas, et al., 2021). This is an open-access article published under the Creative Commons Attribution 4.0 International (CC BY 4.0) https://creativecommons.org/licenses/by/4.0/.

IdU-CldU-IdU, and CldU-IdU-CldU), whereas stalled forks are seen as CIdU only tracts. Fork restart efficiency can be evaluated by determining the percentage of restarted (ClDU + IdU) or stalled (CldU only) forks among the total CldU-labeled tracts (Only CldU and CldU + IdU tracts). A defect in fork restart results in a decreased percentage of restarting forks with a concomitant increase in percentage of stalled forks compared to control cells. Alternatively, to test fork restart delay or impaired fork progression following replication restart, mean IdU tract length or IdU/CldU ratio of the restarted forks can be determined (Datta, Biswas, et al., 2021; Mukherjee et al., 2019). A reduction in IdU tract length or IdU/CldU ratio compared to control sample suggests delayed fork restart or slow fork progression following replication resumption. In this approach, a relatively longer 2nd pulse (IdU) is generally required to accurately examine the ability of fork progression following replication restart.

2.3.3 Nucleolytic degradation of stalled forks

Stalled forks require limited nucleolytic processing of DNA ends to restart DNA synthesis (Petermann & Helleday, 2010; Thangavel et al., 2015). However, uncontrolled nucleolytic resection of stalled forks in cells genetically deficient in fork protection function (e.g., *BRCA1*- or *BRCA2*-mutated cancer cells) drives genomic instability (Schlacher et al., 2011). Nucleolytic degradation of nascent DNA at stalled forks can be precisely determined by replication tract analysis using DNA fiber technique. In a typical fork stability assay, replication forks are first sequentially labeled with 1st (CldU) and 2nd (IdU) thymidine analogs for an equal amount of time followed by incubation with replication stalling agents such as HU (Fig. 6). Because the 2nd label was incorporated just prior to the fork stalling, nucleolytic degradation of the nascent DNA strands is evaluated by the relative shortening of the 2nd label as compared to the 1st label (Fig. 6). For quantitative comparison of stalled fork stability between biological samples, it is important to normalize 2nd label tract length to that of the 1st label to ensure that the shorter tracts reflect nucleolytic resection, not the defects in fork progression. IdU/CldU ratios quantified that are significantly less than 1 represent cellular conditions where fork stabilization is impaired.

3. Expected outcomes

The DNA fiber experiments described in this chapter are designed to provide molecular insight to the functional requirements of specified gene products for normal cellular DNA replication and the response to replication

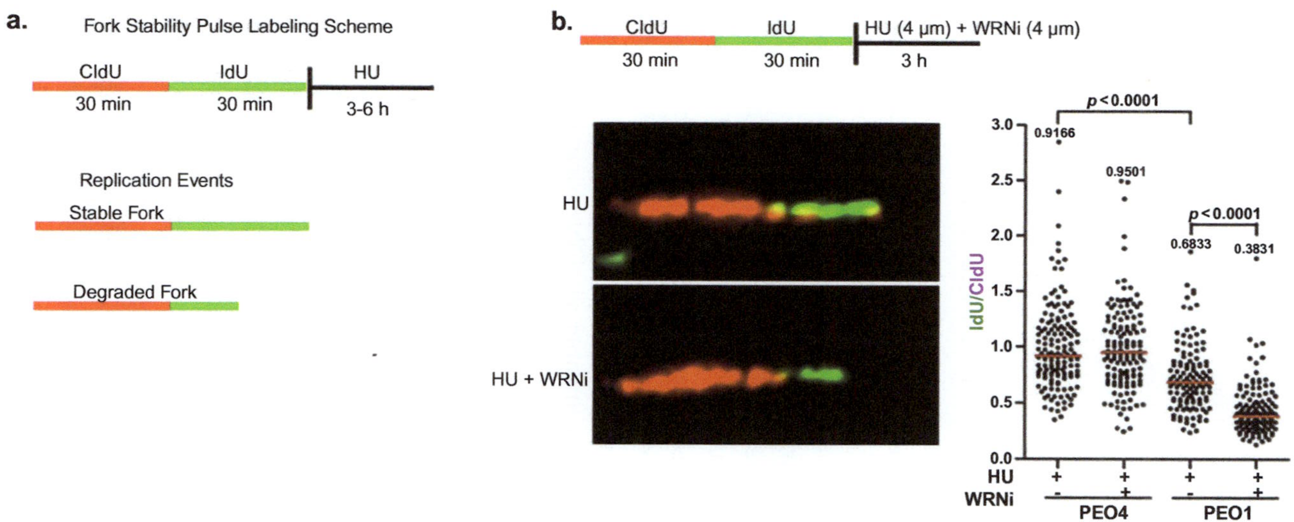

Fig. 6 DNA fiber experiments to detect fork instability under conditions of replication stress. (A) Scheme of fiber pulse-labeling for fork stability assay to examine nucleolytic degradation of nascent DNA at stalled replication forks. Relative shortening of the IdU-labeled strand following HU-induced replication stalling is indicative of fork degradation. (B) Fiber assay showing enhanced fork degradation induced by WRNi treatment of *BRCA2*-mutated PEO1 cells. Graphical quantification of stalled fork stability in *BRCA2*-wild type PEO4 and *BRCA2*-mutated PEO1 cells exposed to WRNi. Reduced IdU/CldU ratios in PEO1 cells upon WRNi treatment suggest enhanced fork degradation. The graph was published in (Datta, Biswas, et al., 2021).

stress induced by a pharmacological treatment. These studies can be performed with a minimal set of key reagents, providing some flexibility dependent on the specific research questions and experimental system being employed. The key reagents required are: (1) Defined isogenic cell lines, preferably CRISPR KO and WT backgrounds; (2) Gene expression plasmid(s) suitable for transfection into human cells and engineered with site-directed mutations corresponding to catalytic domains of interest, e.g., WRN ATPase/helicase mutants; (3) Suitable RNAi oligonucleotides that achieve a high efficiency (>90%) depletion of the target gene; (4) Small molecule inhibitors of catalytic enzymes implicated in fork stability mechanisms, e.g., Mre11 or Dna2 nuclease inhibitor, WRN helicase inhibitor. With these tools in hand and access to epifluorescence microscopy designed for immunofluorescence detection of specific fluorophores dictated by nucleoside analogs, and imaging software, the researcher can design a constellation of cell-based experimental scenarios that address fork stalling, fork degradation, fork asymmetry, dormant origin activation, nascent strand degradation at the fork, and fork restart.

This collection of experimental outcomes permits a wholistic assessment of replication fork dynamics that can be altered by the loss of individual fork protection factors in a synergistic manner. Moreover, mechanistic insights can be gleaned from carefully designed experiments in which forks are subjected to replication stress by timely addition of a selected drug and monitoring the outcome in a kinetic manner via DNA strand lengthening or shortening as detected by fluorescence of chromophores in the nascently synthesized or degraded DNA strands.

4. Notes

1. Depending on cell type, pulse labeling time should be adjusted to optimize the labeling efficiency and read-out
2. Always use a high concentration of the thymidine analog for the second labeling step to dilute out any residual label from the first analog. Try not to waste any time for the addition of second label for the two-step labeling to ensure continuity in the experiment
3. Nice, stretched DNA fibers along the length of the slide are needed for good fiber scoring with microscope. It is important to optimize the angle at which the slide should be tilted so that the liquid gradually spreads down the slide and does not flow too quickly
4. Make sure to avoid trapping bubbles in this step

5. Advantages

DNA fiber studies provide a very sensitive and direct assessment of perturbations to replication forks in living cells that can be used to investigate genetic- or chemical-induced SL. By characterizing the cellular outcomes of genetic ablation, modification or depletion, as well as catalytic inhibition of the gene product target by a small molecule, the researcher can assemble a working model for pathway crosstalk between genetically interacting factors suspected to play a role in fork stability. By doing so, mechanistic questions can be hypothesized and further studied. We found that modifications of the DNA fiber assay enabled us to address the consequences of either WRN loss or pharmacological inhibition of WRN helicase activity in the setting of BRCA2 deficiency in cancer cells.

6. Limitations

Although DNA fiber experiments provide the opportunity to study fork dynamics in a well-defined cell-based system, the translational value of elucidated SL interactions requires significantly further validation as mitigating factors may come into play. The genetic complexity of specific cancer cell types and in situ environmental factors in cancerous tissues potentially complicate the efficacious application of genetic- or chemical-induced SL, revealed by DNA fiber experiments and other cell-based assays, to apply even in pre-clinical model organisms. For example, the chemical scaffold or pharmacokinetic parameters of a drug-like compound that has bioactivity in a cell-based system may not have suitable properties for application in model genetic organisms. For the WRN helicase inhibitor NSC617145, we were able to show its utility in a nude mouse xenograft model whereby the intraperitoneal administration of the drug cause tumor shrinkage and histological detection of DNA damage biomarkers in the cancerous tissue (Datta, Biswas, et al., 2021).

7. Conclusions

DNA fiber experiments to study replication fork dynamics in living cells has moved to the forefront of experimental strategies to investigate the consequences of replication stress as a function of genetic loss or small molecule inhibition of cellular DNA nucleases and helicases.

The experiments described in this work should enable a laboratory to address basic but highly relevant questions in this realm. The DNA fiber experiments are relatively straightforward in design, but care in their execution must be performed. Depending on experimental design, quantitative analyses of DNA fiber lengths can provide a wealth of information to characterize fork progression, nascent strand degradation, dormant origin firing, and replication fork restart. Such experimental results can be used to identify unique genetic interactions that underlie SL and help to build pathway models for replication fork protection mechanisms in human cells.

Acknowledgments

This work was supported by the Intramural Research Program, National Institute on Aging, NIH (1ZIAAG000741-19).

References

Berti, M., Cortez, D., & Lopes, M. (2020). The plasticity of DNA replication forks in response to clinically relevant genotoxic stress. *Nature Reviews. Molecular Cell Biology, 21*, 633–651.

Bianco, J. N., Poli, J., Saksouk, J., Bacal, J., Silva, M. J., Yoshida, K., et al. (2012). Analysis of DNA replication profiles in budding yeast and mammalian cells using DNA combing. *Methods (San Diego, California), 57*, 149–157.

Cortez, D. (2015). Preventing replication fork collapse to maintain genome integrity. *DNA Repair, 32*, 149–157.

Datta, A., Biswas, K., Sommers, J. A., Thompson, H., Awate, S., Nicolae, C. M., et al. (2021). WRN helicase safeguards deprotected replication forks in BRCA2-mutated cancer cells. *Nature Communications, 12*, 6561.

Datta, A., Dhar, S., Awate, S., & Brosh, R. M., Jr. (2021). Synthetic lethal interactions of RECQ helicases. *Trends Cancer*, 7, 146–161.

Jensen, R. B., Carreira, A., & Kowalczykowski, S. C. (2010). Purified human BRCA2 stimulates RAD51-mediated recombination. *Nature, 467*, 678–683.

Liao, H., Ji, F., Helleday, T., & Ying, S. (2018). Mechanisms for stalled replication fork stabilization: New targets for synthetic lethality strategies in cancer treatments. *EMBO Reports, 19*.

Moynahan, M. E., Pierce, A. J., & Jasin, M. (2001). BRCA2 is required for homology-directed repair of chromosomal breaks. *Molecular Cell*, 7, 263–272.

Mukherjee, C., Tripathi, V., Manolika, E. M., Heijink, A. M., Ricci, G., Merzouk, S., et al. (2019). RIF1 promotes replication fork protection and efficient restart to maintain genome stability. *Nature Communications, 10*, 3287.

Oshima, J., Sidorova, J. M., & Monnat, R. J., Jr. (2017). Werner syndrome: Clinical features, pathogenesis and potential therapeutic interventions. *Ageing Research Reviews, 33*, 105–114.

Panzarino, N. J., Krais, J. J., Cong, K., Peng, M., Mosqueda, M., Nayak, S. U., et al. (2021). Replication gaps underlie BRCA deficiency and therapy response. *Cancer Research, 81*, 1388–1397.

Parra, I., & Windle, B. (1993). High resolution visual mapping of stretched DNA by fluorescent hybridization. *Nature Genetics, 5*, 17–21.

Petermann, E., & Helleday, T. (2010). Pathways of mammalian replication fork restart. *Nature Reviews. Molecular Cell Biology*, *11*, 683–687.

Quinet, A., Lemaçon, D., & Vindigni, A. (2017). Replication fork reversal: Players and guardians. *Molecular Cell*, *68*, 830–833.

Ray Chaudhuri, A., Callen, E., Ding, X., Gogola, E., Duarte, A. A., Lee, J. E., et al. (2016). Replication fork stability confers chemoresistance in BRCA-deficient cells. *Nature*, *535*, 382–387.

Rickman, K., & Smogorzewska, A. (2019). Advances in understanding DNA processing and protection at stalled replication forks. *The Journal of Cell Biology*, *218*, 1096–1107.

Schlacher, K., Christ, N., Siaud, N., Egashira, A., Wu, H., & Jasin, M. (2011). Double-strand break repair-independent role for BRCA2 in blocking stalled replication fork degradation by MRE11. *Cell*, *145*, 529–542.

Scully, R., Elango, R., Panday, A., & Willis, N. A. (2021). Recombination and restart at blocked replication forks. *Current Opinion in Genetics & Development*, *71*, 154–162.

Setton, J., Zinda, M., Riaz, N., Durocher, D., Zimmermann, M., Koehler, M., et al. (2021). Synthetic lethality in cancer therapeutics: The next generation. *Cancer Discovery*, *11*, 1626–1635.

Thakar, T., & Moldovan, G. L. (2021). The emerging determinants of replication fork stability. *Nucleic Acids Research*, *49*, 7224–7238.

Thangavel, S., Berti, M., Levikova, M., Pinto, C., Gomathinayagam, S., Vujanovic, M., et al. (2015). DNA2 drives processing and restart of reversed replication forks in human cells. *The Journal of Cell Biology*, *208*, 545–562.

Ying, S., Hamdy, F. C., & Helleday, T. (2012). Mre11-dependent degradation of stalled DNA replication forks is prevented by BRCA2 and PARP1. *Cancer Research*, *72*, 2814–2821.

Yu, C. E., Oshima, J., Fu, Y. H., Wijsman, E. M., Hisama, F., Alisch, R., et al. (1996). Positional cloning of the Werner's syndrome gene. *Science (New York, N.Y.)*, *272*, 258–262.

CHAPTER NINE

Optimizing CMG helicase and CMG-dependent replication assays by designing DNA fork substrates and choosing nucleotide analogues for helicase preloading

Nina Y. Yao[a] and Mike E. O'Donnell[a,b,*]
[a]Laboratory of DNA Replication, The Rockefeller University, New York, NY, United States
[b]Howard Hughes Medical Institute at Rockefeller University, New York, NY, United States
[*]Corresponding author: e-mail address: odonnel@rockefeller.edu

Contents

Abstract

The replication machinery that synthesizes new copies of chromosomal DNA is located at the junction where double-stranded DNA is separated into its two strands. This replication fork DNA structure is at the heart of most assays involving DNA helicases. The helicase enzyme unwinds the replication fork structure into two single-stranded

Methods in Enzymology, Volume 672
ISSN 0076-6879
https://doi.org/10.1016/bs.mie.2022.02.023

templates which are converted into two daughter duplexes by other proteins, including DNA polymerases. In eukaryotes, the CMG (Cdc45/Mcm2-7/GINS) helicase plays the pivotal role of unwinding the parental duplex DNA and at the same time interacts with numerous other proteins, including the leading strand polymerase, Pol **ε**. This chapter first describes how we design and prepare synthetic replication forks used in our CMG-related assays. Then we describe how to load CMG onto the fork. The Mcm2-7 motor subunits of CMG form a closed ring, as do all cellular replicative helicases, that encircles ssDNA for helicase function. Thus, the first step in these assays is the loading of CMG onto the fork DNA, followed by DNA unwinding and replication. We explain protocols for different strategies of preloading CMG onto the DNA fork using different ATP analogues. Additionally, the presence of Mcm10, an intimate partner of CMG, affects how CMG is preloaded onto a fork substrate.

1. Introduction

DNA replication of a cellular genome starts at unique sites along the chromosome known as origins. An origin binding protein, and other factors, assemble two helicases onto an origin. The site of replisome movement is a Y-shaped structure referred to as a replication fork. The helicase is the "point of the spear," acting at the fork to separate the parental duplex (Fig. 1A).

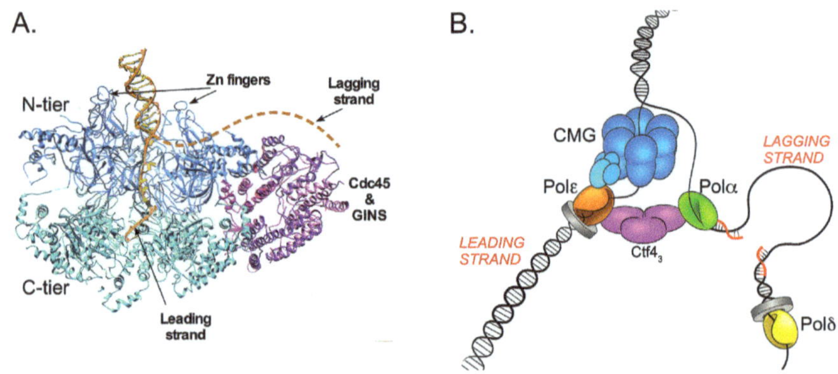

Fig. 1 Structure of CMG at a replication fork. (A) The Mcm2–7 hexamer is shown as a blue/turquoise double tier ring. The N-terminal domain and C-terminal domain of each Mcm subunit form the N (blue)- and C (turquoise)-tiers in the hexamer, respectively. The Cdc45 and GINS tetramer are shown in purple, and act to bind other proteins. (B) CMG is located at the DNA fork junction where it unwinds the duplex DNA into the leading and lagging strand templates. The CMG-interacting Pol ε replicates the leading strand while Pol α synthesizes primers for Pol δ to extend replication on the lagging strand. *Langston, L. D., Mayle, R., Schauer, G. D., Yurieva, O., Zhang, D., Yao, N. Y., et al. (2017). Mcm10 promotes rapid isomerization of CMG-DNA for replisome bypass of lagging strand DNA blocks.* eLife, 6, *e29118. doi:https://doi.org/10.7554/eLife.29118.*

The helicases attract other factors (e.g., primase/polymerases) to form a "replisome machine" that duplicates the DNA (Fig. 1B). In eukaryotes the helicase is composed of 11 different proteins called CMG (an abbreviation for its three main components: Cdc45, the Mcm2-7 heterohexamer and the GINS heterotetramer) (Ilves, Petojevic, Pesavento, & Botchan, 2010; Moyer, Lewis, & Botchan, 2006). The Mcm2-7 heterohexamer is the motor of CMG and forms a ring that must encircle ssDNA to act as a helicase.

CMG utilizes ATP (or dATP) hydrolysis to translocate along ssDNA in the 3′ to 5′ direction while unwinding the duplex at the forked junction. CMG, like other ring shaped hexameric helicases, functions by steric exclusion, in which the hexameric ring encircles one strand of ssDNA and excludes the other strand as it translocates along the encircled strand (Fu et al., 2011). In this fashion, it acts as a moving wedge to drive strands apart. These helicases typically function with DNA polymerases and single-strand (ss) DNA-binding protein, thereby preventing the unwound strands from simply reannealing upon separation (Bell & Labib, 2016).

Many studies of CMG helicase activity utilize DNA forks assembled from synthetic oligonucleotides. One strand of the forked substrate is radiolabeled so that unwinding activity can be monitored by visualizing single stranded products by native polyacrylamide gel electrophoresis (PAGE). If one is also examining the function of CMG with a DNA polymerase, the synthetic fork can be "primed" using a 5′ ^{32}P-primer that anneals to the leading strand. Synthetic DNA forks have been an important tool in structural studies of CMG. The cryo-EM structure of CMG at a replication fork was the first structure of a replicative helicase unwinding a fork, and it revealed the unexpected finding that CMG tracks N-tier first (Georgescu et al., 2017). Before this finding, CMG was thought to track on ssDNA C-tier first (Bell & Labib, 2016). The N-first direction of CMG on ssDNA has important implications for how hexameric helicases may open dsDNA at origins (Langston & O'Donnell, 2019). Through the use of synthetic DNA forks, it was shown that an associated protein, Mcm10, stimulates CMG to unwind longer stretches of DNA and bypass DNA barriers (Langston et al., 2017). This finding was also shown to occur for CMGs formed at origins (Lõoke, Maloney, & Bell, 2017).

While CMG is assembled at origins from its component parts by the actions of many factors (Yeeles, Janska, Early, & Diffley, 2017), co-expression of the 11 subunits has been used to make recombinant CMG of human, Drosophila and yeast (Georgescu et al., 2014; Ilves et al., 2010;

Kang, Galal, Farina, Tappin, & Hurwitz, 2012). The purified recombinant CMG can be assembled onto the 3′ ssDNA arm of a synthetic fork to study helicase activity. If one uses a non-hydrolysable (AMPpNp) or a slowly hydrolysable (ATPγS) analog, one can preload CMG onto DNA before initiating efficient unwinding upon adding ATP (Abid Ali et al., 2016; Burnham, Kose, Hoyle, & Yardimci, 2019; Georgescu et al., 2017; Kose, Larsen, Duxin, & Yardimci, 2019; Kose, Xie, Cameron, Strycharska, & Yardimci, 2020; Langston et al., 2017; Petojevic et al., 2015; Yao, Zhang, Yurieva, & O'Donnell, 2022; Yuan et al., 2019).

Through trial and error and by incorporating the ever-evolving knowledge of CMG mechanism and structure along with its associated proteins, we have found ways to improve our CMG helicase assays and also CMG-dependent replication assays. In this chapter, we describe the design of various DNA fork substrates and offer examples of alternative strategies for CMG preloading using different ATP analogues as the first step in both CMG helicase and replication assays.

2. Designing DNA fork substrates for studying CMG structure and function

2.1 Basic DNA fork for a CMG helicase assay

A DNA fork substrate for a CMG helicase assay is comprised of two oligonucleotides that represent the leading strand and the lagging strand. They are partially complimentary and anneal to each other at one end forming a duplex region adjacent to two separate noncomplementary ssDNA arms (Fig. 2). In a helicase assay, the fork substrate is radiolabeled with ^{32}P on the 5′ end of one of the strands. After a CMG unwinding assay is quenched, the unwound single-stranded product can be separated from the radiolabeled fork by electrophoresis on a native polyacrylamide gel and visualized by either autoradiography, or use of a phosphor imaging system. Examples of DNA sequences used for CMG helicase assays in our lab are presented in Table 1.

2.1.1 Leading strand arm

CMG translocates on ssDNA in the 3′–5′ direction, therefore it must load onto the 3′ leading strand arm (Bell & Labib, 2016). The Hurwitz lab showed that human CMG is more active in unwinding when the 3′ tail is longer, up to a point (80 > 40 > 20 nts long) (Kang et al., 2012). They also showed that a stretch of only dT's was a better substrate than an equal length of random nucleotide sequence. For yeast CMG we typically use a leading strand arm of 40 dTs.

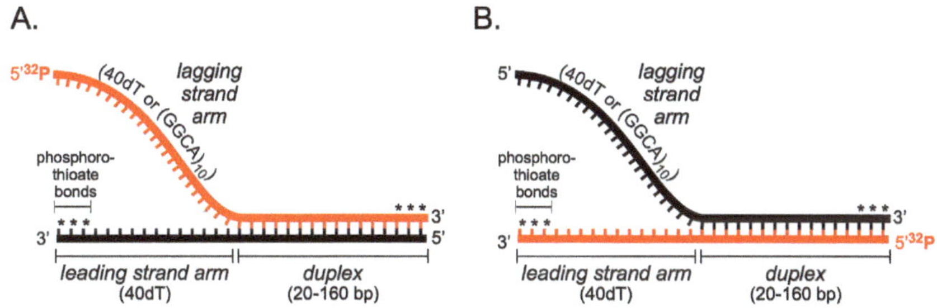

Fig. 2 A basic DNA fork for helicase assays. The forked DNA is made by annealing two synthetic oligos that are partially complementary. The ssDNA tails are often composed of approximately 40 deoxythymidines (dT). The lagging strand arm can also be made of 10 (GGCA) repeats. Either the lagging (A) or the leading (B) strand can be radiolabeled with ^{32}P by T4 polynucleotide kinase (PNK) (shown in red) for monitoring helicase unwinding of the fork substrate into its single-stranded DNA products by native polyacrylamide gel electrophoresis (PAGE). The asterisks represent phosphorothioate bonds between the last 3 nucleotides of the 3′ ends to protect against the 3′–5′ exonuclease of Pol ε, especially if used in the context of a replication experiment.

Table 1 Sequence of oligonucleotides used for making DNA forks.

NAME	size	OLIGONUCLEOTIDE SEQUENCE	source
30bp helicase fork LEAD	70 nt	5'GACGCTGCCGAATTCTGGCTTGCTAGGACATTTTTTTTTTTTTTTTT TTTTTTTTTTTTTTTTTTTT*T*T*T 3'	IDT
30bp helicase fork LAG	70 nt	5'GGCAGGCAGGCAGGCAGGCAGGCAGGCAGGCAGGCAGGCTTGT CCTAGCAAGCCAGAATTCGGCAGC*G*T*C 3'	IDT
60bp helicase fork LEAD	99 nt	5'CTCGTTTTACAACGTCGTGCTGAGGTGATATCTGCTGAGGCAATG GGAATTCGCCAACCTTTTTTTTTTTTTTTTTTTTTTTTTTTTTTTTTTTTT *T*T*T3'	IDT
60bp helicase fork LAG	99 nt	5'GGCAGGCAGGCAGGCAGGCAGGCAGGCAGGCAGGCAGGCAGGT TGGCGAATTCCCATTGCCTCAGCAGATATCACCTCAGCACGACGTT GTAAAAC*G*A*G 3'	IDT
160bp helicase fork LEAD	200 nt	5'AGAGAGTAGAGTTGAGTTGTGATGTGTAGAGTTGTTGTAGAGAAG AGTTGTGAAGTGTTGAGTAGAGAAGAGAAGAGAAGTGTTGTGATGT GTTGAGTAGTGTAGAGTTGAGAAGTAGAGATGTGTTGAGATGAGAA GAGTTGTAGTTGAGTTGAAGTGGTTTTTTTTTTTTTTTTTTTTTTTTTTT TTTTTTTT*T*T*T*T*T3'	IDT
160bp helicase fork LAG	200 nt	5'TTTTTTTTTTTTTTTTTTTTTTTTTTTTTTTTTTTTTTTTCCACTTCAAC TCAACTACAACTCTTCTCATCTCAACACATCTCTACTTCTCAACTCTA CACTACTCAACACATCACAACACTTCTCTTCTCTTCTCTACTCAACAC TTCACAACTCTTCTCTACAACAACTCTACACATCACAACTCAACTCTA CTCTCT3'	IDT
61bp replication fork LEAD	180 nt	5'AGGTGTAGATTAATGTGGTTAGAATAGGATGTGGTAGGAAGTGA GAATTGGAGAGTGTGTTTTTTTTTTTTTTTTTTTTTTTTTTTTTTTTTTT TTTTAAAGGTGAGGGTTGGGAAGTGGAAGGATGGGCTCGAGAGGTT TTTTTTTTTTTTTTTTTTTTTTTTTTTTTTTTTTTT*T*T*T-3	IDT
61bp replication fork LAG	100 nt	5'GGCAGGCAGGCAGGCAGGCAGGCAGGCAGGCAGGCAGGCACAC ACTCTCCAATTCTCACTTCCTACCACATCCCTATTCTAACCACATTAA TCTACA*C*C*T3'	IDT
61bp replication fork PRIMER	37 nt	5' CCTCTCGAGCCCATCCTTCCACTTCCCAACCCTCACC 3'	IDT
61bp replication fork "junction" size marker	80 nt	5'CCTCTCGAGCCCATCCTTCCACTTCCCAACCCTCACCTTTAAAAAA AAAAAAAAAAAAAAAAAAAAAAAAAAAAAAAAAA 3'	IDT
61bp replication fork "full length" size marker	140 nt	5'CCTCTCGAGCCCATCCTTCCACTTCCCAACCCTCACCTTTAAAAAA AAAAAAAAAAAAAAAAAAAAAAAAAAAAAAAAAACACACTCTCCAATT CTCACTTCCTACCACATCCCTATTCTAACCACATTAATCTACACCT 3'	IDT

The nucleotides in red represent the leading strand 3′ arm and the nucleotides in green represent the lagging strand 5′ arm. Asterisks denote phosphorothioate bonds.

2.1.2 Lagging strand arm

The lagging strand arm can be designed using any sequence. Often, it is a string of approximately 40 dTs to prevent pairing to the leading strand dT arm. But a recent study using an electrophoretic mobility shift assay (EMSA) found that CMG bound both leading and lagging dT arms (Petojevic et al., 2015). This same study showed that CMG will not bind to a series of (GGCA) repeats. So, we often use a lagging strand arm composed of 10 (GGCA) repeats or use 40 dTs. Note however, that GGCA repeats are a better substrate for radiolabeling by T4 Polynucleotide kinase (PNK). Refer to Section 3.1 for more details about PNK's specificity for the type of 5′ nucleotide on an oligo during the radiolabeling reaction.

2.1.3 Duplex region of the DNA replication fork

For CMG helicase assays, the duplex region of the DNA fork is typically shorter compared to the replication forks that will be discussed in Section 2.2. Its length usually ranges from 20 to 60 basepairs (bp). CMG can easily unwind this length of duplex but be aware that if ATPγS is used during the CMG preloading step (see protocol in Section 4.1), CMG can use this slowly hydrolysable ATP analog to unwind short forks (especially duplexes shorter than 40 bp), depending on the time given (Yao et al., 2022). When preloading CMG onto "short" forks, incubation with AMPpNp may be a safer choice rather than ATPγS. The duplex region should not be shorter than 20 bp because spontaneous melting may occur either during the reaction incubation at 30 °C or by the heat generated during subsequent gel electrophoresis. Longer duplexes have been used in our lab in the past. Indeed, yeast CMG alone cannot unwind a 160 bp fork, however upon addition of Mcm10, a protein that binds CMG, enables the efficient unwinding of this longer duplex (Langston et al., 2017).

2.2 Primed mini-fork for a CMG-dependent replication assay

The basic DNA fork substrate used in the CMG helicase assay can be converted into a fork substrate that is functional in a CMG-dependent DNA replication assay by including a primer-binding site on the leading strand arm and then annealing a ^{32}P-radiolabeled complimentary primer oligonucleotide for PCNA clamp loading and DNA polymerase extension (Fig. 3). The "mini" replication fork described here is the foundation on which longer replication forks can be made, and the sequences used in many of our CMG dependent DNA replication assays are presented in Table 1.

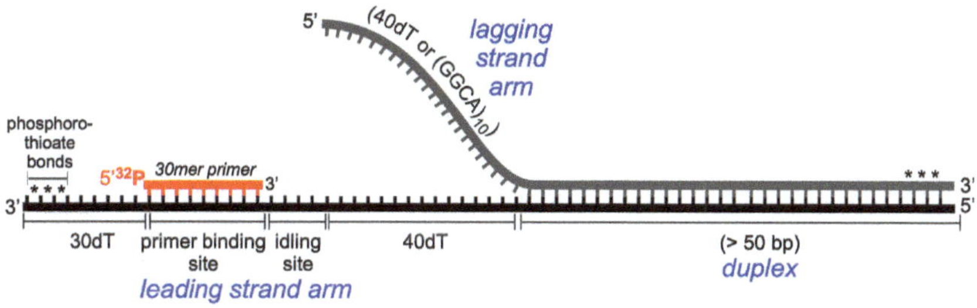

Fig. 3 A primed mini-fork for CMG-dependent replication assays. The typical forked DNA is modified by use of a long leading ssDNA arm to which a primer oligonucleotide can be annealed. The 3′ end contains some phosphorothioate bonds to prevent polymerase (Pol ε or δ) proofreading exonuclease action. See Table 1 for an example of specific sequences.

2.2.1 Modifying the leading strand arm to accommodate primer extension for DNA polymerases

The primer oligonucleotide anneals to a complimentary sequence located at the center of the leading strand arm and is flanked on either side by stretches of single-stranded dTs. We find that the CMG helicase can load at the 3′ ssDNA end, traverse across the primer-template duplex, and continue sliding over to the fork junction where it will begin to unwind the fork duplex. We use primers of 25–40 ntds. This provides plenty of room for RFC to load a PCNA clamp onto the primer (Bowman, O'Donnell, & Kuriyan, 2004).

Immediately upstream of the primer-binding site is a unique sequence referred to as the "polymerase idling site" (Fig. 3). The actual sequence can vary. It can be as simple as a few dAs (e.g., 5′-AAA-3′) (Yuan et al., 2019) where in the presence of only dTTP, the polymerase can advance forward 3 nucleotides but be stalled once it reaches the long stretch of dT's before the fork junction. A more complicated sequence such as 5′-TTGTAAGAAAG
GAG-3′ (Georgescu et al., 2014) would give the added advantage of being able to load PCNA using dATP by RFC while idling the polymerase. In between the "idling site" and the fork junction is a stretch of 40 dTs that can easily accommodate CMG (Georgescu et al., 2017; Petojevic et al., 2015; Sun et al., 2015).

2.2.2 Mini replication fork duplex is a versatile foundation for building much longer forks

The duplex region at a natural replication fork is much longer than the synthetic duplex of a fork made for a helicase assay. The processivity of CMG unwinding a fork by itself is far less, when compared to CMG function in combination with replisomal proteins because CMG forms a complex with Pol ε and several other factors, and together they unwind and replicate several kilobases of duplex DNA (Bell & Labib, 2016; Langston et al., 2017; Lewis et al., 2020; Yeeles, Deegan, Janska, Early, & Diffley, 2015).

Longer forked DNAs can be made by simply ligating the mini fork discussed here to the end of a linearized plasmid. For a detailed protocol on the construction of these long forks, refer to the BioProtocol publication on these assays (Schauer, Finkelstein, & O'Donnell, 2017). Our lab often uses a 3 kb duplex attached to the synthetic DNA fork in our biochemical studies of the eukaryotic replisome (Georgescu et al., 2014, 2015; Langston et al., 2014). For example, this 3 kb fork was used to demonstrate the asymmetric nature of the Pol ε and Pol δ replicating the leading strand and lagging strand respectively (Burgers & Kunkel, 2017; Kunkel & Burgers, 2014). However, even longer forks, like the 19 kb fork used by the van Oijen lab in single-molecule studies of the dynamic exchange of polymerases at a eukaryotic replication fork (Lewis et al., 2020) was initially built from a mini-fork like the one discussed here.

3. Protocols for DNA fork preparation

Materials and equipment

- 1.5 and 0.5 mL Protein LB microfuge tube (Sarstedt, 72.706.600, 72.704.600)
- Heated water bath
- Nalgene™ Acrylic Benchtop Beta Radiation Shield (Thermo Scientific; 6700-1812)
- Metal heating block (VWR, catalog number: 12621-084)
- 16 mm × 125 mm borosilicate glass test tube (Fisher Scientific: 14-961-30)
- Electrophoresis power supply EPS3501XL (GE, GE18-1130-05)
- Standard Dual Vertical Protein Electrophoresis Unit (Hoefer; SE600)
- Regular Glass Plates, 18 × 16 cm, 2 pcs (Hoefer, SE6102)
- Standard Comb, 15 wells, 1 mm thickness, 5.7 mm width (Hoefer, SE511-15-1.0)
- Spacers, 1 mm thickness, 16 cm length, 2 cm width, 2 pcs (Hoefer, SE6119-2-1.0)
- Variable mode gel scanner (GE Typhoon FLA9500)
- Storage Phosphor screen (Imaging Screen-K (Kodak, 20 × 25 cm) and Exposure cassette-K (Bio-Rad, 170-7843 and 170-7861))

- Microfuge 18 centrifuge (Beckman Coulter)
- Liquid scintillation counter (PerkinElmer Tri-Carb2910TR)
- 7 mL scintillation vials with caps (Fisherbrand, 03-337-26)
- 1.5 cm Whatman DE81 ion exchange filter discs (discontinued) (Alternative: Amersham Hybond-N+ sheet, RPN119B or Amersham Hybond XL, sheet, RPN1520S cut into circles)
- 18 mini vial Sample cassette (PerkinElmer)
- Portable geiger counter (Technical Associates, TBM-3S)
- QuantaSmart™

  Note: Safety Considerations and Standards

  When handling radioactive substances, follow all appropriate lab safety guidelines. For example, work behind a plexiglass shield, check for radioactive contamination with a Geiger counter and dispose of radioactive waste in the assigned waste containers and dispose accordingly.

3.1 Preparation of DNA fork substrate for CMG helicase assay

3.1.1 Before you begin

1. Design your oligonucleotides and check with the IDT Oligonucleotide Analyzer tool (or comparable):
   - **a.** To verify that all of the oligos anneal to each other as expected
   - **b.** To confirm that annealed duplexes do not have melting temperatures less than 40 °C
   - **c.** To check for unexpected homodimerizations that could prevent efficient hybridization of the leading and lagging strand
   - **d.** To check for unexpected secondary structures that may affect strand hybridization or cause unwanted replication stall sites
2. Order oligonucleotides at appropriate scale and purity:

We often order ssDNA oligos from Integrated DNA Technologies (IDT). For oligonucleotides less than 100 nucleotides long, a 250 nmol scale batch with PAGE purification will ensure that the oligonucleotides are of uniform size and plentiful enough for many fork preps. For template strands that are between 100 and 200 nucleotides long, IDT can synthesize PAGE purified Ultramer DNA oligos. Yield is decreased as the oligo's length increases (the minimum guaranteed yield for a 200mer is 0.25 nmol). So, for these extra-long oligonucleotides, ordering an extra batch is sometimes required, depending on how much fork you need for your studies.

3. Dissolve DNA oligonucleotides in TE buffer to make
   - **a.** A stock concentration of 100 μM
   - **b.** approximately 100 μL of 10 μM working dilution

Table 2 List of reagents used for making DNA fork substrates.

Reagent	Source	Identifier
T4 Polynucleotide Kinase (10,000 Units/mL) 10× T4 Polynucleotide Kinase Reaction buffer	New England Biolabs	M0201S or M0201L B0201S
^{32}P-Y-ATP, 3000 Ci/mmol, 3.3 μM	PerkinElmer	BLU002A
Tris base	Fisher	BP152-5
NaCl	US Biological	S5000
EDTA, tetrasodium salt	Amresco	0245-2.5kg
Sodium citrate dihydrate	Fisher	S279-3
Ammonium formate	Sigma-Aldrich	156264-2.5kg
Sodium pyrophosphate anhydrous	EMD	S074-1
10× TBE buffer	Amresco	EW-21853-84
Acrylamide:bis-acrylamide (29:1) 40% solution	Fisher Scientific	BP1408-1
Ammonium persulfate	Sigma	A3678-100G
TEMED	Bio-Rad	161-0801
Bromophenol Blue	Bio-Rad	161-0404
Xylene Cyanol FF	Bio-Rad	161-0423
Glycerol	Sigma-Aldrich	G5516
EcoLume	MPbio	882470

4. Prepare the following buffers (see Table 2).
   a. **5× Hybridization buffer:** 10 mM Tris-Cl, pH 8.5, 300 mM NaCl, 30 mM sodium citrate
   b. **6× DNA loading dye:** 30% (*v*/v) glycerol, 0.25% (*w*/*v*) bromophenol blue, 0.25% (w/v) xylene cyanol FF
   c. **TE buffer:** 10 mM Tris·Cl, pH 8.0; 1 mM EDTA
   d. **Filter disc wash buffer:** 0.3 M ammonium formate (pH 7.8), 10 mM sodium pyrophosphate

3.1.2 Radiolabel either the leading or the lagging strand oligonucleotide

Note: It is important to note that T4 Polynucleotide kinase greatly favors oligonucleotides ending in dG (van Houten, Denkers, van Dijk, van den

Brekel, & Brakenhoff, 1998). A 5′ dT or dA is also satisfactory for radiolabeling, but a dC can sometimes yield a very poor signal. You can label either the leading or lagging strand for helicase assays. But if you plan on comparing helicase activity using several different forks, it is advisable to design the strand you plan to radiolabel with a common 5′ nucleotide that is present on all your forks and verify that the T4 PNK radiolabels the different oligonucleotides with similar specific radioactivity.

1. In a 1.5 mL microfuge tube

10 μM Lagging (or Leading) strand oligonucleotide	2 μL (20 pmol)
10 × T4 PNK reaction buffer	5 μL
3.3 μM γ-^{32}P-ATP	5 μL (16.5 pmol)
T4 Polynucleotide Kinase (10 units/μL)	5 μL
ddH_2O	33 μL
	50 μL
	Concentration of ^{32}P-leading strand oligonucleotide = 400 nM

2. Incubate tube at 37 °C in a water bath for 40 min

3. Inactivate the PNK by incubating the tube on a 65 °C metal heating block for 20 min

3.1.3 Anneal the leading and lagging strand oligonucleotides together

Note: 40 μL of the radiolabeled oligo (out of the total 50 μL) is used for the following annealing reaction. The remaining 10 μL sample is saved at 4 °C and is used in Section 3.1.4 for quantification.

1. In a 1.5 mL microfuge tube

400 nM ^{32}P-Lagging (or Leading) strand oligo	40 μL (16 pmol)
10 mM complimentary Leading (or Lagging) strand oligo	2 μL (20 pmol)
5 × Hybridization buffer	5 μL
ddH_2O	3 μL
	50 μL

2. Incubate tube in a glass test tube having near boiling water and allow to cool to room temp over 30 min
3. Add 10 μL of 6 × DNA loading dye and keep tube on ice until ready to load onto a native polyacrylamide gel

3.1.4 Gel purify the radiolabeled DNA fork

Note: The purpose of gel purifying the ^{32}P-fork is to ensure a pristine starting material with no excess single-stranded oligos that might bind helicase non-productively. Also, PAGE purification ensures there is no background of radiolabeled strand to be confused with the identical CMG unwound product on a native PAGE gel.

1. Prepare a native 10% polyacrylamide (14 cm × 16 cm) gel using 1 mm spacers and a 10 well comb

40% Acrylamide:bis-acrylamide solution (29:1)	7.5 mL
10 × TBE	3.0 mL
ddH_2O	19.5 mL
10% ammonium persulfate	300 μL
TEMED	30 μL

2. Load the radiolabeled fork reaction mix into 1 lane of the gel

   Note: If you are purifying more than one type of radiolabeled fork, it is advisable to load each reaction as far away from each other in case there is cross-contamination of the products during electrophoresis.
3. Run the gel in 1 × TBE (4 L capacity) at 35 mAmps (~250 V) until the bromophenol blue dye migrates 1 in. from the bottom of the gel (approximately 2 h).
4. Carefully separate one of the pieces of glass off the gel and cover the gel with plastic wrap
5. Expose the gel to a phosphor imaging screen for approximately 10 min. and scan with a Typhoon Scanner to confirm the presence of the radiolabeled fork. Sometimes you may notice a lower band that represents unannealed oligonucleotides if the annealing reaction was incomplete

6. Remove the plastic wrap, and using a razor blade, cut out the lane of interest in the gel. Then chop this lane into 14 pieces
7. Carefully cover these gel pieces with plastic wrap again and expose the phosphor imaging screen for another 10 min
8. Upon scanning the screen with the Typhoon scanner, you will be able to determine which gel piece(s) contain the radiolabeled fork DNA
9. Mince the gel piece(s) of interest and transfer the tiny gel pieces into a 1.5 mL microfuge tube containing 200 μL TE
10. Leave the tube in the cold room overnight to allow the radiolabeled fork DNA to "self-elute" out of the gel pieces

3.1.5 Determine the concentration of the purified ^{32}P-fork

1. Spin the tube of gel pieces at 15,000 rpm in a tabletop microfuge for 5 min
2. Collect the radioactive DNA in the supernatant and determine the concentration by spotting 1 μL of this purified ^{32}P-fork onto a DE81 filter disc. As a control, dilute 1 μL of the 400 nM ^{32}P-lagging strand from Section 3.1.1 to 40 nM with 9 μL TE. Then spot 1 μL of 40 nM ^{32}P-lagging strand (equivalent to 40 fmol) onto 2 DE81 filter discs. One of these 2 control filter discs is washed 6 times with filter wash buffer, 2 times with deionized water and once with ethanol and allowed to air dry under a heat lamp

   Note: The DE81 filter discs used here are discontinued. Amersham Hybond-N+ or Hybond XL sheet are alternative materials that can be used but they need to be cut into small squares.
3. The radioactivity of the three filters is then quantitated separately by liquid scintillation counting
4. The specific activity of the ^{32}P-lagging strand = cpm/40 fmol
5. The concentration of the radiolabeled fork = (cpm/specific activity of the ^{32}P-lagging strand)/1 μL
6. Efficiency of radiolabeling in Section 3.1.1 = (cpm of the washed filter)/(cpm of the unwashed filter)

   Note: The yield of purified ^{32}P-DNA fork generated from this protocol is typically 3–7 pmol DNA in a volume of 120 μL at a concentration of 30–60 nM and a specific activity of 1000–2000 cpm/fmol.

3.2 Preparation of primed-DNA fork for CMG-dependent replication assay

3.2.1 Radiolabeling the primer oligonucleotide

1. In a 1.5 mL microfuge tube

10 μM primer oligonucleotide	2 μL (20 pmol)
10 × T4 PNK reaction buffer	5 μL
3.3 μM γ-^{32}P-ATP	5 μL (16.5 pmol)
T4 PNK (10 units/μL)	5 μL
ddH_2O	33 μL
	50 μL
Concentration of ^{32}P-primer = 400 nM	

2. Incubate tube in 37 °C water bath for 40 min
3. Inactivate the PNK by incubating the tube on a 65 °C metal heating block for 20 min

3.2.2 Annealing the primer, leading and lagging strand oligonucleotides

1. In a 1.5 mL microfuge tube

400 nM ^{32}P-primer (above)	40 μL (16 pmol)
10 μM Leading strand oligonucleotide	2 μL (20 pmol)
10 μM Lagging strand oligonucleotide	2.4 μL (24 pmol)
5 × Hybridization buffer	5 μL
ddH_2O	3 μL
	50 μL

2. Incubate tube in a test tube of near boiling water and allow to cool to room temp

Note: The ^{32}P-primed DNA replication fork can be gel purified in the same way as DNA forks for the helicase assay in Section 3.1.

4. Protocols for CMG helicase assays

Materials and equipment

- BIO-RAD MINI-PROTEAN Tetra Cell, 2 Gel ELECTROPHORESIS System (older model: BZA652089 or newer model: 1658003EDU/1658022EDU)
- Outer and inner glass plates, Bio-Rad Mini-Protean II (The Gel Company, GBS07S, GBS07L)
- Mini-Protean II Comb, 15 wells, 1.5 mm thickness, 80 mm width (BIO-RAD,165–2926)
- Spacers, 1.5 mm thickness, 8.7 cm length, 1 cm width, 2 pcs (BIO-RAD,165–2933)
- 3 M Whatman filter paper
- Pre Proteins and reagents (Tables 3 and 4)

4.1 Helicase assays

4.1.1 Before you begin

1. Dissolve 25 mg AMPpNp powder in 937.5 μL ddH_2O to make

Table 3 Purified *S. cerevisiae* proteins used in CMG helicase and CMG-dependent DNA replication assays (Sections 4 and 5).

Name	Working conc.	Reference for purification
CMG (Cdc45 Mcm2-7 GINS) (*Helicase*)	0.2 ∞M	Georgescu et al. (2014)
Mcm10 (Mini Chromosome Maintenance) (*Helicase accessory factor*)	0.4 ∞M	Langston et al. (2017)
RFC (Replication Factor C) (*Clamp loader*)	1 ∞M	Georgescu et al. (2014)
PCNA (Proliferating Cell Nuclear Antigen) (*Clamp*)	5 ∞M	Georgescu et al. (2014)
Pol ε (*Leading strand polymerase*)	2 ∞M	Georgescu et al. (2014)
Pol δ (*Lagging strand polymerase*)	2 ∞M	Georgescu et al. (2014)
Pol α (*Primase/polymerase*)	2 ∞M	Georgescu et al. (2014)
RPA (Replication Protein A) (*Single-strand DNA-binding protein*)	20 ∞M	Georgescu et al. (2014)

Each protein used in the assays was purified separately according to the methods described in the references. The protein concentrations were then adjusted by diluting to the following working concentrations.

Table 4 List of reagents used for CMG helicase assays.

Reagent	Source	Identifier
Purified ^{32}P-helicase assay fork	Refer to Section 3.1	
ATPγS (adenosine-5′-o-(3-thio-triphosphate) (100 mM)	Roche	11162306001
AMPpNp (adenylyl-imidodiphosphate)	Roche	10102547001
100 mM ATP	Cytiva	GE27-2056-01
Magnesium sulfate	Sigma-Aldrich	M7506
DTT (1,4-dithiothreitol)	Roche	11583786001
3MM Whatman paper	Whatman	WHA3030917
Other reagents found in Section 3		

   a. Stock concentration of 50 mM (store at −80 °C as 100 μL aliquots)
   b. Approximately 100 μL of 10 mM working dilution
2. Prepare approximately 100 μL of 10 mM ATPγS mM by diluting the 100 mM stock with TE
3. Prepare the following buffers

(a) 10× Helicase buffer		1× conc
1 M Tris-acetate, pH 7.6	2.0 mL	20 mM
1 M DTT	0.5 mL	5 mM
0.5 M EDTA	20 μL	0.1 mM
ddH$_2$O	7.48 mL	
	10 mL Store at −20 °C as 1 mL aliquots	
(b) 5× Helicase Stop Solution		1× conc
0.5 M EDTA	2.0 mL	20 mM
60% glycerol	4.17 mL	5%
1% Bromophenol blue	665 μL	
1% Xylene Cyanol	665 μL	
10% SDS	500 μL	0.1%
ddH$_2$O	2.0 mL	
	10 mL	

(c) Helicase Stop + Trap oligo solution		1 × conc
5 × Helicase Stop solution (above)	500 μL	1 ×
10 μM cold oligo (the same oligo that was radiolabeled in the fork)	1 μL	6.7 nM
ddH_2O	999 μL	

Note: The trap oligo was added to the helicase stop solution to prevent the radiolabeled unwound product from reannealing back to its complementary oligo after the CMG has been inactivated by the EDTA and SDS. At the concentration of ^{32}P-fork that we work with (i.e., 0.5 nM) we do not see any significant reannealing of products, but include this trap oligo just to be sure.

4.1.2 Helicase assay with preloading CMG onto a DNA fork with either ATPγS or AMPpNp

Note: If you want to take rapid time points during your helicase assay then maximize the number of CMG's preloaded on the fork substrate so that you get a large burst of helicase unwinding upon addition of ATP. For the CMG loading reaction, you can either use ATPγS or AMPpNp. Since CMG is capable of slowly hydrolyzing ATPγS leading to a gradual unwinding of the fork duplex during the preincubation loading step, it is important that the duplex region on your DNA forks not be too short if you use ATPγS during your preloading step, or do not preincubate too long. Studies done in our lab have shown that forks that are 60 bp, generally do not exhibit much unwinding using ATPγS during preincubation times up to 2 h. Helicase assays with forks shorter than 40 bp would probably benefit from using AMPpNp for CMG loading instead of ATPγS, unless preincubation time is kept short. Since AMPpNp is a totally nonhydrolyzable analog of ATP, it allows the CMG helicase to bind DNA but not unwind it. Choosing which ATP analog to use for CMG preloading will depend on how long your fork is and for how long you plan to use for CMG preloading.

In the following example protocol, CMG is first preloaded onto a "short duplex" fork (e.g., 30 bp long) using AMPpNp in a 30-min preincubation step. After loading the CMG onto the fork, ATP is added to the reaction to initiate CMG unwinding activity. Over the course of 20 min, 10 aliquots of the reaction were withdrawn and quenched with EDTA every 2 min (Fig. 4).

1. Aliquot 15 μL of Helicase Stop + trap oligo solution into ten 0.5 mL microfuge tubes

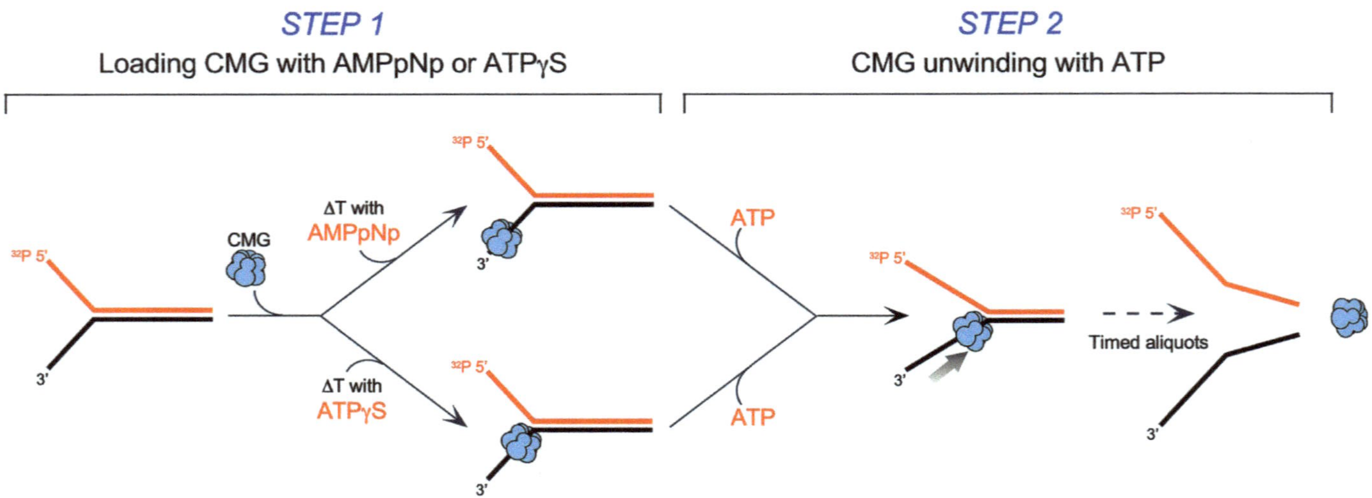

Fig. 4 Scheme for CMG helicase assays that include a CMG preloading step. CMG can be preloaded onto the fork substrate with the use of AMPpNp or ATPγS before initiating unwinding with ATP.

2. In a 1.5 mL microfuge tube, mix

	Per 10 μL reaction	(10 rxns + 10% excess)	Final conc
		×11	
10× Helicase buffer	1.0 μL	11.0 μL	1×
0.5 M $MgSO_4$	0.2 μL	2.2 μL	10 mM
Purified ^{32}P-30bp helicase fork (50 nM)	0.1 μL	1.1 μL	0.5 nM
10 mM AMPpNp	0.2 μL	2.2 μL	200 μM
200 nM CMG	1.0 μL	11.0 μL	20 nM
ddH_2O	7.0 μL	77.0 μL	
	9.5 μL	104.5 μL	

3. Incubate the tube (containing the equivalent of 11 reactions) in a 30 °C water bath for 30 min
4. Add 5.5 μL of 100 mM ATP (final ATP concentration = 5 mM)
5. Take a 10 μL aliquot of the reaction mix every 2 min (up to 20 min) and quench it with the pre-aliquoted 15 μL of Helicase Stop + trap oligo solution in the smaller tubes
6. Keep the reaction tubes on ice until ready for the gel electrophoresis step in Section 4.2 or store at −20 °C overnight

4.1.3 Helicase assay without preloading CMG onto a DNA fork

Note: If your helicase assay does not involve rapid time points, a separate CMG preloading step may not be necessary. The reaction can simply be initiated by adding the CMG to the ^{32}P-fork and ATP.

In the following example, CMG is allowed to unwind the 60 bp fork for 30 min. Under 5 mM ATP conditions, we find that the level of CMG unwinding of this fork has reached close to its maximum by this time. The reaction mix was also scaled up for the purpose of easier pipetting and allows for multiple reactions (Fig. 5).

1. In a 1.5 mL microfuge tube, mix

	Per 10 μL reaction	×10	Final conc.
10× Helicase buffer	1.0 μL	10.0 μL	1×
0.5M $MgSO_4$	0.2 μL	2.0 μL	10mM
Purified ^{32}P-helicase fork (50nM)	0.1 μL	1.0 μL	0.5nM
100mM ATP	0.5 μL	5.0 μL	5mM
ddH_2O	7.2 μL	72.0 μL	
	9.0 μL	90.0 μL	

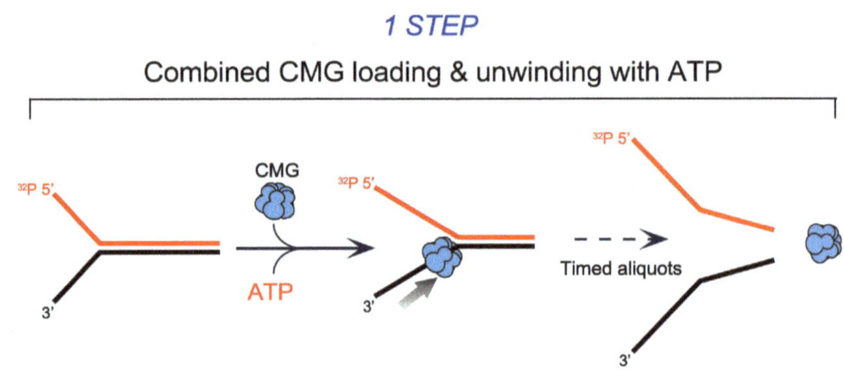

Fig. 5 Scheme for a 1 step CMG helicase assay. CMG loading and unwinding take place together in the presence of ATP.

2. Aliquot 9.0 μL of reaction mix to a 0.5mL microfuge tube
3. Initiate the reaction by adding 1.0 μL of 200nM CMG (final CMG concentration = 20nM)
4. Incubate the reaction in a 30 °C water bath for 30min
5. Quench reaction with 15 μL of Helicase Stop + trap oligo solution
6. Keep the reaction tubes on ice until ready for the gel electrophoresis step in Section 4.2 or store at −20 °C overnight

4.1.4 Helicase assay with the stimulation of Mcm10 preloading CMG onto a DNA fork

Note: CMG has difficulty unwinding long duplex forks. Addition of Mcm10 greatly stimulates the processivity of CMG unwinding (Langston et al., 2017). However, Mcm10 also reanneals ssDNA very efficiently (Mayle et al., 2019). Therefore, helicase assays with Mcm10 must have a trap oligo added with the ATP, or shortly thereafter to prevent the newly unwound products from being reannealed together by Mcm10. The trap is in excess and complementary to the unlabeled strand. It is the same trap oligo that is also present in the quenching solution.

In this example, a long 160 bp fork is unwound by CMG in the presence of ATP and Mcm10. One minute after initiating the unwinding, an oligo trap is added to prevent Mcm10 from reannealing the unwound products (Fig. 6).

1. Prepare some 250 nM trap oligo solution:

   1 μL of 10 μM unradiolabeled oligo (the same strand that was labeled in Section 3.1.1)

   +39 μL TE

   40 μL of 250 nM trap oligo

2. In a 1.5 mL microfuge tube, mix

	Per 10 μL reaction	×10	Final conc
10× Helicase buffer	1.0 μL	10.0 μL	1×
0.5 M $MgSO_4$	0.2 μL	2.0 μL	10 mM
Purified ^{32}P-160bp helicase fork (50 nM)	0.1 μL	1.0 μL	0.5 nM
200 nM CMG	1.0 μL	10.0 μL	20 nM
400 nM Mcm10	1.0 μL	10.0 μL	40 nM
ddH_2O	5.2 μL	52.0 μL	
	8.5 μL	85.0 μL	

3. Aliquot 8.5 μL of reaction mix to a 0.5 mL microfuge tube
4. Initiate the reaction by adding 0.5 μL of 100 mM ATP (final ATP concentration = 5 mM)

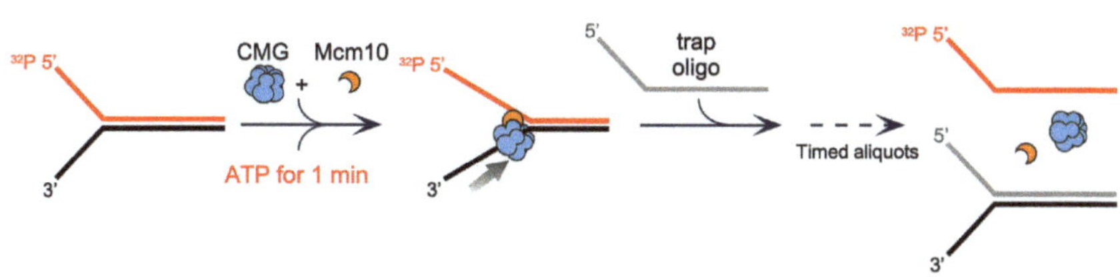

Fig. 6 Scheme for CMG helicase assay involving long forks. Mcm10 is included in this reaction to stimulate the processivity of the CMG helicase, thus allowing it to unwind longer duplexes. Since Mcm10 is also capable of reannealing the unwound ssDNA, a trap oligo that is complementary to the nonradiolabeled strand of the fork is added during the unwinding reaction.

5. Incubate the reaction in a 30 °C water bath for 1 min before adding 1.0 μL of 250 nM trap oligo
6. Quench reaction with 15 μL of Helicase Stop + trap oligo solution
7. Keep the reaction tubes on ice until ready for the gel electrophoresis step in Section 4.2 or store at −20 °C overnight

4.2 Native gel electrophoresis to analyze fork unwinding by CMG helicase

1. Prepare a mini 7 cm(H) × 8 cm(W) native 10% polyacrylamide gel using 1.5 mm spacers and a 15 well comb. This recipe is sufficient for 2 gels

40% acrylamide:bis-acrylamide solution (29:1)	7.5 mL
10 × TBE	3.0 mL
ddH_2O	19.5 mL
10% ammonium persulfate	300 μL
TEMED	30 μL

2. Load the helicase reactions onto the mini-gel along with a lane for a ^{32}P-fork marker and a ^{32}P-lagging strand marker
3. Run the gel in 1 × TBE buffer at 125 V until the bromophenol blue dye is near the bottom of the gel (approximately 50 min)
4. Mount the gel on a piece of 3MM Whatman paper and cover with plastic wrap
5. Expose the gel to a phosphor imaging screen for approximately 30 min
6. Scan the screen on the Typhoon scanner and use the manufacturer's computer program ImageQuant to quantify the volume of radioactivity in each band
7. The percent of ^{32}P-forks unwound by CMG is equal to the volume of the "unwound ^{32}P-lagging strand" band divided by the volume of the "^{32}P-fork" band

5. Protocols for CMG-dependent replication assays

Before you begin

The primed mini- fork substrate described in this chapter (Section 3.2) is the basis for our lab's other more longer and elaborate replication forks.

In these CMG-dependent replication assays, the primed fork substrate can be made either using a radiolabeled ^{32}P-primer, or constructed entirely of nonradiolabeled oligonucleotides. With this latter "cold" fork substrate, ^{32}P-dNTPs are used to follow DNA synthesis. If no lagging strand priming is present, only the leading strand will be observed. But if Pol α-primase is present, both strands will be made in the same reaction. In this event, we use a synthetic 3 kb DNA fork in which there is no dC on one strand of the duplex and no dG on the other (Georgescu et al., 2015). Hence, replication of leading and lagging strands are distinguished by which radionucleotide is used for the replication phase. A detailed protocol for this 3 kb-fork replication was previously published by Schauer et al. (2017).

In this section, we present two alternative ways to perform CMG-dependent replication assays using the primed mini-fork. Both methods require a two-step process. First, the CMG is preloaded onto the DNA fork. Second, the polymerase is added to extend the primer. In one protocol, CMG is loaded onto the DNA fork using a brief preincubation with ATP and in the second protocol, CMG is loaded onto the fork using a much longer preincubation with ATPγS. See Table 5 additional details about the reagents.

Note: Both methods run the risk of CMG unwinding the fork before the polymerase has the chance to "catch up." Therefore, it is imperative to first test the CMG loading conditions with your primed fork in a

Table 5 List of reagents used for CMG-dependent replication assays.

Reagent	Source	Identifier
Purified ^{32}P-primed replication assay fork	Refer to Section 3.2	
ATPγS (adenosine-5′-o′(3-thio-triphosphate)) (100 mM)	Roche	11162306001
100 mM ATP	Cytiva	GE27-2056-01
dATP, dCTP, dGTP, dTTP (100 mM)	Cytiva	
Magnesium sulfate	Sigma-Aldrich	M7506
BSA (bovine serum albumin)	NEB	B9000S
Formamide (deionized)	Ambion	AM9342
Urea	Fisher Chemical	U15-3
Other reagents found in Section 3 and 4		

helicase assay (refer to Section 4). If any of the fork is unwound during the time of CMG loading, then either the choice of nucleotide has to be changed (i.e., ATP to ATPγS or AMPpNp), or the incubation time has to be shortened.

5.1 CMG-dependent DNA replication assays

5.1.1 Before you begin

1. Dilute each dNTP (dATP, dCTP, dGTP and dTTP) to 3 mM: 3 μL 100 mM dNTP +97 μL TE
2. Prepare the following buffers

(a) 10× Replication buffer		1× conc
1 M Tris-acetate, pH 7.5	2.5 mL	25 mM
1 M DTT	200 μL	2 mM
0.5 M EDTA	20 μL	0.1 mM
20 mg/mL BSA	200 μL	40 μg/mL
70% Glycerol	7.08 mL	5%
	10.0 mL	Store at −20 °C as 1 mL aliquots

(b) 2× Replication Stop Solution		1× conc
Formamide (deionized)	7.7 mL	38.5%
0.5 M EDTA, pH 8.0	800 μL	20 mM
10% SDS	1.0 mL	0.5%
1% Bromophenol blue	250 μL	0.0125%
1% Xylene cyanol FF	250 μL	0.0125%
	10 mL	

5.1.2 Replication assay with preloading CMG onto a DNA fork with ATP

In this example assay, CMG is preloaded onto a primed fork with 200 μM ATP for 3 min before the replication reaction is initiated with Pol ε, RFC, PCNA, 5 mM ATP and the dNTPs. The reaction mix is scaled up so that multiple time points can be tested (Fig. 7).

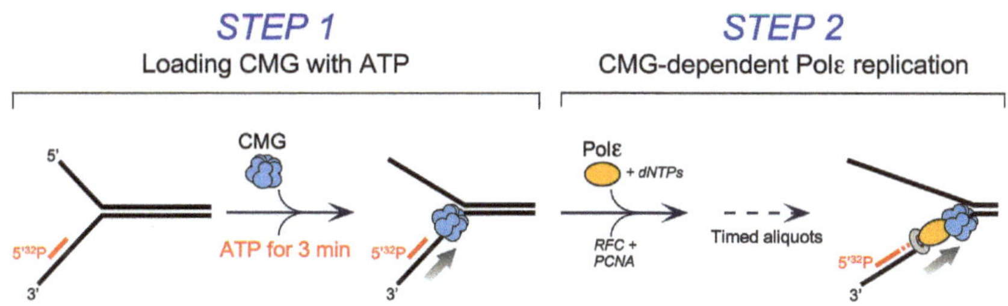

Fig. 7 Scheme for a replication assay with a quick CMG preloading step with ATP. CMG is first preloaded onto the primed mini-fork in the presence of ATP. After a brief 3-min preloading step, RFC, PCNA and Pol ε with the dNTPs are added to form a processive replisome that extends the ^{32}P-primer along the leading strand template.

1. In a 1.5 mL microfuge tube, make mix A

	Per 20 μL reaction	×10	Final conc
10× Replication buffer	2.0 μL	20.0 μL	1×
0.5 M $MgSO_4$	0.2 μL	2.0 μL	10 mM
Purified ^{32}P-60bp primed fork (50 nM)	0.4 μL	4.0 μL	1 nM
10 mM ATP	0.4 μL	4.0 μL	200 μM
ddH_2O	10.92 μL	109.2 μL	
	13.92 μL	139.2 μL	

In a 1.5 mL microfuge tube, make mix B			
1 μM RFC	0.1 μL	1.0 μL	5 nM
5 μM PCNA	0.1 μL	1.0 μL	25 nM
2 μM Pol ε	0.2 μL	2.0 μL	20 nM
100 mM ATP	1.0 μL	10.0 μL	5 mM
3 mM dATP	0.67 μL	6.7 μL	100 μM
3 mM dCTP	0.67 μL	6.7 μL	100 μM
3 mM dGTP	0.67 μL	6.7 μL	100 μM
3 mM dTTP	0.67 μL	6.7 μL	100 μM
	4.08 μL	40.8 μL	

2. Add 20 μL of 200 nM CMG (final CMG concentration = 20 nM) to Mix A (139.2 μL)
3. Incubate the reaction in a 30 °C water bath for 3 min
4. Initiate replication with Mix B (40.8 μL)
5. Take a 20 μL aliquot of the reaction mix every 3 min (up to 30 min) and quench it with 20 μL of 2× Replication Stop solution
6. Keep the reaction tubes on ice until ready for the gel electrophoresis step in Section 5.2 or store at −20 °C overnight

5.1.3 Replication assay with preloading CMG onto a DNA fork with ATPγS

In this next example assay, CMG is preloaded onto a primed fork with ATPγS for a long 2 h incubation. A previous helicase assay performed with this fork template verified that CMG was not unwinding the DNA during this 2 h time frame (Langston et al., 2017). In this replication reaction, Pol α was used and therefore RFC and PCNA were not needed. The reaction mix is scaled up so that multiple time points can be tested (Fig. 8).

1. In a 1.5 mL microfuge tube, make mix A

	Per 20 μL Reaction	×10	final conc
10× Replication buffer	2.0 μL	20.0 μL	1×
0.5 M $MgSO_4$	0.2 μL	2.0 μL	10 mM
Purified ^{32}P-60bp primed fork (50 nM)	0.2 μL	2.0 μL	0.5 nM
10 mM ATPγS	0.2 μL	2.0 μL	100 μM
ddH_2O	11.52 μL	115.2 μL	
	14.12 μL	141.2 μL	

In a 1.5 mL microfuge tube, make mix B			
2 μM Pol α	0.2 μL	2.0 μL	20 nM
100 mM ATP	1.0 μL	10.0 μL	5 mM
3 mM dATP	0.67 μL	6.7 μL	100 μM
3 mM dCTP	0.67 μL	6.7 μL	100 μM
3 mM dGTP	0.67 μL	6.7 μL	100 μM
3 mM dTTP	0.67 μL	6.7 μL	100 μM
	3.88 μL	38.8 μL	

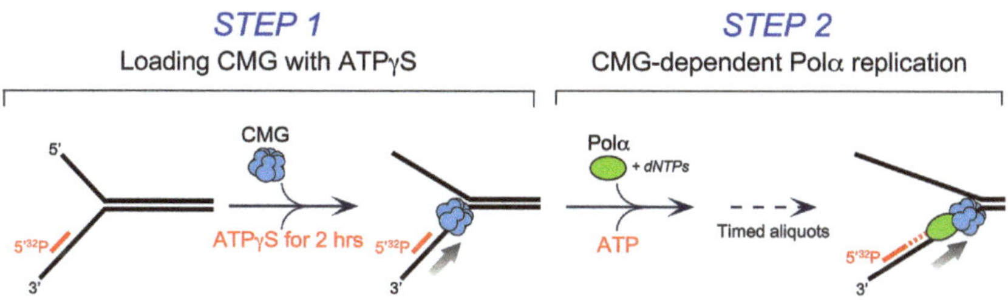

Fig. 8 Scheme for a replication assay with a long CMG preloading step with ATPγS. CMG can be loaded onto a primed mini-fork using ATPγS. Prior to this assay, CMG was demonstrated to exhibit no noticeable unwinding in a helicase assay, therefore it was possible to extend this CMG preloading step up to 2 h. CMG-dependent replication was then initiated with Polα, ATP and dNTPs.

2. Add 20 μL of 200 nM CMG (final CMG concentration = 20 nM) to Mix A (141.2 μL)
3. Incubate the reaction in a 30 °C water bath for 120 min
4. Initiate replication with Mix B (38.8 μL)
5. Take a 20 μL aliquot of the reaction mix every 3 min (up to 30 min) and quench it with 20 μL of 2 × Replication Stop solution
6. Keep the reaction tubes on ice until ready for the gel electrophoresis step in Section 5.2 or store at −20 °C overnight

5.2 Denaturing gel electrophoresis to analyze replication products synthesized by DNA polymerases in the presence of CMG

1. Prepare radiolabeled size markers. Oligonucleotides with the sequences corresponding to the expected "full length" product and "up to the fork junction" product (refer to Table 1) are radiolabeled according to the protocol in Section 3.1.1
2. Prepare a 14 cm(W) × 16 cm(H) 10% 8 M Urea denaturing polyacrylamide gel using 1 mm spacers and a 15 well comb

Urea	28.8 g
40% acrylamide solution (29:1)	15 mL
10 × TBE	6 mL
Bring volume to 60 mL with ddH$_2$O and gently dissolve the urea by mixing on a heated stir plate.	
Cool the solution to room temperature and add 360 μL 10% ammonium persulfate and 36 μL TEMED. Pour gel solution into glass casting plates with spacers then insert comb.	

3. Boil the tubes from the replication assays in Section 5.1.1 or Section 5.1.2 for 2 min in a hot pot before loading onto the gel. Remember to include lanes for the size markers (radiolabeled oligos from step 1).
4. Run the gels in 1 × TBE (4 L) at 38 mAmps (maximum voltage and wattage) until the bromophenol blue dye is approximately 1 in. from the bottom of the gel (approximately 1 h).
5. Carefully mount the gel onto a piece of 3MM Whatman filter paper and cover with plastic wrap
6. Expose gel to a phosphor imaging screen for approximately 1 h before scanning with the Typhoon scanner
7. The lengths of the DNA products synthesized in the replication assays can be determined by comparing the position of those bands with the positions of the size markers

6. Summary

The protocols described in this report can be used in various combinations for either helicase or replication-associated reactions. In addition, there is plenty of room for diverging from these exact protocols. Even within our own lab, we still vary parameters to search for optimal conditions for any given assay. Specifically, the behavior and activity of combinations of proteins/enzymes will take on new and unpredictable biochemical characteristics that may require a different regime. Therefore, the protocols outlined here are only to be taken as a starting guide for developing an assay and can be customized to suit your enzymes whether they are from different organisms, or reactions that only partially utilize the particular proteins discussed in these protocols.

Acknowledgments

We thank the entire O'Donnell lab for their advice and expertise in DNA-related assays. We are also grateful for support by the US National Institutes of Health grants GM115809 and the Howard Hughes Medical Institute.

References

Abid Ali, F., Renault, L., Gannon, J., Gahlon, H. L., Kotecha, A., Zhou, J. C., et al. (2016). Cryo-EM structures of the eukaryotic replicative helicase bound to a translocation substrate. *Nature Communications*, 7, 10708. https://doi.org/10.1038/ncomms10708.

Bell, S. P., & Labib, K. (2016). Chromosome duplication in Saccharomyces cerevisiae. *Genetics*, *203*(3), 1027–1067. https://doi.org/10.1534/genetics.115.186452.

Bowman, G. D., O'Donnell, M., & Kuriyan, J. (2004). Structural analysis of a eukaryotic sliding DNA clamp-clamp loader complex. *Nature*, *429*(6993), 724–730. https://doi.org/10.1038/nature02585.

Burgers, P., & Kunkel, T. A. (2017). Eukaryotic DNA replication fork. *Annual Review of Biochemistry*, *86*, 417–438. https://doi.org/10.1146/annurev-biochem-061516-044709.

Burnham, D. R., Kose, H. B., Hoyle, R. B., & Yardimci, H. (2019). The mechanism of DNA unwinding by the eukaryotic replicative helicase. *Nature Communications*, *10*(1), 2159. https://doi.org/10.1038/s41467-019-09896-2.

Fu, Y. V., Yardimci, H., Long, D. T., Ho, T. V., Guainazzi, A., Bermudez, V. P., et al. (2011). Selective bypass of a lagging strand roadblock by the eukaryotic replicative DNA helicase. *Cell*, *146*(6), 931–941. https://doi.org/10.1016/j.cell.2011.07.045.

Georgescu, R. E., Langston, L., Yao, N. Y., Yurieva, O., Zhang, D., Finkelstein, J., et al. (2014). Mechanism of asymmetric polymerase assembly at the eukaryotic replication fork. *Nature Structural & Molecular Biology*, *21*(8), 664–670. https://doi.org/10.1038/nsmb.2851.

Georgescu, R. E., Schauer, G. D., Yao, N. Y., Langston, L. D., Yurieva, O., Zhang, D., et al. (2015). Reconstitution of a eukaryotic replisome reveals suppression mechanisms that define leading/lagging strand operation. *eLife*, *4*, e04988. https://doi.org/10.7554/eLife.04988.

Georgescu, R., Yuan, Z., Bai, L., de Luna Almeida Santos, R., Sun, J., Zhang, D., et al. (2017). Structure of eukaryotic CMG helicase at a replication fork and implications to replisome architecture and origin initiation. *Proceedings of the National Academy of Sciences of the United States of America*, *114*(5), E697–E706. https://doi.org/10.1073/pnas.1620500114.

Ilves, I., Petojevic, T., Pesavento, J. J., & Botchan, M. R. (2010). Activation of the MCM2-7 helicase by association with Cdc45 and GINS proteins. *Molecular Cell*, *37*(2), 247–258. https://doi.org/10.1016/j.molcel.2009.12.030.

Kang, Y. H., Galal, W. C., Farina, A., Tappin, I., & Hurwitz, J. (2012). Properties of the human Cdc45/Mcm2-7/GINS helicase complex and its action with DNA polymerase epsilon in rolling circle DNA synthesis. *Proceedings of the National Academy of Sciences of the United States of America*, *109*(16), 6042–6047. https://doi.org/10.1073/pnas.1203734109.

Kose, H. B., Larsen, N. B., Duxin, J. P., & Yardimci, H. (2019). Dynamics of the eukaryotic replicative helicase at lagging-Strand protein barriers support the steric exclusion model. *Cell Reports*, *26*(8), 2113–2125.e6. https://doi.org/10.1016/j.celrep.2019.01.086.

Kose, H. B., Xie, S., Cameron, G., Strycharska, M. S., & Yardimci, H. (2020). Duplex DNA engagement and RPA oppositely regulate the DNA-unwinding rate of CMG helicase. *Nature Communications*, *11*(1), 3713. https://doi.org/10.1038/s41467-020-17443-7.

Kunkel, T. A., & Burgers, P. M. (2014). Delivering nonidentical twins. *Nature Structural & Molecular Biology*, *21*(8), 649–651. https://doi.org/10.1038/nsmb.2852.

Langston, L. D., Mayle, R., Schauer, G. D., Yurieva, O., Zhang, D., Yao, N. Y., et al. (2017). Mcm10 promotes rapid isomerization of CMG-DNA for replisome bypass of lagging strand DNA blocks. *eLife*, *6*, e29118. https://doi.org/10.7554/eLife.29118.

Langston, L. D., & O'Donnell, M. E. (2019). An explanation for origin unwinding in eukaryotes. *eLife*, *8*, e46515. https://doi.org/10.7554/eLife.46515.

Langston, L. D., Zhang, D., Yurieva, O., Georgescu, R. E., Finkelstein, J., Yao, N. Y., et al. (2014). CMG helicase and DNA polymerase ε form a functional 15-subunit holoenzyme for eukaryotic leading-strand DNA replication. *Proceedings of the National Academy of Sciences of the United States of America*, *111*(43), 15390–15395. https://doi.org/10.1073/pnas.1418334111.

Lewis, J. S., Spenkelink, L. M., Schauer, G. D., Yurieva, O., Mueller, S. H., Natarajan, V., et al. (2020). Tunability of DNA polymerase stability during eukaryotic DNA replication. *Molecular Cell*, 77(1), 17–25.e5. https://doi.org/10.1016/j.molcel.2019.10.005.

Lõoke, M., Maloney, M. F., & Bell, S. P. (2017). Mcm10 regulates DNA replication elongation by stimulating the CMG replicative helicase. *Genes & Development, 31*(3), 291–305. https://doi.org/10.1101/gad.291336.116.

Mayle, R., Langston, L., Molloy, K. R., Zhang, D., Chait, B. T., & O'Donnell, M. E. (2019). Mcm10 has potent strand-annealing activity and limits translocase-mediated fork regression. *Proceedings of the National Academy of Sciences of the United States of America, 116*(3), 798–803. https://doi.org/10.1073/pnas.1819107116.

Moyer, S. E., Lewis, P. W., & Botchan, M. R. (2006). Isolation of the Cdc45/Mcm2-7/GINS (CMG) complex, a candidate for the eukaryotic DNA replication fork helicase. *Proceedings of the National Academy of Sciences of the United States of America, 103*(27), 10236–10241. https://doi.org/10.1073/pnas.0602400103.

Petojevic, T., Pesavento, J. J., Costa, A., Liang, J., Wang, Z., Berger, J. M., et al. (2015). Cdc45 (cell division cycle protein 45) guards the gate of the eukaryote replisome helicase stabilizing leading strand engagement. *Proceedings of the National Academy of Sciences of the United States of America, 112*(3), E249–E258. https://doi.org/10.1073/pnas.1422003112.

Schauer, G., Finkelstein, J., & O'Donnell, M. (2017). In vitro assays for eukaryotic leading/lagging strand DNA replication. *Bio-Protocol,* 7(18), e2548. https://doi.org/10.21769/BioProtoc.2548.

Sun, J., Shi, Y., Georgescu, R. E., Yuan, Z., Chait, B. T., Li, H., et al. (2015). The architecture of a eukaryotic replisome. *Nature Structural & Molecular Biology, 22*(12), 976–982. https://doi.org/10.1038/nsmb.3113.

van Houten, V., Denkers, F., van Dijk, M., van den Brekel, M., & Brakenhoff, R. (1998). Labeling efficiency of oligonucleotides by T4 polynucleotide kinase depends on 5′-nucleotide. *Analytical Biochemistry, 265*(2), 386–389. https://doi.org/10.1006/abio.1998.2900.

Yao, N. Y., Zhang, D., Yurieva, O., & O'Donnell, M. E. (2022). CMG helicase can use ATPγS to unwind DNA: Implications for the rate-limiting step in the reaction mechanism. *Proceedings of the National Academy of Sciences of the United States of America, 119*(4), e2119580119. https://doi.org/10.1073/pnas.2119580119.

Yeeles, J. T., Deegan, T. D., Janska, A., Early, A., & Diffley, J. F. (2015). Regulated eukaryotic DNA replication origin firing with purified proteins. *Nature, 519*(7544), 431–435. https://doi.org/10.1038/nature14285.

Yeeles, J., Janska, A., Early, A., & Diffley, J. (2017). How the eukaryotic replisome achieves rapid and efficient DNA replication. *Molecular Cell, 65*(1), 105–116. https://doi.org/10.1016/j.molcel.2016.11.017.

Yuan, Z., Georgescu, R., Santos, R., Zhang, D., Bai, L., Yao, N. Y., et al. (2019). Ctf4 organizes sister replisomes and Pol α into a replication factory. *eLife, 8*, e47405. https://doi.org/10.7554/eLife.47405.001.

CHAPTER TEN

ReconSil: An electron microscopy toolbox to study helicase function at an origin of replication

Thomas Pühringer[a], Julia F. Greiwe[a], Thomas C.R. Miller[b,*], and Alessandro Costa[a,*]

[a]Macromolecular Machines Laboratory, The Francis Crick Institute, London, United Kingdom
[b]Center for Chromosome Stability, Department of Cellular and Molecular Medicine, University of Copenhagen, Copenhagen, Denmark
[*]Corresponding authors: e-mail address: tmiller@sund.ku.dk; alessandro.costa@crick.ac.uk

Contents

Methods in Enzymology, Volume 672
ISSN 0076-6879
https://doi.org/10.1016/bs.mie.2022.03.016

Abstract

The loading of the MCM replicative helicase onto eukaryotic origins of replication occurs via a sequential, symmetric mechanism. Here, we describe a method to study this multistep reaction using electron microscopy. Tools presented include protein expression and purification protocols, methods to produce asymmetric replication origin substrates and bespoke image processing strategies. DNA templates include recognisable protein roadblocks that help to orient DNA replication factors along a specific origin sequence. Detailed electron microscopy image processing protocols are provided to reposition 2D averages onto the original micrograph for the in silico reconstitution of fully occupied origins of replication. Using these tools, a chemically trapped helicase loading intermediate is observed sliding along origin DNA, showcasing a key feature of the MCM loading mechanism. Although developed to study replicative helicase loading, this method can be employed to investigate the mechanism of other multicomponent biochemical reactions, occurring on a flexible polymeric substrate.

Abbreviations

ACS	ARS consensus sequence
ARS	autonomously replicating sequence
ATP	adenosine triphosphate
CMG	Cdc45-Mcm2-7-GINS
CTF	contrast transfer function
EM	electron microscopy
GINS	Go-Ichi-Ni-San
GUI	graphical user interface
MCM	minichromosome maintenance
M.HpaII	Haemophilus parainfluenzae Methyltransferase
OCCM	ORC-Cdc6-Cdt1-MCM
ORC	origin recognition complex
PAGE	polyacrylamide gel electrophoresis
Pfu	pyrococcus furiosus
YP	yeast peptone

1. Introduction

DNA replication is essential for the propagation of life. In all replicating cells, an initiator recognizes a site on the genome named an origin of replication and recruits a pair of hexameric helicases that unwind the DNA bidirectionally, providing the single stranded DNA template for the replicative polymerases to duplicate. DNA unwinding and synthesis are coordinated through the assembly of a multiprotein complex, named the replisome (Attali, Botchan, & Berger, 2021). DNA replication in eukaryotic cells initiates at multiple origins and requires tight coordination to ensure that the genome is replicated only once per cell division cycle. To prevent re-replication, the loading and activation of eukaryotic helicases are temporally separated and under stringent cell cycle control. Misregulation leads to genomic instability and can lead to malignant cellular transformations (Lewis & Costa, 2020).

From the end of mitosis and throughout the G1 phase of the cell cycle, two copies of a hexameric ring shaped MCM (MiniChromosome Maintenance) helicase are recruited onto origin DNA. DNA-loaded MCMs form a double hexamer, which is catalytically inactive yet contains the symmetry to support bidirectional replication (Abid Ali et al., 2017; Evrin et al., 2009; Noguchi et al., 2017; Remus et al., 2009). Only upon entry into S phase is the DNA unwinding function of the MCM ATPase motor switched on. This occurs upon recruitment of two firing factors named Cdc45 and GINS, which, together with the MCM motor, form the so-called CMG holohelicase (Ilves, Petojevic, Pesavento, & Botchan, 2010; Moyer, Lewis, & Botchan, 2006). The mechanism for loading a double hexamer onto replication origins has been the focus of investigations from several research groups for the last decade, with most of the work focused on the *S. cerevisiae* replication origin, ARS1 (Autonomously Replicating Sequence 1). This contains a linear ~140 bp stretch of AT-rich duplex DNA that is prone to melting and is flanked by two nucleosomes. ARS1 also contains two inverted binding sites for the Origin Recognition Complex (ORC), i.e., the initiator that recruits the MCM helicase. The first site, ACS (ARS Consensus Sequence), is bound by ORC with high affinity. The second site, B2, maps 42 bp downstream of ACS and is bound by ORC with inverted polarity and lower affinity (Fig. 1A) (Bell & Stillman, 1992; Lipford & Bell, 2001; Rao & Stillman, 1995; Rowley, Cocker, Harwood, & Diffley, 1995). The linear architecture of the ARS1 origin suggests a symmetric mechanism for MCM double hexamer loading, whereby two inverted ORCs each recruit an MCM hexamer, leading to the topological loading of the double hexamer. Bulk biochemistry and genetic evidence strongly support this model (Coster & Diffley, 2017). However, single-molecule work that investigated the kinetics of the MCM loading reaction revealed that helicase loading is sequential. According to colocalization single-molecule spectroscopy measurements, ORC recruits one MCM hexamer, while recruitment of the second MCM is dictated by the first loaded hexamer. This observation invited a model whereby the N-terminal domain of MCM, which will become the dimerization interface of the double hexamer, plays a direct role in the recruitment of the second hexameric ring (Ticau, Friedman, Ivica, Gelles, & Bell, 2015).

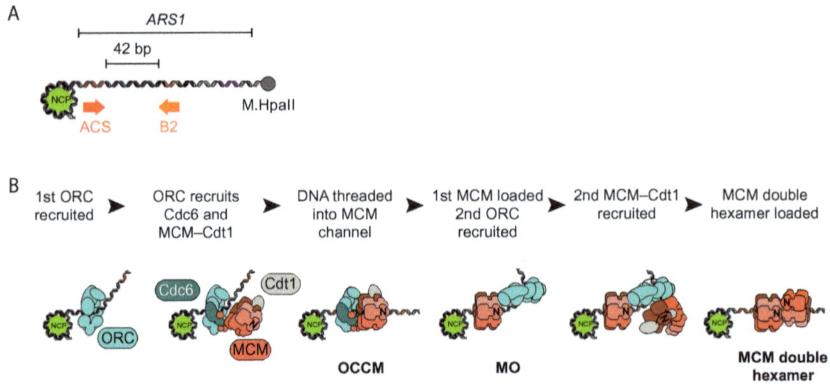

Fig. 1 MCM helicase loading at the ARS1 origin of replication. (A) Architecture of the ARS1 origin of replication. ARS1 contains two inverted binding sites for ORC, a high affinity site named ACS and a lower affinity site named B2. The ARS1 origin sequence used in this study is flanked at the ACS site by a Widom strong positioning sequence used for nucleosome reconstitution. Downstream of the B2 site, a suicide substrate to covalently link a M.*Hpa*II methyltransferase has been engineered. (B) MCM helicase loading mechanism. ORC binds and bends DNA. Together with Cdc6, ORC recruits an MCM-Cdt1 complex and threads DNA through the central channel of MCM, forming the OCCM complex. ORC then disengages from the C-terminal face of MCM and engages the N-terminal face of MCM binding the B2 DNA site in an inverted configuration, forming the MO intermediate. In this new position, ORC is competent to recruit a new MCM-Cdt1 complex via the same mechanism as the first loaded MCM. The end result is the formation of an MCM double hexamer that encircles duplex DNA.

How can helicase loading be symmetric but at the same time sequential? To address this question, we previously employed a time resolved electron microscopy strategy to visualize the structural intermediates of helicase loading, by imaging the reaction as it occurs in vitro. We found that ORC binds at the ACS site and recruits a first MCM ring with the help of additional loading factors Cdc6 and Cdt1. After loading of the first hexamer, we observed that ORC lets go of the C-terminal face of the MCM and binds to the opposed face, i.e., the N-terminal face that will become the dimerization interface in the double hexamer. In this alternate helicase-interacting configuration, ORC can bind to the inverted B2 site on the origin and can recruit a second MCM via the same mechanism as the first. These events lead to the full topological loading of the symmetric MCM double hexamer (Miller, Locke, Greiwe, Diffley, & Costa, 2019). Recent single-molecule fluorescence observations support our time-resolved EM results (Gupta, Friedman, Gelles, & Bell, 2021), explaining how helicase loading can be sequential (first loaded MCM dictates loading of the second MCM) and, at the same time, symmetric (the two helicases are loaded via the same molecular mechanism by two inverted ORC binding events, Fig. 1B).

A separate issue, related to MCM loading, is the spacing between the two inverted ORC binding sites (42 bp in ARS1), which is shorter than the length of DNA trapped inside an MCM double hexamer (~60 bp). While biochemical experiments already support the notion that MCM slides along DNA to create enough space for double hexamer formation (Warner, Azmi, Kang, Zhao, & Bell, 2017), visualizing sliding with EM can help establish at what stage in the pathway sliding occurs. Aside from the time-resolved approach to enrich for certain helicase loading intermediates and monitor DNA sliding, chemical strategies to halt the loading reaction can also be employed. For example, use of a slowly hydrolysable ATP analog (ATPγS) stabilizes a complex between ORC and MCM bound to loading factors Cdc6 and Cdt1 (Yuan et al., 2017). This intermediate, dubbed "OCCM" and representing the first helicase recruitment intermediate, will be used in this chapter as an example to study helicase loading and sliding, in the context of a full origin.

Accurate tracking of the orientation and relative position of different initiator factors along the DNA is essential for developing a full understanding of multicomponent, multistep reactions that occur on a single segment of the DNA double helix. A key development in this context are nucleic-acid substrates containing recognizable asymmetric markers (Miller et al., 2019). These markers can help establish the binding polarity of distinct replication factors as well as their location relative to the DNA sequence. In this chapter, we will detail protocols to express and purify all factors required for loading the *S. cerevisiae* MCM helicase onto an origin of replication. We will also describe protocols to produce the ARS1 origin of replication, asymmetrically capped with either a nucleosome or a covalently linked methyltransferase adduct, which can be used to orient replication factors bound to the same DNA substrate. Image processing protocols are also described to track the relative position and orientation of ORC and MCM along the origin substrate. In particular, due to the limited persistence length of duplex DNA, traditional averaging approaches fail to describe the entire ARS1 origin of replication including MCMs, ORCs and the DNA capping roadblocks. To circumvent this issue, we describe a method to overlay 2D averages onto raw images that contributed to their formation, in the context of the original raw micrographs. This approach yields a view of the entire origin nucleoprotein complex, providing not only the molecular details of isolated protein complexes, but an exhaustive description of the full context of the MCM loading reaction. In our results section, we analyze the ability of the chemically trapped OCCM intermediate to slide along DNA, explaining how enough space is created between two ORC binding sites to support double hexamer loading (Warner et al., 2017). MCM sliding is only one aspect of the helicase loading mechanism, while time resolved approaches combined with ReconSil have been used to describe the entire double-hexamer loading reaction in our previously published study (Miller et al., 2019). In the Discussion section, we describe future applications of the ReconSil toolbox, ranging from origin activation studies to completely different nucleic acid transactions.

2. Protein expressions and purifications

Detailed protocols are also available in Posse, Johansson, and Diffley (2021). *S. cerevisiae* expression strains are listed in Table 1.

Table 1 Yeast overexpression strains.

Strain	Genotype (all in W303 background)	Reference
yAM33	*MATa* *bar1::Hyg* *pep4::KanMx* *his3::pRS303-Cdt1/Gal4 (HIS3)* *ura3::pRS306-Mcm2/Mcm3 CBP-Mcm3 (URA3)* *trp1::pRS304-Mcm4/5 (TRP1)* *leu2::pRS305-Mcm6/7 (LEU2)*	Coster, Frigola, Beuron, Morris, and Diffley (2014)
ySD-ORC	*MATa ade2–1 ura3–1 his3–11,15 trp1–1 leu2–3112 can1–100* *bar1::Hyg* *pep4::KanMX* *his3::pRS303-ORC3/4 (HIS3)* *ura3::pRS306-ORC1/2 CBP-Orc1 (URA3)* *trp1::pRS304-ORC5/6 (TRP1)*	Frigola, Remus, Mehanna, and Diffley (2013)

2.1 Expression of ORC and Mcm2-7–Cdt1 in *S. cerevisiae*

Materials:

YP medium (supplemented with 4 mg/L adenine and 100 μg/mL ampicillin (optional))
Glucose (20%, sterile)
Raffinose (20%, sterile)
Galactose (20%, sterile)
Alpha factor (5 mg/mL)
Lysis buffer (see protein purification protocols in Sections 2.2 and 2.3)
Dithiothreitol (DTT, 1 M)
Protease inhibitors (e.g., complete EDTA free (Roche))

1. Inoculate 50 mL YP + 2% glucose with the protein expression strain (ySD-ORC for ORC; yAM33 for Mcm2-7–Cdt1) and grow at 30 °C and shaking at 200 rpm to saturation (roughly 24 h)
2. Inoculate 12 L YP + 2% raffinose with 2×10^8 cells/L and grow at 30 °C and 200 rpm to 2–3 $\times 10^7$ cells/mL overnight
3. Add alpha factor to a final concentration of 100 ng/mL to arrest cells in the G1 phase of the cell cycle
4. After 2–3 h, when all cells are arrested, induce protein expression by addition of 2% galactose and grow for 3 h at 30 °C and 200 rpm
5. Harvest cells by centrifugation for 10 min at 4500 × *g*
6. Transfer cells to a 50 mL Falcon tube and wash once with the respective lysis buffer (see protein purification protocols). Pellet cells by centrifugation
7. Resuspend cells in 0.4 volumes of lysis buffer supplemented with 2 mM DTT and protease inhibitors
8. Freeze cells dropwise in liquid nitrogen
9. Crush cells in a freezer mill (6875D Freezer/Mill® Dual Chamber Cryogenic Grinderfreezer mill (SPEX SamplePrep)) with 6 cycles of 2 min at a crushing rate of 15 and 1 min rest. Store the frozen cell powder at −80 °C

This protocol is adapted from Coster et al. (2014) and Frigola et al. (2013).

2.2 Purification of CBP-tagged ORC

Materials:

Buffer 1 (25 mM HEPES-KOH pH 7.6, 0.05% NP-40, 10% glycerol, 2 mM β-mercaptoethanol)
Protease inhibitors (e.g., complete EDTA free (Roche))
Potassium chloride (3 M)
Calcium chloride (1 M)

Ethylenediaminetetraacetic acid (EDTA, 0.5 M)
Ethylene glycol-bis(β-aminoethylether)-N,N,N′,N′-tetraacetic acid (EGTA, 0.5 M)
Calmodulin affinity resin (e.g., Agilent 214303-52)
Empty gravity chromatography column
3–8% Tris-acetate gels (Criterion)
Amicon Ultra centrifugal filter units (Millipore), 100,000 MWCO
Superdex 200 16/600 gel filtration column (Cytiva)

1. Thaw cell powder on ice and resuspend in 1 volume of Buffer 1 supplemented with 0.1 M potassium chloride (= lysis buffer) and protease inhibitors
2. Increase potassium chloride concentration to 0.5 M before subjecting the lysate to ultracentrifugation for 60 min at 235,418 × *g* and 4 °C
3. Collect the middle layer without transferring the lipids at the top or the pelleted material
4. Supplement cleared lysate with 2 mM calcium chloride and incubate for 2 h at 4 °C and constant rotation with 5 mL Calmodulin Affinity Resin (bed volume) pre-equilibrated in Buffer 1 + 0.4 M potassium chloride +2 mM calcium chloride
5. Collect beads in an empty gravity chromatography column and wash beads with 80 mL Buffer 1 + 0.4 M potassium chloride +2 mM calcium chloride
6. Elute protein by incubating the resin with Buffer 1 + 1 mM EDTA +2 mM EGTA, collecting 2 mL fractions
7. Analyse fractions by SDS polyacrylamide gel electrophoresis (SDS-PAGE).
8. Pool peak fractions and concentrate to ~2.5 mL using a 100,000 MWCO centrifugal filter unit
9. Remove any precipitate by centrifugation at full speed and 4 °C for 5 min in a benchtop centrifuge
10. Load the sample on a Superdex 200 16/600 gel filtration column, which is equilibrated with Buffer 1 + 0.15 M potassium chloride, using a 5 mL capillary loop
11. Analyse fractions on an SDS-PAGE
12. Pool peak fractions and concentrate to ~1 mg/mL using a 100,000 MWCO centrifugal filter unit
13. Freeze aliquots of 5–20 μL in liquid nitrogen and store at −80 °C

Note: All protein purification steps should be carried out at 4 °C.

This protocol was adapted from Coster et al. (2014).

2.3 Purification of CBP-tagged Mcm2-7–Cdt1

Materials:

Buffer 2 (45 mM HEPES-KOH pH7.6, 100 mM potassium acetate, 5 mM magnesium acetate, 0.02% NP-40, 10% glycerol, 2 mM β-mercaptoethanol)
Protease inhibitors (e.g., complete EDTA free (Roche))
Calcium chloride (1 M)
EDTA (0.5 M)
EGTA (0.5 M)
Calmodulin affinity resin (e.g., Agilent 214303-52)
Empty gravity chromatography column
3–8% Tris-acetate gels (Criterion)
Amicon Ultra centrifugal filter units (Millipore), 100,000 MWCO
Superdex 200 16/600 gel filtration column (Cytiva)

1. Thaw cell powder on ice and resuspend in 1 volume of Buffer 2 supplemented with protease inhibitors
2. Subject the lysate to ultracentrifugation at 235,418 × *g* for 60 min at 4 °C
3. Collect the middle layer without transferring the lipids at the top or pelleted material
4. Supplement cleared lysate with 2 mM calcium chloride and incubate for 2 h at 4 °C and constant rotation with 2 mL Calmodulin Affinity Resin (bed volume) pre-equilibrated in Buffer 2 + 2 mM calcium chloride
5. Collect beads in an empty gravity chromatography column and wash beads with 80 mL Buffer 2 + 2 mM calcium chloride
6. Elute protein by incubating the resin with Buffer 2 + 1 mM EDTA +2 mM EGTA, collecting 1 mL fractions
7. Analyse fractions on an SDS-PAGE
8. Pool peak fractions with stoichiometric subunit composition and concentrate to ~2.5 mL using a 100,000 MWCO centrifugal filter unit
9. Remove any precipitate by centrifugation at full speed and 4 °C for 5 min in a benchtop centrifuge
10. Load the sample on a Superdex 200 16/600 gel filtration column, which is equilibrated with Buffer 2, using a 5 mL capillary loop
11. Elute the protein and collect 1 mL fractions
12. Analyse fractions on an SDS-PAGE
13. Pool peak fractions and concentrate to 2 mg/mL using a 100,000 MWCO centrifugal filter unit
14. Freeze aliquots of 5–20 μL in liquid nitrogen and store at −80 °C

This protocol is adapted from Frigola et al. (2013).

2.4 Expression of Cdc6 in *E. coli*

Materials:

LB-agar plate supplemented with 100 μg/mL ampicillin and 30 μg/mL chloramphenicol
LB medium
Ampicillin (100 mg/mL)
Chloramphenicol (30 mg/mL)
Isopropyl β-D-1-thiogalactopyranoside (IPTG, 1 M)
Protease inhibitors (e.g., cOmplete EDTA free (Roche))

1. Transform *E. coli* expression strain of choice, (BL21 (DE3) RIL is a suitable option), with the plasmid containing the Cdc6 sequence (pAM3) and grow at 37 °C on a LB-agar plate supplemented with 100 μg/mL ampicillin and 30 μg/mL chloramphenicol
2. Inoculate 50 mL LB + ampicillin + chloramphenicol with one colony and grow at 37 °C and 200 rpm overnight
3. Inoculate 3 L LB + ampicillin + chloramphenicol with the overnight starter and grow to an optical density 600 (OD_{600}) of 0.6 at 37 °C
4. Cool the culture to 18 °C
5. Induce expression by adding IPTG to a final concentration of 0.5 mM and grow at 18 °C and 200 rpm for 5 h
6. Harvest cells by centrifugation for 10 min at 4500 × *g* and freeze in liquid nitrogen. Store the frozen cell pellet at −80 °C

2.5 Purification of GST-tagged Cdc6

Materials:

Buffer 3 (50 mM potassium phosphate pH 7.6, 150 mM potassium acetate, 5 mM magnesium chloride, 2 mM ATP, 1% triton X-100, 1 mM DTT)
Buffer 4 (50 mM potassium phosphate pH 7.6, 5 mM magnesium chloride, 2 mM ATP, 0.1% triton X-100, 1 mM DTT)
Buffer 5 (25 mM HEPES-KOH pH 7.6, 10 mM magnesium acetate, 0.02% NP-40, 100 mM potassium acetate, 10% glycerol, 5 mM β-mercaptoethanol)
Protease inhibitors (e.g., cOmplete EDTA free (Roche))
Glutathione Sepharose® 4B (GE)
Empty gravity chromatography column
PreScission protease
Bio-Gel HTP Hydroxyapatite powder (Bio-Rad)
4–15% Criterion TGX gel (Criterion)

Amicon Ultra centrifugal filter units (Millipore), 30,000 MWCO

1. Thaw cell pellet on ice and resuspend in 30 mL Buffer 3 supplemented with protease inhibitors
2. Lyse cells by sonication
3. Centrifuge at 58,545 × *g* and 4 °C for 20 min and collect the soluble phase
4. Apply the soluble phase twice to pre-equilibrated 2 mL Glutathione Sepharose® 4B beads (bed volume) on a gravity chromatography column
5. Wash beads with 40 mL Buffer 3 + protease inhibitors and 40 mL Buffer 3
6. Resuspend beads in Buffer 3 to generate a 50% slurry and incubate at 4 °C overnight with 50 μL PreScission protease
7. Collect eluate and mix with Buffer 4 to bring potassium acetate concentration to 75 mM
8. Hydrate 2 g Bio-Gel HTP Hydroxyapatite (Bio-Rad) powder with 12 mL Buffer 4 + 75 mM potassium acetate and incubate at room temperature for 5 min with constant rotation
9. Let resin settle for 10 s and remove the suspension of fine material at the top. Replace buffer and repeat this procedure two more times
10. Prepare a 50% slurry of the resin and add 4 mL of the slurry to the diluted eluate
11. Incubate the diluted eluate for 15 min with the hydroxyapatite
12. Wash beads with 10 mL buffer 4 + 75 mM potassium acetate +15% glycerol and subsequently with 10 mL buffer 4 + 150 mM potassium acetate +15% glycerol
13. Elute the protein with ~10 mL buffer 4 + 400 mM potassium acetate +15% glycerol, collecting 1 mL fractions
14. Analyse fractions on an SDS-PAGE and pool fractions containing Cdc6
15. Dialyse sample against buffer 5 overnight
16. Concentrate to 0.5 mg/mL using a 30,000 MWCO centrifugal filter unit
17. Freeze aliquots of 5–20 μL in liquid nitrogen and store at −80 °C

The expression and purification protocols of Cdc6 are adapted from Frigola et al. (2013).

2.6 Expression of histone octamers in *E. coli*

Materials:

LB-agar plate supplemented with 100 μg/mL ampicillin, 30 μg/mL chloramphenicol and 50 μg/mL spectinomycin

LB medium
Ampicillin (100 mg/mL)
Chloramphenicol (30 mg/mL)
Spectinomycin (50 mg/mL)
IPTG (1 M)
Protease inhibitors (e.g., complete EDTA free (Roche)

1. Transform *E. coli* expression strain of choice, (BL21 (DE3) RIL is a suitable option), simultaneously with plasmids pCDFDuet containing *S. cerevisiae* H2A (*Nco*I *Not*I), H2B (*Nde*I *Xho*I) and pETDuet containing H3 (NcoI NotI), H4 (NdeI XhoI) (Saravanan et al., 2012).
2. Grow at 37 °C on a LB-agar plate supplemented with 100 μg/mL ampicillin, 30 μg/mL chloramphenicol and 50 μg/mL spectinomycin
3. Inoculate 50 mL LB + ampicillin + chloramphenicol + spectinomycin with one colony and grow at 37 °C and 200 rpm overnight
4. Inoculate 12 L LB + ampicillin + chloramphenicol + spectinomycin with the overnight starter and grow to an OD_{600} of 0.6 at 37 °C
5. Induce expression by adding IPTG to a final concentration of 0.5 mM and grow at 37 °C and 200 rpm for 4 h
6. Harvest cells by centrifugation for 10 min at 4500 × *g* and freeze in liquid nitrogen. Store the frozen cell pellet at −80 °C or proceed with the purification

2.7 Purification of histone octamers

Materials:

Buffer 6 (20 mM HEPES pH 8.0, 0.1 mM EDTA, 10 mM β-mercaptoethanol)
Sodium chloride (5 M)
Protease inhibitors (e.g., complete EDTA free (Roche))
Lysozyme
Syringe filter (0.45 μm)
HiTrap heparin HP column (5 mL; Cytiva)
Superdex 200 16/600 gel filtration column (Cytiva)
10–20% Criterion TGX gel (Criterion)
Amicon Ultra centrifugal filter units (Millipore), 30,000 MWCO

1. Thaw cell pellet on ice and resuspend in 30 mL Buffer 6 + 100 mM sodium chloride + protease inhibitors + 0.2 mg/mL lysozyme
2. Lyse cells by sonication

3. Centrifuge at 58,545 × *g* and 4 °C for 20 min and collect the soluble phase
4. Filter lysate with a syringe filter
5. Apply the soluble phase to a 5 mL heparin column. Wash column with Buffer 6 + 500 mM sodium chloride until UV reading reaches baseline. Elute histones using a salt gradient to Buffer 6 + 2 M sodium chloride in 16 CV
6. Analyse fractions by SDS-PAGE
7. Pool fractions containing all four histones and concentrate using a 30,000 MWCO centrifugal filter
8. Remove any precipitates by centrifugation at full speed and 4 °C for 5 min in a benchtop centrifuge
9. Load the sample on a Superdex 200 16/600 gel filtration column equilibrated with Buffer 6 + 2 M sodium chloride, using a 5 mL capillary loop
10. Elute the protein and collect 2 mL fractions
11. Analyse fractions by SDS-PAGE
12. Pool peak fractions and concentrate to 2 mg/mL using a 30,000 MWCO centrifugal filter unit
13. Keep at 4 °C for direct usage
14. For long-term storage, add glycerol (50% final concentration), freeze aliquots in liquid nitrogen and store at −80 °C. Remove glycerol by gel filtration on a Superdex 200 gel filtration column with Buffer 6 + 2 M sodium chloride before nucleosome reconstitution

3. DNA template, OCCM complex preparation and staining

In *S. cerevisiae*, ORC recognises specific sequences found at replication origins. In our reconstitution experiment, we used the *ARS1* origin sequence, which contains a high-affinity ORC binding site named an ACS. 42 bp downstream of the ACS, an inverted, lower-affinity ORC binding site is found (B2) (Coster & Diffley, 2017). ORC loads MCM double hexamers that passively slide along duplex DNA (Abid Ali et al., 2017; Evrin et al., 2009; Gros et al., 2015; Noguchi, Vassilev, Ghosh, Yates, & DePamphilis, 2006; Remus et al., 2009). DNA roadblocks can limit the linear diffusion of double hexamers along the double helix. Examples of roadblocks are nucleosomes that naturally flank origin DNA in cells, or the methyl transferase M.*Hpa*II, which can be covalently trapped on duplex DNA using a modified recognition sequence (CC*GG) containing a 5-fluoro-2′-deoxycytosine (C*) that acts as a suicide substrate (Miller et al., 2019). Distinct roadblocks can be added on each end of an origin DNA sequence. This strategy produces asymmetric origins, ensuring that the polarity of DNA binding for different replication factors, as well as their relative position along the origin substrate, can be tracked. To study the mechanism of MCM loading onto ARS1, we adopted this approach and introduced Widom nucleosome positioning sequences (Lowary & Widom, 1998) for nucleosome reconstitution at both ends of an ARS1 origin DNA sequence. Amplification of this origin with different primers allowed us to generate nucleosome-capped, MH-capped or asymmetric nucleosome-MH substrates that trap MCM loading intermediates on the DNA for visualisation by EM. Desthiobiotin-conjugated primers can be employed, which function as handles for tethering to streptavidin-coated paramagnetic beads. Desthiobiotin has lower affinity for streptavidin compared to biotin, which allows efficient elution using biotin (Eickhoff et al., 2019).

Sequence of the ARS1 origin DNA construct flanked by Widom 601 at the 5′ and Widom 603 at the 3′ end of DNA:

5′ – CCGCCCTGGAGAATCGCGGTGCCGAGGCCGCTCAATT GGTCGTAGACAGCTCTAGCACCGCTTAAACGCACGTACGCG CTGTCCCCCGCGTTTTAACCGCCAAGGGGATTACTCCCTAG TCTCCAGGCACGTGTCAGATATATACATCCTGTATTTTACAG ATTTTATGTTTAGATCTTTTATGCTTGCTTTTCAAAAGGCCT GCAGGCAAGTGCACAAACAATACTTAAATAAATACTACTCAG TAATAACCTATTTCTTAGCATTTTTGACGAAATTTGCTATTT TCCCAGTTCGCGCGCCCACCTACCGTGTGAAGTCGTCACTC GGGCTTCTAAGTACGCTTAGCGCACGGTAGAGCGCAATCCA AGGCTAACCACCGTGCATCGATGTTGAAAGAGGCCCTCCGT CCTTATTACTTCAAGTCCCTGGGGTGCAGG – 3′.

Primers used for PCR amplification. To amplify the different symmetric and asymmetric constructs following primers are used:

1. NCP F – 5′ - (Des)-CGATAGAACTCGGGCCGCCCTGGAGAA TCGCGGTGCCG – 3′.
2. NCP R – 5′ - CCTGCACCCCAGGGACTTGAAGTAATAAGG AC – 3′.
3. MTRB F – 5′ - (Des)- ATATATCC*GGCCTGTATTTTACAGA TTTTATGTTTAGA;TCTTTTATGC – 3′.
4. MTRB R – 5′ - AGGGCGCC*GGAACTGGGAAAATAGCAAA TTTCGTCAAAAATGC – 3′.

(Des) – Desthiobiotin TEG.

C* – 5-fluoro-2′-deoxycytosine. The modified recognition sequence (CC*GG) functions as the suicide substrate, establishing a covalent interaction between MH and DNA (Yardimci et al., 2012).

Amplification of the Widom-flanked ARS1 sequence with different combination of primers allows the reconstitution of the following substrates:

The nucleosome-M.HpaII capped origin has been used as an example in the work detailed below.

3.1 Amplification and purification of DNA templates

Materials:

Buffer 7 (10×; 200 mM Tris pH 8.6, 100 mM potassium chloride, 100 mM $(NH_4)_2SO_4$, 20 mM magnesium chloride, 1% triton X-100, 1 mg/mL BSA)
Buffer 8 (50 mM Tris pH 8.0, 5 mM beta-mercaptoethanol)
TE buffer (10 mM Tris pH 8.0, 1 mM EDTA)
TAE buffer (50×)
HpaII methyltransferase (NEB)
ARS1 DNA construct
Primers (see Table 2)
dNTP mix (25 mM each)
Pfu polymerase
Resource Q column (1 mL; Cytiva)
Ethanol (100%)
2% agarose gel (prepared with 1× TAE buffer)

1. Amplify the linear origin DNA sequences by PCR:
   a. Set up 40 × 100 μL reactions by mixing (Table 3)
   b. Run PCR with the following settings (Table 4):
2. Pool all reactions

Table 2 Combination of primers to produce the differently roadblocked DNA constructs.

Construct	Primers
Nucleosome-Nucleosome	1 and 2
Nucleosome-M.*Hpa*II	1 and 4
M.HpaII-Nucleosome	2 and 3
M.HpaII-M.*Hpa*II	3 and 4

Table 3 Recipe of PCR reaction.

Component	Final concentration	Volume
Water		As required to bring total volume to 100 μL
Buffer 7	1×	10 μL
dNTP mix	0.25 mM	1 μL of 25 mM stock
Primer 1	0.5 μM	0.5 μL of 100 mM stock
Primer 2	0.5 μM	0.5 μL of 100 mM stock
Template	10 ng / 100 μL reaction	1 μL of 10 ng/μL stock
Pfu polymerase	2.5 u/100 μL reaction	

Table 4 PCR program.

Segment	Cycles	Temperature [°C]	Time
1	1	95	2 min
2	35	95 56 70	25 s 20 s 10 s
3	1	70	30 s
4	1	4	Hold

3. Remove free oligonucleotides by anion exchange chromatography using a 1 mL Resource Q column (Cytiva) with a 24 CV gradient from Buffer 8 to Buffer 8 + 2 M sodium chloride
4. Analyse fractions on a 2% agarose gel
5. Pool peak fractions and perform a buffer exchange using one of the following two options:
   a. Option 1: Ethanol precipitation
      i. Add 2.5 volumes of ice-cold ethanol and mix thoroughly
      ii. Incubate for 30 min at −80 °C, mix regularly, or incubate overnight at −20 °C
      iii. Pellet DNA by centrifugation for 30 min at 4 °C and maximum speed in a benchtop centrifuge
      iv. Remove supernatant and carefully wash the pellet with 70% ice-cold ethanol without disrupting the pellet
      v. Centrifuge for 30 min at 4 °C and maximum speed in a benchtop centrifuge
      vi. Remove supernatant and air-dry pellet

   - **vii.** Resuspend pellet in 50 µL TE buffer
- **b.** Option 2: Centrifugal filter unit
   - **i.** Concentrate the pooled sample to half the volume using a 30,000 MWCO centrifugal filter unit
   - **ii.** Dilute sample by adding 1 volume Buffer 8
   - **iii.** Repeat steps 1 and 2 four times
   - **iv.** Concentrate to ~50 µL

3.2 Nucleosome reconstitution

Materials:

Buffer 9 (20 mM HEPES pH 7.6, 5 mM β-mercaptoethanol)
Buffer 10 (20 mM HEPES pH 7.6, 80 mM potassium chloride, 100 mM sodium acetate, 1 mM DTT)
Sodium chloride (5 M)
100 µL dialysis cups (10,000 MWCO, Thermo Scientific)
TAE buffer (50 ×)
4% native PAGE

1. Mix soluble histone octamers with origin DNA (1:1 molar ratio of histone octamers and Widom sites, optimize ratio on a small scale) in Buffer 9 + 2 M sodium chloride at a final DNA concentration of 0.5–0.7 mg/mL
2. Dialyse sample overnight against 100 mL Buffer 9 + 2 M sodium chloride that is gradually supplemented with 1.9 L Buffer 9 + 50 mM sodium chloride
3. Carry out a final dialysis step against Buffer 10 for 2 h
4. Check correct assembly of nucleosomes by 4% native PAGE in 0.5 × TAE buffer

This protocol is adapted from Dyer et al. (2004).

3.3 Methyltransferase conjugation

Materials:

Buffer 8 (composition see Section 3.1)
Buffer 11 (50 mM potassium acetate, 25 mM Tris pH 7.5, 10 mM magnesium acetate, 1 mg/mL BSA, 150 mM S-adenosyl-methionine (NEB))
Sodium chloride (5 M)
Resource Q column (1 mL; Cytiva)

Methyltransferase conjugation is performed after nucleosome reconstitution for origin templates containing nucleosomes.

1. Mix M.*Hpa*II and 100 µg of the DNA template in a 3:1 molar ratio and incubate in a total volume of 200 µL Buffer 11 at 30 °C overnight
2. For templates without nucleosomes: separate products on a 1 mL Resource Q column using a 40 CV gradient from Buffer 8 to Buffer 8 + 2 M sodium chloride. For chromatinized templates do not use salt gradients to avoid perturbing reconstituted nucleosomes. Instead, purify nucleosome templates via size exclusion chromatography using a Superose 6 column (see Section 3.4)
3. Concentrate to 1 mg/mL using a 30,000 MWCO centrifugal filter unit
4. Templates without nucleosomes can be frozen in aliquots of 5–20 µL in liquid nitrogen and stored at −80 °C

3.4 Purification of assembled origins

Materials:

Buffer 10 (composition see Section 3.2)
Superose 6 Increase 3.2/300 column (Cytiva)
Amicon Ultra centrifugal filter units (0.5 mL, 100,000 MWCO; Millipore)
TAE buffer (50 ×)
4% native PAGE

1. Subject nucleosome origin templates to size exclusion chromatography using a Superose 6 Increase 3.2/300 column, which is equilibrated with Buffer 10. Collect 60 µL fractions
2. Analyse fractions on a 4% native PAGE in 0.5 × TAE buffer
3. Pool peak fractions and concentrate using a centrifugal filter unit
4. Store reconstituted origins at 4 °C

3.5 OCCM assembly

Materials:

EM buffer (25 mM HEPES pH 7.6, 10 mM magnesium acetate, 100 mM potassium acetate, 1 mM DTT)
Glycerol
100 mM ATPγS

1. Incubate 6 nM ORC with 5 nM nucleosome-M.*Hpa*II origin DNA for 5 min at 30 °C with mixing at 1,250 rpm in EM buffer +10% glycerol +0.5 mM ATPγS. ATPγS is a slowly hydrolysable ATP analog that blocks helicase loading at the stage of the first MCM ring recruitment
2. Add 6 nM Cdc6 to the reaction and incubate for further 5 min

3. Add 6 nM Mcm2-7–Cdt1, bringing the total reaction volume to 30 μL, and incubate for further 30 min
4. Proceed to grid preparation

3.6 Negative stain electron microscopy

Materials:

Sample to be analysed
300-mesh copper grids with carbon film (EM Resolutions)
2% uranyl acetate
Filter paper

1. Glow discharge grids for 30 s at 45 mA
2. Apply 3 μL sample to each grid
3. Incubate for 1 min
4. Stain on four drops of 50 μL 2% uranyl acetate with gentle stirring for 10 s each
5. Remove excess stain with filter paper and dry grids completely before imaging
6. Collect micrographs on a transmission electron microscope at a nominal magnification of 30,000 × (yielding a pixel size of 3.45 Å at the specimen level) within a −0.5 to −2.5 μm defocus range. A representative micrograph is shown in Fig. 2A

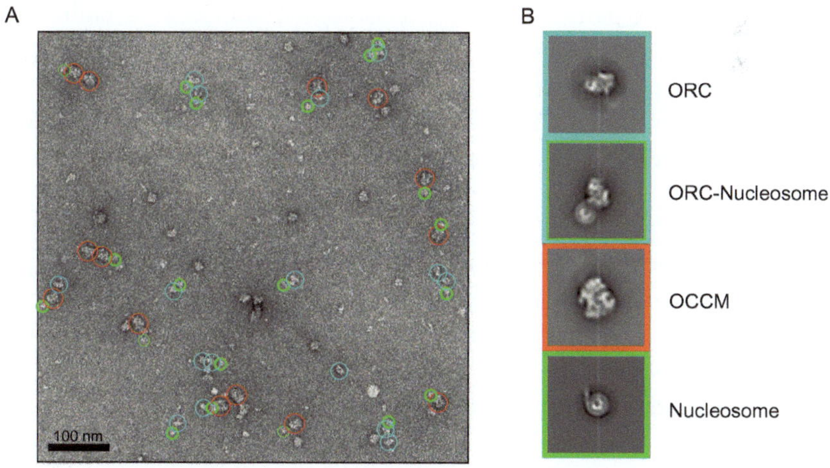

Fig. 2 Raw data and 2D averages for the OCCM recruitment on a capped ARS1 origin of replication. (A) Representative raw micrograph with highlighted particles corresponding to nucleosomes (green), isolated ORC (orange) and OCCM (cyan). (B) Representative 2D averages of Nucleosomes, ORC, nucleosome-flanked ORC and OCCM.

4. Image processing

In this section, we describe the image processing pipeline, from raw micrographs to the generation of signal-enhanced views of OCCM on origins of replication. This requires basic familiarity with the command line and the EM image processing software package RELION (Zivanov et al., 2018). Please refer to the RELION tutorial pages:

https://www3.mrc-lmb.cam.ac.uk/RELION/index.php/Main_Page.

4.1 Data import and CTF estimation

1. Open Terminal, create a "Micrographs" folder in your processing directory and move the recorded micrographs (.mrc format) into this folder
2. Load the RELION 3.1 GUI and import the Micrographs specifying pixel size, voltage, spherical aberration, amplitude contrast. In our experiment, the pixel size was 3.45 Å/px, voltage 120 kV, spherical aberration 6.2 mm and amplitude contrast 0.15
3. Perform CTF estimation using Gctf (here, version 1.06 was used).

4.2 Particle picking and extraction

1. Create an External Topaz picking job using the CTF estimation output as Input Micrographs. Our Topaz model required the following parameters:
   a. number_of_particles = 250, scalefactor = 1, device = 0, radius = 40 (pixel), selectthreshold = −3
   b. model = path/model. sav
2. Extract boxes with a box size of 128 pixels, do not invert the contrast. This step yielded 14,011 particles from 96 micrographs in our experiment

4.2.1 Notes

1. To inspect the results of the picking job with Topaz, first save the display settings for a ManualPick Job in RELION, before viewing the Topaz picked particles contained in the coords_suffix_topazpick.star file in the RELION GUI
2. The parameters of our Topaz picking model were optimized for our dataset using training data containing images of nucleosomes, ORC, OCCM and other MCM-containing complexes and may not yield optimal picking for other samples. For detailed guidance on how to perform Topaz particle picking, refer to (Bepler et al., 2019).

4.3 2D classification and class selection

1. Perform 2D classification of the extracted particles with 60 classes and a mask diameter of 280 Å. Limit resolution *E*-step to 15 Å and add the flag "--only_flip_phases" in the Running tab. This is preferred when averaging negatively stained particles
2. Manually curate the resulting 2D classes by running a Subset Selection job and exclude particles that do not average into distinct nucleosome, ORC, nucleosome-ORC or OCCM classes (compare to our final 2D classifications in Fig. 2B).
3. Perform further 2D classification of the individual molecular complexes as required to remove noise and obtain clear 2D classes of nucleosomes, MCM-Cdt1, ORC, nucleosome-ORC and OCCM complexes
   a. Re-extract the nucleosomes with a box size of 96 pixels and re-classify with a mask diameter of 200 Å. The mask diameter should be increased to 350 Å for optimal nucleosome-ORC classification

4.4 In silico reconstitution

The output of 2D classification includes image files (.mrcs files) of nucleosome, ORC, nucleosome-ORC and OCCM class averages, in our case (see Section 3.6). It also includes optimiser.star and data.star files that contain information on class composition, micrographs, picking coordinates and alignment parameters for individual particles that contribute to the class averages. The in silico reconstitution was initially developed (Miller et al., 2019) using scripts created in our labs and tools from Xmipp (Sorzano et al., 2004), RELION (Zivanov et al., 2018), and EMAN2 (Tang et al., 2007). The same results can now be achieved by running the RELION_particle_reposition command in RELION3.1 and later versions (Zivanov et al., 2018). Particle repositioning modifies the raw micrographs by overlaying 2D class averages onto the raw particles that contributed to those classes, in their original orientation.

4.4.1 Particle repositioning

1. Use the command line in your processing directory
2. Reposition the class averages resulting from the desired 2D classification job into the micrographs from which particles were extracted. To do this, use the command:

   > relion_particle_reposition –i Import/job001/micrographs.star –opt Class2D/jobXXX/run_it025_optimiser.star –o repos –odir RECON_JXXX

   a. Change the XXX placeholder to the job number of your chosen 2D classification
   b. This command creates a new set of reconstituted micrographs, with overlayed 2D class averages in an output folder called: RECON_JXXX/Micrographs/.
3. It is possible to reposition class averages from a general 2D classification containing multiple complexes of interest and directly proceed with extracting reconstituted origins. However, we find that classifying different complexes separately returns cleaner results (e.g., higher quality of class averages due to individualized mask diameter, box size, number of classes). This requires iterative particle reposition, starting with the most abundant classified species, and repositioning the second-most populated complex into the ReconSil micrographs of the previous round
4. Continuing from step 2, create a subdirectory called "originals" in your Micrograph folder and move the original micrographs into this directory
5. Create additional subdirectories for every species that you intend to reposition
6. Soft link the reconstituted micrographs into the "Micrographs" directory:
   ln -s ../RECON_JXXX/Micrographs/*.mrc
7. Rename the links to the ReconSil micrographs to match the name of original micrographs:
   rename _repos.mrc .mrc *.mrc
   a. This will allow working with the original micrograph .star files as an input to reposition class averages into ReconSil micrographs
8. Repeat steps 2–7 until you have repositioned all classes of interest
9. If one only performs a single round of 2D classification before repositioning, it is still necessary to carry out steps 4–7 once, so you can use the original micrograph star files as input to extract images from Reconsil micrographs

4.5 Extraction of entire reconstituted origins

In our experiment, helicase loading onto asymmetric, road-blocked DNA has been halted using ATPγS. This blocks the double hexamer formation at the first helicase recruitment step with ORC, Cdc6, Cdt1 and MCM forming the OCCM intermediate. OCCM particles are the largest particles used in the particle-reposition process, and they are expected to map close to the center of the reconstituted origins. Re-extracting OCCMs from ReconSil micrographs using a large-enough box size will therefore yield images of full OCCM-bound origins. To re-extract OCCMs from the ReconSil micrographs, first re-arrange the directory architecture and nomenclature, with the aim of swapping original micrographs with raw micrographs. To this end:

1. In the RELION GUI perform a subset selection to choose the OCCM averages to be re-extracted. For comparison, a representative OCCM class, which we used as input for our ReconSil analysis, is shown in Fig. 2B
2. Perform a particle extraction job to extract the selected OCCM particles using a box size that encompasses the entire origin length. In our case a 300 × 300 pixel box works. For the "OR re-extract refined particles?" option in the I/O tab, select "yes." Input files should be the micrographs_ctf.star file corresponding to the CTF corrected micrographs and the subset selected particles.star file. In our case:

   CtfFind/job002/micrographs_ctf.star ("micrograph STAR file") and Select/jobXXX/particles.star ("Refined particles STAR file").

   a. To speed up the initial analysis and save memory, we recommend binning the data by 2 (choose "Yes" for the "Rescale particles?" option in the "extract" tab and e.g., "150" for "Re-scaled size (pixels)," if using an original box size of 300 pixels

4.6 Origin selection

Before delving into the analysis of the relative position of helicase complexes along origin DNA, it is critical to identify which of the extracted particle images represent fully reconstituted origins and discriminate them from groups of particles that fortuitously map closely to one another but are not bound to the same DNA molecule. Given our origin-substrate design, the slowly-hydrolysable nucleotide analog added, and the extraction strategy employed, the origins analysed here are expected to contain a single nucleosome capping one DNA end, a methyltransferase adduct at the other end, at least one OCCM particle trapped in between, and, in some cases an ORC complex. Non-DNA-bound MCM–Cdt1 particles can also be identified in the 2D classes and re-extracted ReconSiled images but can be ignored for this analysis. Finally, our Topaz model, and therefore class averages do not include the methyltransferase adduct (~45 kDa in size and featureless at negative stain EM resolution). For this reason, the DNA substrate is only expected to be capped by a single nucleosome in the ReconSiled images. Methyltransferase adducts may, however, be discernible in the background raw image.

4.6.1 Rules for origin selection

We apply four rules to select successfully reconstituted origins, with the goal of excluding instances where origin composition cannot be unambiguously defined.

1. If one OCCM particle maps in proximity to multiple nucleosome averages, discard the image, as the two nucleosomes likely belong to different DNA molecules
2. If particles are visible in the background of the ReconSil micrograph, which are not overlayed with a 2D average, the reconstituted origin should be discarded
3. Do not include any origins in crowded areas or aggregates
4. Exclude origins that are close to the edge of the micrograph, because particles mapping outside of the field of view might be bound to the same DNA

4.7 Extracting reconstituted origins that comply with the origin selection rules

1. For the initial origin selection, create a subset selection job with the re-extracted OCCM particles.star as an input and choose fully reconstituted origins that comply with the origin selection rules described above
2. Extract the selected origins from the ReconSil micrographs skipping particle rescaling
3. In the command line, copy the resulting *.mrcs files corresponding to selected reconciled origins into a new folder in your processing directory, which can be named "selected_origins_reconstituted"
4. Remove the symbolic link to the ReconSil micrographs in your "Micrographs" directory and replace it with a symbolic link to the original micrographs in the "originals" subdirectory

   ln -s originals/*.mrc)
5. Extract the selected origins from the original, raw micrographs
6. For downstream analysis, a stack of .mrcs files can be generated using:

   relion_stack_create –i Extract/jobXXX/particles.star –output Extract/jobXXX/selected_origins
7. These .mrcs stacks can be directly opened and further processed in ImageJ

4.7.1 Notes

To facilitate the selection and rejection of origins for downstream data analysis, we recommend applying a combination of the following strategies:

1. When preparing grids for data analysis, work with dilute sample concentrations to maximize spacing between distinct origins. MCM helicase loading reactions in the presence of ATP or ATPγS are efficient at low nanomolar concentrations, permitting grid preparation with dispersed particles. For nucleic acid-bound particles that require higher concentrations for assembly or stability, crosslinking before dilution can help to retain intact complexes
2. To maximise particle number from low concentration grids, record micrographs at the lowest possible magnification that yields high-quality particle alignment and 2D class averages. To have the opportunity to identify rare events (1–5% of the total population), we recommend using datasets that contain at least 200 reconstituted origins. In our dataset, we recorded 96 micrographs, resulting in a final selection of 851 OCCMs with an average of 8.9 per micrograph, yielding 263 fully reconstituted origins
3. To avoid hypothesis-based bias when different sample conditions are to be compared, data analysis should be performed blind. We routinely have a colleague codify datasets with non-descriptive identifiers prior to analysis
4. When in doubt, reject candidate origins where the components cannot be unambiguously identified

5. Results

One application of the ReconSil approach is presented here, where the MCM helicase loading reaction onto roadblocked ARS1 DNA is stalled at the OCCM intermediate by using ATPγS. A representative micrograph is shown in Fig. 2A. Averages of the isolated nucleosome, isolated ORC, nucleosome-proximal ORC and OCCM are shown in Fig. 2B. These particles are repositioned onto the raw micrographs and reconstituted OCCM-containing origins are then extracted. The resulting origins are classified in eight categories, depending on their relative orientation along the ARS1 DNA sequence (Fig. 3A). As expected from the presence of a high-affinity ORC binding site (ACS) mapping next to the Widom positioning sequence, most origins (172) show a single OCCM particle with ORC mapping in proximity to the nucleosome. A category of interest (ORC-OCCM tandem, 24 origins) shows one ORC at the ACS site and an OCCM in the same orientation mapping further downstream. Such a configuration could derive from an OCCM recruitment event outside of the ACS sequence, however maintaining the same binding polarity as ACS. Alternatively, an initial OCCM recruitment

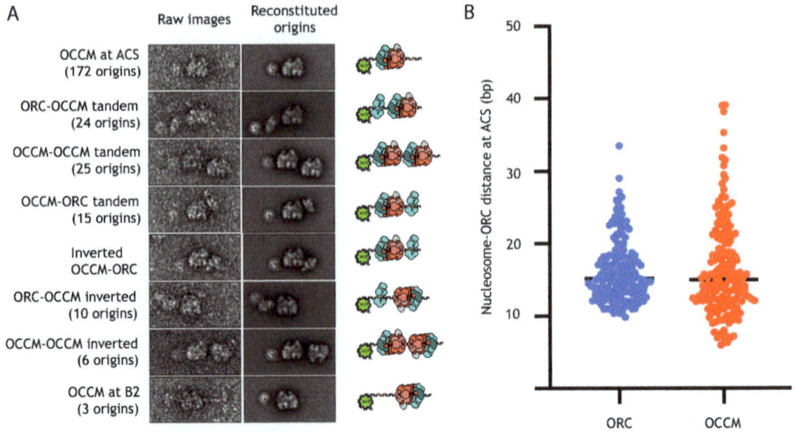

Fig. 3 In silico reconstituted origins and inter-particle distance measurements. (A) Origins reconstituted in silico display distinct binding polarity and levels of saturation on ARS1 DNA. (B) Distance between nucleosome and ORC (blue) or nucleosome and OCCM (red) expressed in base pairs. While the mean distance does not change indicating ACS occupancy, the OCCM distribution becomes broader.

at ACS could be followed by OCCM sliding along the ARS1 origin and a new ORC-binding event at ACS. Notably, an OCCM-ORC tandem configuration, with OCCM at ACS and ORC binding downstream, occurs less frequently (15 origins) than the ORC-OCCM tandem. This observation supports the notion that the ORC-OCCM tandem category is the product of OCCM sliding. Tandem OCCM binding events are also observed (25 origins), supporting the notion that ORC incorporated in OCCM is capable of sliding. Further evidence, suggesting that ORC becomes mobilized when OCCM is formed, can be obtained by plotting OCCM distances from the nucleosome and comparing them to a control experiment where MCM is omitted. When inspected in isolation, ORC has a tight distribution centered around ACS, i.e., 15 bp downstream of the nucleosome, in full agreement with the Widom-flanked ARS1 substrate design. When ACS occupancy is analysed for the full OCCM recruitment experiment, the mean nucleosome-OCCM distance is unchanged, however, the distribution appears broader (Fig. 3B). Other qualitative observations informing the loading mechanism of the MCM double hexamers have been discussed extensively in (Miller et al., 2019). These include OCCM at ACS mapping next to an inverted ORC at the B2 site, an inverted OCCM at the B2 site, alone or next to an ACS-bound ORC and two OCCM assemblies facing one another (Fig. 3A).

6. Summary

In this chapter, we described a collection of biochemical reconstitution and single-particle electron microscopy methods to study the process of MCM helicase loading at a yeast origin of replication. Using particle repositioning to obtain signal enhanced views of helicase loading intermediates, we showed how key biological information can be retrieved on the relative position and orientation of different loading intermediate particles bound to the same origin DNA molecule (Miller et al., 2019). Helicase loading at an origin is only the first event toward the initiation of DNA replication. Upon entry into S phase, firing factors Cdc45 and GINS bind to MCM, forming the CMG holohelicase, and Mcm10 activates its ATP hydrolysis function. At this stage the two helicase particles cross paths, to ensure that the ejected strand from one helicase ring becomes the translocations strand of the symmetry related ring (Douglas, Ali, Costa, & Diffley, 2018; Georgescu et al., 2017). We envisage that ReconSil approaches are going to be important to describe the origin activation reaction at the molecular level (Greiwe, Zanetti, Miller, & Costa, 2022). Likewise, studies on the dynamics of the replication fork might benefit from the in silico reconstitution approaches described in this chapter. One example is the elusive, transient association between the lagging strand polymerase Pol delta with the replisome core (Lewis et al., 2020). Replication fork progression through nucleosomes could be studied by reconstituting the entire context of the chromatin substrate captured as it is being uncoiled by the advancing replisome (Willhoft & Costa, 2021). At the same time, ReconSil approaches might prove useful in understanding the molecular basis of other nucleic acid transactions.

Acknowledgments

We would like to thank the members of the Costa and Miller laboratories for helpful discussion. T.P. is the recipient of a Boehringer Ingelheim Fonds PhD fellowship. Work in A.C.'s lab is funded jointly by the Wellcome Trust, MRC, and CRUK at the Francis Crick Institute (FC001065) and by the European Research Council (ERC) under the European Union's Horizon 2020 research and innovation programme (grant agreement no. 820102). Work in T.C.R.M's lab is funded by the Danish National Research Foundation (DNRF115).

References

Abid Ali, F., Douglas, M. E., Locke, J., Pye, V. E., Nans, A., Diffley, J. F. X., et al. (2017). Cryo-EM structure of a licensed DNA replication origin. *Nature Communications*, *8*(1), 2241.

Attali, I., Botchan, M. R., & Berger, J. M. (2021). Structural mechanisms for replicating DNA in eukaryotes. *Annual Review of Biochemistry*, *90*, 77–106.

Bell, S. P., & Stillman, B. (1992). ATP-dependent recognition of eukaryotic origins of DNA replication by a multiprotein complex. *Nature*, *357*(6374), 128–134.

Bepler, T., Morin, A., Rapp, M., Brasch, J., Shapiro, L., Noble, A. J., et al. (2019). Positive-unlabeled convolutional neural networks for particle picking in cryo-electron micrographs. *Nature Methods*, *16*(11), 1153–1160.

Coster, G., & Diffley, J. F. X. (2017). Bidirectional eukaryotic DNA replication is established by quasi-symmetrical helicase loading. *Science*, *357*(6348), 314–318.

Coster, G., Frigola, J., Beuron, F., Morris, E. P., & Diffley, J. F. (2014). Origin licensing requires ATP binding and hydrolysis by the MCM replicative helicase. *Molecular Cell*, *55*(5), 666–677.

Douglas, M. E., Ali, F. A., Costa, A., & Diffley, J. F. X. (2018). The mechanism of eukaryotic CMG helicase activation. *Nature*, *555*(7695), 265–268.

Dyer, P. N., Edayathumangalam, R. S., White, C. L., Bao, Y., Chakravarthy, S., Muthurajan, U. M., et al. (2004). Reconstitution of nucleosome core particles from recombinant histones and DNA. *Methods in Enzymology*, *375*, 23–44.

Eickhoff, P., Kose, H. B., Martino, F., Petojevic, T., Abid Ali, F., Locke, J., et al. (2019). Molecular basis for ATP-hydrolysis-driven DNA translocation by the CMG helicase of the eukaryotic replisome. *Cell Reports*, *28*(10), 2673–2688 (e8).

Evrin, C., Clarke, P., Zech, J., Lurz, R., Sun, J., Uhle, S., et al. (2009). A double-hexameric MCM2-7 complex is loaded onto origin DNA during licensing of eukaryotic DNA replication. *Proceedings of the National Academy of Sciences of the United States of America*, *106*(48), 20240–20245.

Frigola, J., Remus, D., Mehanna, A., & Diffley, J. F. (2013). ATPase-dependent quality control of DNA replication origin licensing. *Nature*, *495*(7441), 339–343.

Georgescu, R., Yuan, Z., Bai, L., de Luna Almeida, R., Santos, J., Sun, D., et al. (2017). O'Donnell, structure of eukaryotic CMG helicase at a replication fork and implications to replisome architecture and origin initiation. *Proceedings of the National Academy of Sciences of the United States of America*, *114*(5), E697–E706.

Greiwe, J. F., Zanetti, G., Miller, T. C. R., & Costa, A. (2022). In silico reconstitution of DNA replication. Lessons from single-molecule imaging and cryo-tomography applied to single-particle cryo-EM. *Current Opinion in Structural Biology*, *72*, 279–286.

Gros, J., Kumar, C., Lynch, G., Yadav, T., Whitehouse, I., & Remus, D. (2015). Post-licensing specification of eukaryotic replication origins by facilitated Mcm2-7 sliding along DNA. *Molecular Cell*, *60*(5), 797–807.

Gupta, S., Friedman, L. J., Gelles, J., & Bell, S. P. (2021). A helicase-tethered ORC flip enables bidirectional helicase loading. *eLife*, *10*.

Ilves, I., Petojevic, T., Pesavento, J. J., & Botchan, M. R. (2010). Activation of the MCM2-7 helicase by association with Cdc45 and GINS proteins. *Molecular Cell*, *37*(2), 247–258.

Lewis, J. S., & Costa, A. (2020). Caught in the act: Structural dynamics of replication origin activation and fork progression., biochem society. *Transactions*, *48*(3), 1057–1066.

Lewis, J. S., Spenkelink, L. M., Schauer, G. D., Yurieva, O., Mueller, S. H., Natarajan, V., et al. (2020). Tunability of DNA polymerase stability during eukaryotic DNA replication. *Molecular Cell*, 77(1), 17–25 (e5).

Lipford, J. R., & Bell, S. P. (2001). Nucleosomes positioned by ORC facilitate the initiation of DNA replication. *Molecular Cell*, 7(1), 21–30.

Lowary, P. T., & Widom, J. (1998). New DNA sequence rules for high affinity binding to histone octamer and sequence-directed nucleosome positioning. *Journal of Molecular Biology*, *276*(1), 19–42.

Miller, T. C. R., Locke, J., Greiwe, J. F., Diffley, J. F. X., & Costa, A. (2019). Mechanism of head-to-head MCM double-hexamer formation revealed by cryo-EM. *Nature*, *575*(7784), 704–710.

Moyer, S. E., Lewis, P. W., & Botchan, M. R. (2006). Isolation of the Cdc45/Mcm2-7/GINS (CMG) complex, a candidate for the eukaryotic DNA replication fork helicase. *Proceedings of the National Academy of Sciences of the United States of America, 103*(27), 10236–10241.

Noguchi, K., Vassilev, A., Ghosh, S., Yates, J. L., & DePamphilis, M. L. (2006). The BAH domain facilitates the ability of human Orc1 protein to activate replication origins in vivo. *The EMBO Journal, 25*(22), 5372–5382.

Noguchi, Y., Yuan, Z., Bai, L., Schneider, S., Zhao, G., Stillman, B., et al. (2017). Cryo-EM structure of Mcm2-7 double hexamer on DNA suggests a lagging-strand DNA extrusion model. *Proceedings of the National Academy of Sciences of the United States of America, 114*(45), E9529–E9538.

Posse, V., Johansson, E., & Diffley, J. F. X. (2021). Eukaryotic DNA replication with purified budding yeast proteins. *Methods in Enzymology, 661*, 1–33.

Rao, H., & Stillman, B. (1995). The origin recognition complex interacts with a bipartite DNA binding site within yeast replicators. *Proceedings of the National Academy of Sciences of the United States of America, 92*(6), 2224–2228.

Remus, D., Beuron, F., Tolun, G., Griffith, J. D., Morris, E. P., & Diffley, J. F. (2009). Concerted loading of Mcm2-7 double hexamers around DNA during DNA replication origin licensing. *Cell, 139*(4), 719–730.

Rowley, A., Cocker, J. H., Harwood, J., & Diffley, J. F. (1995). Initiation complex assembly at budding yeast replication origins begins with the recognition of a bipartite sequence by limiting amounts of the initiator, ORC. *The EMBO Journal, 14*(11), 2631–2641.

Saravanan, M., Wuerges, J., Bose, D., McCormack, E. A., Cook, N. J., Zhang, X., et al. (2012). Interactions between the nucleosome histone core and Arp8 in the INO80 chromatin remodeling complex. *Proceedings of the National Academy of Sciences of the United States of America, 109*(51), 20883–20888.

Sorzano, C. O., Marabini, R., Velazquez-Muriel, J., Bilbao-Castro, J. R., Scheres, S. H., Carazo, J. M., et al. (2004). XMIPP: A new generation of an open-source image processing package for electron microscopy. *Journal of Structural Biology, 148*(2), 194–204.

Tang, G., Peng, L., Baldwin, P. R., Mann, D. S., Jiang, W., Rees, I., et al. (2007). EMAN2: An extensible image processing suite for electron microscopy. *Journal of Structural Biology, 157*(1), 38–46.

Ticau, S., Friedman, L. J., Ivica, N. A., Gelles, J., & Bell, S. P. (2015). Single-molecule studies of origin licensing reveal mechanisms ensuring bidirectional helicase loading. *Cell, 161*(3), 513–525.

Warner, M. D., Azmi, I. F., Kang, S., Zhao, Y., & Bell, S. P. (2017). Replication origin-flanking roadblocks reveal origin-licensing dynamics and altered sequence dependence. *The Journal of Biological Chemistry, 292*(52), 21417–21430.

Willhoft, O., & Costa, A. (2021). A structural framework for DNA replication and transcription through chromatin. *Current Opinion in Structural Biology, 71*, 51–58.

Yardimci, H., Wang, X., Loveland, A. B., Tappin, I., Rudner, D. Z., Hurwitz, J., et al. (2012). Bypass of a protein barrier by a replicative DNA helicase. *Nature, 492*(7428), 205–209.

Yuan, Z., Riera, A., Bai, L., Sun, J., Nandi, S., Spanos, C., et al. (2017). Structural basis of Mcm2-7 replicative helicase loading by ORC-Cdc6 and Cdt1. *Nature Structural & Molecular Biology, 24*(3), 316–324.

Zivanov, J., Nakane, T., Forsberg, B. O., Kimanius, D., Hagen, W. J., Lindahl, E., et al. (2018). New tools for automated high-resolution cryo-EM structure determination in RELION-3. *eLife, 7*.

CHAPTER ELEVEN

CMG helicase activity on G4-containing DNA templates

Sahil Batra, Sujan Devbhandari, and Dirk Remus*

Molecular Biology Program, Memorial Sloan Kettering Cancer Center, New York, NY, United States

*Corresponding author: e-mail address: remusd@mskcc.org

Contents

Abstract

G-quadruplexes (G4s) are non-canonical nucleic acid structures that form in G-rich regions of the genome and threaten genome stability by interfering with DNA replication. However, the underlying mechanisms are poorly understood. We have recently found that G4s can stall eukaryotic replication forks by blocking the progression of replicative DNA helicase, CMG. In this paper, we detail the methodology of DNA unwinding assays to investigate the impact of G4s on CMG progression. The method details the purification of recombinantly expressed CMG from the budding yeast, *Saccharomyces cerevisiae*, purification of synthetic oligonucleotides, and covers various aspects of DNA substrate preparation, reaction setup and result interpretation. The use of synthetic

Methods in Enzymology, Volume 672
ISSN 0076-6879
https://doi.org/10.1016/bs.mie.2022.02.020

oligonucleotides provides the advantage of allowing to control the formation of G4 structures in DNA substrates. The methods discussed here can be adapted for the study of other DNA helicases and provide a general template for the assembly of DNA substrates with distinct G4 structures.

1. Introduction

G-quadruplexes (G4 structures/G4s) are stable DNA or RNA secondary structures that are prevalent in characteristic G-rich sequences. Importantly, due to their potential to impede the progression of DNA polymerases, G-quadruplexes in chromosomal DNA have been implicated in genome instability (Lerner & Sale, 2019). Moreover, in a recent study we have determined that G-quadruplexes can also cause replication fork stalling by impeding the progression of the eukaryotic replicative DNA helicase, CMG (Cdc45-MCM-GINS), when present on the translocation strand ahead of the CMG (Kumar, Batra, Griffith, & Remus, 2021). Here, we will describe methods to assess the impact of G4s on DNA unwinding by CMG from the budding yeast, *Saccharomyces cerevisiae*, using oligonucleotide-based DNA templates and recombinant CMG.

G4s are formed by stacking of co-planar guanine tetrads or quartets that are hydrogen bonded through Hoogsteen base pairing (Fig. 1) (Bochman, Paeschke, & Zakian, 2012; Spiegel, Adhikari, & Balasubramanian, 2020).

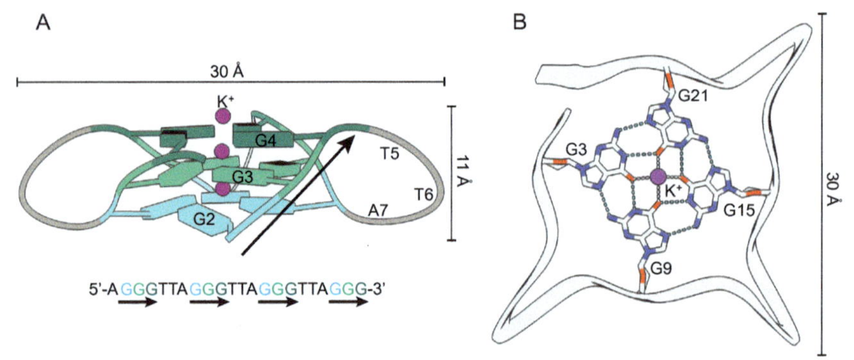

Fig. 1 G-quadruplex structure. (A) Structural representation of a G4 from human telomeric DNA (PDB: 1KF1) that folds into a parallel G4 structure (Parkinson, Lee, & Neidle, 2002). The guanines are represented as slabs and colored differently per quartet. G-strands are marked with black arrows in both the structural model and the corresponding DNA sequence. (B) Interactions between the guanines of one quartet and with K^+ ion (purple). Hydrogen bonds and ionic interactions are depicted as blue and gray dashed lines, respectively.

The stacks are stabilized by coordination of monovalent ions to the centrally facing carbonyl oxygens of guanine bases, with K^+ ions providing the maximum stabilization (Bhattacharyya, Arachchilage, & Basu, 2016). The four G-strands of a G4 can be provided by multiple strands (intermolecular), including hybrid RNA-DNA structures (Wanrooij et al., 2012), or a single nucleic acid strand (intramolecular). G4s can further adopt various topologies (Bochman et al., 2012; Spiegel et al., 2020), classified as parallel, antiparallel or hybrid depending on the orientation of G-strands with respect to each other.

Intramolecular G4s in chromosomal DNA that may impact DNA replication or gene expression were thought to predominantly fold from four contiguous G-stretches that are interspersed by 1-7 random nucleotides ($G_{\geq 3}N_{1\text{-}7}G_{\geq 3}N_{1\text{-}7}G_{\geq 3}N_{1\text{-}7}G_{\geq 3}$) (Huppert & Balasubramanian, 2005; Todd, Johnston, & Neidle, 2005). Each G-stretch in the G4 sequence forms a G-strand, while the interspersed sequences form peripheral loops (Fig. 1A). However, G4s with longer loops and discontinuous G- stretches have also been described, significantly broadening the G4 potential of genomic regions (Bedrat, Lacroix, & Mergny, 2016; Mukundan & Phan, 2013).

G4-forming sequences are widely distributed across genomes (Chambers et al., 2015; Marsico et al., 2019) and impact several physiological processes, including transcription regulation and telomere maintenance (Spiegel et al., 2020). Irrespective of their physiological significance, G4s can potentially interfere with DNA replication (Kumar et al., 2021; Sato, Martin-Pintado, Post, Altelaar, & Knipscheer, 2021) and must be resolved to provide the template for DNA synthesis; failure in doing so threatens genetic and epigenetic stability (De & Michor, 2011; Paeschke et al., 2013; Ribeyre et al., 2009; Sarkies, Reams, Simpson, & Sale, 2010). G4 DNA structures increase during S phase of the cell cycle and subsequently decline upon transition to G2 phase, indicating that G4s are effectively managed during DNA replication (Biffi, Tannahill, McCafferty, & Balasubramanian, 2013). Indeed, several monomeric DNA helicases that can unwind G4 DNA structures are important for genetic stability, suggesting that they facilitate replisome progression through G4-containing genomic regions (Bosch et al., 2014; Dahan et al., 2018; Vannier et al., 2013). Examples include helicases belonging to superfamily 1 (DNA2, Pif1) or superfamily 2 (RTEL1, DDX11, FANCJ, RecQ, WRN, BLM, DHX36, DHX9) (Lerner & Sale, 2019; Singleton, Dillingham, & Wigley, 2007). Structural studies on RecQ and DHX36 provided some mechanistic insights into G4 DNA unwinding (Chen et al., 2018; Voter, Qiu, Tippana, Myong, & Keck, 2018). However, no general principles for DNA helicase-mediated G4 unwinding could be derived from these studies. RecQ has a guanine- specific pocket (GSP) that flips out a guanine from the 3' end of the G4, suggesting that sequential sequestration of the terminal guanines could unfold the G4 structure (Voter et al., 2018). DHX36, on the other hand, has a DHX36 specific motif (DSM) which upon binding to G4 transduces structural changes in the helicase core and C-terminal domain that destabilizes the G4 structure even without requiring ATP; ATP hydrolysis is thought to be important for the translocation step (Chen et al., 2018).

Some of these helicases also exhibit distinct substrate preferences. For example, DHX36 is more active on parallel G4s (Heddi, Cheong, Martadinata, & Phan, 2015), while Pif1 may prefer non-parallel G4s (Byrd, Bell, & Raney, 2018). This diversity in features and mechanisms may hint at a functional specialization of accessory DNA helicases to mediate replisome progression at distinct genomic regions.

The eukaryotic replisome is a large multi-subunit complex structurally organized around the replicative DNA helicase, CMG. (Attali, Botchan, & Berger, 2021). CMG is composed of 11 subunits comprising a hetero-hexamer of catalytic subunits, Mcm2-7 (MCM), and two essential non-catalytic subunits, Cdc45 and the hetero-tetrameric GINS (Go, Ichi, Ni, San, an acronym for Sld5 and subunits Psf1-3) complex (Attali et al., 2021). The C-terminal AAA^+ ATPase domains of the MCM hexamer power the CMG helicase, while the non-catalytic subunits promote CMG processivity (Ilves, Petojevic, Pesavento, & Botchan, 2010). CMG translocates in 3'-5' direction on the leading strand template with the MCM N-terminal face leading at the fork junction (Douglas, Ali, Costa, & Diffley, 2018; Fu et al., 2011; Georgescu et al., 2017; Yardimci, Loveland, Habuchi, van Oijen, & Walter, 2010). Other replisome components that directly contact the CMG helicase include the trimeric complex of Csm3, Tof1 and Mrc1, known as fork protection complex (FPC), which enhances fork rates and maintains normal replisome progression under replication stress (Bando et al., 2009; Szyjka, Viggiani, & Aparicio, 2005; Tourriere, Versini, Cordon-Preciado, Alabert, & Pasero, 2005; Yeeles, Janska, Early, & Diffley, 2017). Of note, Csm3-Tof1 form a heterodimer that associates with the CMG through interactions with the N-terminal domains of Mcm2, -4, -6 and -7, contacting duplex DNA ahead of the fork junction (Baretic et al., 2020). This places Csm3-Tof1 at the fork apex and thus in prime position to detect G4s ahead of the CMG. Consistent with this notion, the vertebrate homologue of Tof1, Timeless, has been shown to possess a C-terminal DNA binding domain (CTD) with enhanced specificity for G4 DNA that is important for the maintenance of epigenetic stability around G4 sequences (Lerner et al., 2020). In addition, the CTD of Timeless interacts with Ddx11 (Cali, Bharti, Di Perna, Brosh Jr., & Pisani, 2016). It is, therefore, thought that Timeless mediates replisome progression through G4 DNA by recruiting Ddx11 (Cali et al., 2016; Lerner et al., 2020).

We found that replisome stalling occurs at G4s during budding yeast DNA replication *in vitro* independent of the FPC, suggesting that G4s impact CMG directly (Kumar et al., 2021). Moreover, we found that a G4 specifically in the template strand can impede DNA unwinding by purified CMG (Kumar et al., 2021). This is consistent with CMG being generally capable of bypassing bulky groups on the non-template (lagging) strand, since the lagging strand is sterically excluded from the central CMG channel (Fu et al., 2011). However, the specific structure of a lagging strand adduct may affect its potential to block CMG progression (Kose, Larsen, Duxin, & Yardimci, 2019; Langston et al., 2017). Thus, it remains a possibility that certain G4 structures on the lagging strand also impact CMG progression. G4s on the translocation strand could conceivably block the central CMG channel. Although, the central channel is flexible enough to accommodate ~20 Å wide double stranded DNA (Langston & O'Donnell, 2017), it might struggle to accommodate G4s that are generally 23–28 Å in diameter (Do, Lim, Teo, Heddi, & Phan, 2011) (Fig. 1). Moreover, G4s display a variability in structure and size (Spiegel et al., 2020), illustrating the importance of determining the effect of distinct G4s on CMG activity.

Here we present the experimental procedure to perform CMG helicase assays to investigate the impact of G4s on CMG progression. For this we have focused on the sequence d(GGGTGGGTGGGTGGG), which folds into a highly thermostable parallel G4 *in vitro* (Do et al., 2011) and is known to interfere with DNA replication in yeast cells (Piazza et al., 2015). Recombinant *Saccharomyces cerevisiae* CMG is purified from budding yeast cells co-overexpressing all eleven CMG subunits. While CMG is normally formed in a complex and highly cell cycle-regulated sequence of events involving many accessory factors at cellular replication origins (Parker, Botchan, & Berger, 2017), co-overexpression of the eleven CMG subunits in yeast or baculovirus expression systems has been demonstrated previously to allow reconstitution of functional CMG complexes in the absence of replication origin activity (Georgescu et al., 2014; Ilves et al., 2010; Zhou et al., 2017). The assay monitors the unwinding of oligo-nucleotide-based DNA substrates in the presence of CMG and can be readily modified to test the impact of different G4s and accessory factors.

2. CMG purification

Recombinant CMG is purified from a diploid budding yeast strain that conditionally overexpresses all eleven *Saccharomyces cerevisiae* CMG subunits. For this, respective open reading frames (ORFs) are cloned downstream of the GAL1-10 promoter in expression vectors that are integrated at auxotrophic marker loci. Analogously to a previously published approach (Frigola, Remus, Mehanna, & Diffley, 2013; Posse, Johansson, & Diffley, 2021), to maximize expression levels, ORFs have been codon- optimized for expression in budding yeast by converting the endogenous codon usage to that of the twenty most abundant proteins in the budding yeast proteome (Ghaemmaghami et al., 2003). In addition, each ORF is flanked by PGK1 5' and 3' UTR sequences. The strain also harbors a galactose-inducible expression cassette for the Gal4 transcriptional activator to boost GAL1-10 promoter activity. Expression is induced in asynchronous cultures by the addition of galactose to the medium. Mcm3 is N-terminally fused to a 3x FLAG tag and Sld5 is N-terminally fused to a calmodulin binding peptide (CBP)-tag, allowing for enrichment of the CMG holo-complex by tandem affinity purification. CMG is subsequently purified by anion exchange chromatography using an FPLC system designed for micro-preparative columns, which helps to concentrate the sample.

2.1 Equipment

- Sorvall RC 3C Plus centrifuge with Rotor HBB6
- Thermo Scientific Sorvall Lynx 6000 centrifuge with Rotor F12-6x500
- Eppendorf 5810 R centrifuge with Rotor A-4-81
- Freezer/Mill 6875D (Spex)
- Thermo Scientific Sorvall WX+ ultracentrifuge with Rotor T-647.5
- Bio-Rad Econo Pump and Columns 1.0x10cm and 0.7x5cm
- Bio-Rad Mini-PROTEAN Tetra Cell—casting and electrophoresis systems
- MiniQ PC 3.2/3 (GE)
- AKTA micro FPLC system

2.2 Buffers and reagents

- YPD: 1% (w/v) yeast extract and 2% (w/v) peptone in ddH_2O. Autoclave and add 2% (w/v) glucose from a sterile stock of 20% (w/v) glucose before use.

- YPD-Agar: 1% (w/v) yeast extract, 2% (w/v) and 2.5% (w/v) agar in ddH_2O. Autoclave and add 2% (w/v) glucose from a sterile stock of 20% (w/v) glucose before pouring the plates.
- YPLG: For 12L of YPLG, 120g yeast extract and 240g peptone are dissolved in 11.4 L of ddH_2O. 950mL of this YP media is distributed to each of twelve 2L flasks and autoclaved. After the media has cooled down, 40mL of 50% (v/v) glycerol, 23mL of lactic acid (88% concentrated solution) and 15mL of 50% (w/v) NaOH are added in that order.
- Buffer M(+): 40mM HEPES-KOH pH 7.6/10% (v/v) glycerol/ 300mM potassium chloride/2mM magnesium acetate/0.005% TWEEN 20/1mM DTT.
- Buffer M(−): 40mM HEPES-KOH pH 7.6/10% (v/v) glycerol/ 300mM potassium chloride/0.005% TWEEN 20/1mM DTT.
- Buffer K(−): 40mM HEPES-KOH pH 7.6/10% (v/v) glycerol/2mM magnesium acetate/0.005% (v/v) TWEEN 20/1mM DTT.
- Buffer T: 25mM Tris–HCl pH 7.2/10% (v/v) glycerol/2mM magnesium acetate/0.005% (v/v) TWEEN 20/1mM DTT.
- CMG storage buffer: 25mM HEPES-KOH pH 7.6/10% (v/v) glycerol/ 200mM potassium acetate/2mM magnesium acetate/1mM DTT.
- Pierce protease inhibitor tablets–EDTA free (1 tablet per 50mL of buffer).
- Anti-FLAG M2 affinity resin (Sigma).
- 3x FLAG peptide (Sigma).
- GelCode blue stain (Thermo Scientific).
- 4–20% TGX gels (Bio-Rad).
- Calmodulin affinity resin (Agilent).

2.3 Strain genotype

YSD35: MATa/α *ade2-1 ura3-1 his3-11,15 trp1-1 leu2-3,112 can1-100 pep4::kanMX bar::hphNAT1 his3*::Gal-Gal4 (HIS3) *trp1*::GAL-MCM5/ MCM4 (TRP1) *ura3*::GAL- MCM2/FLAG-MCM3 (URA3) *leu2*::GAL-MCM7/MCM6 (LEU2) *trp1*::GAL-CDC45 (TRP1) *ura3*::GAL-PSF2/ PSF3 (URA3) *leu2*::GAL-PSF1/CBP-SDL5 (LEU2)

2.4 Protein expression and cell extract preparation

1. Streak out strain YSD35 from glycerol stock on YPD-agar plate and grow at 30 °C for 3 days.

2. Inoculate 3–4 colonies in 60 mL of YPD media in a 250 mL conical flask and culture at 30 °C and 200 rpm for 20 h. The cell density of this pre-culture should reach ~5×10^7 cells/mL.
3. Inoculate 12 L of YPLG media with 60 mL of pre-culture by adding 5 mL of inoculum to each 2 L Erlenmeyer flask containing 1 L of YPLG. Incubate the cultures at 30 °C and 200 rpm for 20–22 h until they reach a cell density of ~5×10^7 cells/mL.
4. Induce expression by adding 20 g of galactose to each flask and continue incubation at 30 °C and 200 rpm for another 6 h.
5. After induction, harvest cells by centrifugation in 1 L bottles at 4,500 rpm for 10 min at 4 °C, using a Sorvall RC 3C Plus centrifuge or equivalent.
6. Resuspend cell pellets in 40–50 mL of 25 mM HEPES-KOH pH 7.6/1 M sorbitol per centrifuge bottle.
7. Pool cell suspensions in 500 mL centrifuge bottle and centrifuge at 5,000 rpm for 5 min at 4 °C in Sorvall Lynx 6000 centrifuge or equivalent.
8. Resuspend cell pellet from the sorbitol wash in 50 mL of Buffer M(+). Divide the suspension into four 50 mL centrifuge tubes and centrifuge in a swing out rotor at 4,000 rpm for 10 min at 4 °C using an Eppendorf 5810 R centrifuge or equivalent.
9. Note the volumes of the cell pellets, add 0.5x volume of Buffer M(+) (supplemented with protease inhibitor), and resuspend the pellets by stirring with a glass rod to form a thick slurry.
10. Freeze the cell slurry dropwise into liquid nitrogen using a 10 mL serological pipette. The frozen cell pellets are referred as popcorn and can be stored at −80 °C until further processing.
11. Crush the popcorn in a cryogenic mill with a run time of 2 min at 15 CPS for 10 cycles.

    Transfer the crushed powder to a glass beaker and allow to thaw on ice.
12. After the powder liquifies, add an equal volume of the Buffer M(+) (supplemented with protease inhibitor) and mix thoroughly on a magnetic stirrer for 10 min at 4 °C.
13. Transfer the cell extracts to six 60 mL ultra-centrifuge bottles and centrifuge at 41,000 rpm for 1 h at 4 °C.
14. Transfer the clarified soluble phase to fresh tubes and snap-freeze in liquid nitrogen. The frozen soluble phase can be stored at −80 °C until further processing.

2.5 CMG affinity purification

All steps are performed at 4 °C.

FLAG pulldown

1. Transfer 3 mL of anti-FLAG M2 affinity resin (Sigma) to a 1.0 x 10 cm Econo-column and equilibrate with 5 column volumes (CV) of Buffer M(+).
2. Pass the soluble phase over the anti-FLAG resin at a flow rate of 1.5 mL/min using a peristaltic pump (e.g. Bio-Rad Econo pump).
3. Wash the resin with 5 CV of Buffer M(+) to remove unbound proteins.
4. Wash the resin with 25 mL of Buffer M(+)/1 mM ATP, followed by one wash with 5 CV of Buffer M(+).
5. Elute bound proteins by gravity flow in five fractions of 1 CV each. Elute fraction 1 with 3 mL. Buffer M(+)/2 mM $CaCl_2$/0.2 mg/mL 3x FLAG peptide, fractions 2 and 3 with 3 mL of Buffer M(+)/2 mM $CaCl_2$/ 0.1 mg/mL 3x FLAG peptide each, and fractions 4 and 5 with 3 mL of Buffer M(+)/2 mM $CaCl_2$ each. Snap-freeze eluates in liquid nitrogen and store at −80 °C.
6. Analyze 5 μL of each elution fraction by SDS-PAGE on 4–20% precast TGX gels (Bio-Rad).
   Stain the gels with GelCode blue.
7. Repeat steps 1–6 until CMG is depleted from the soluble phase.

CBP pulldown

1. Equilibrate 1 mL calmodulin affinity resin with 5 CV of Buffer M(−)/ 2 mM $CaCl_2$ in a 0.7 x 5 cm Econo-column.
2. Pool the eluates from the FLAG pulldown and pass over the calmodulin affinity resin using a peristaltic pump at a flow rate of 1 mL/min.
3. Wash the resin with 15 CV of Buffer M(−)/2 mM $CaCl_2$.
4. Elute bound proteins by gravity flow with 8 x 1 CV of Buffer M(−)/ 2 mM EGTA/2 mM EDTA.
5. Supplement the eluates with 5 mM magnesium acetate, snap-freeze in liquid nitrogen and store at −80 °C.
6. Analyze 5 μL of each fraction by SDS-PAGE.
7. Repeat steps 1–5 until the CMG is depleted from FLAG elution fractions.

Anion exchange chromatography

1. Pool the eluates from the CBP pulldown and dilute by adding an equal volume of Buffer K(−) to reduce the KCl concentration to 150 mM.
2. Connect the MiniQ column to the FPLC system.

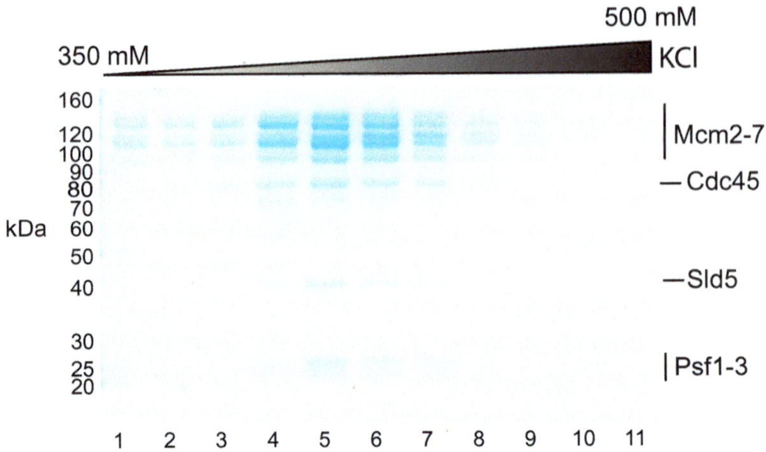

Fig. 2 Purification of CMG. Coomassie-stained SDS-PAGE analysis of MiniQ fractions.

3. Equilibrate the MiniQ column in Buffer T/150 mM KCl.
4. Pass the CBP eluate over the MiniQ column at a flow rate of 0.12 mL/min.
5. Wash the column with 10 CV of Buffer T/150 mM KCl.
6. Elute bound proteins with a KCl gradient of 150 mM–1000 mM over 30 CV in Buffer T.
7. Collect 0.12 mL fractions and analyze 2 μL from each peak fraction by SDS-PAGE (Fig. 2).
8. Pool peak fractions containing stoichiometric amounts of each CMG subunit and dialyze for 4 h against 2 L of CMG storage buffer. Aliquot dialyzed fractions, snap-freeze in liquid nitrogen and store at −80 °C.
9. Determine the concentration of CMG by SDS-PAGE alongside protein standard (e.g. BSA).

   For this, scan the gel and determine band intensities with ImageJ software (https://imagej.nih.gov/ij/).

3. DNA substrates

Our forked DNA substrates contain a 3' $(dT)_{40}$ tail which allows the loading of CMG on the translocating strand. The non-template strand features a secondary structure-forming $(dGdGdCdA)_{10}$ sequence on the 5' end to disfavor CMG loading and aid the steric exclusion of this strand by CMG (Fig. 3A–F) (Georgescu et al., 2017; Kose et al., 2019; Petojevic et al., 2015). A G4-forming sequence [d(TTGGGTGGGTGGGTGGGTT)] is inserted either in the template or non-template strand of the DNA

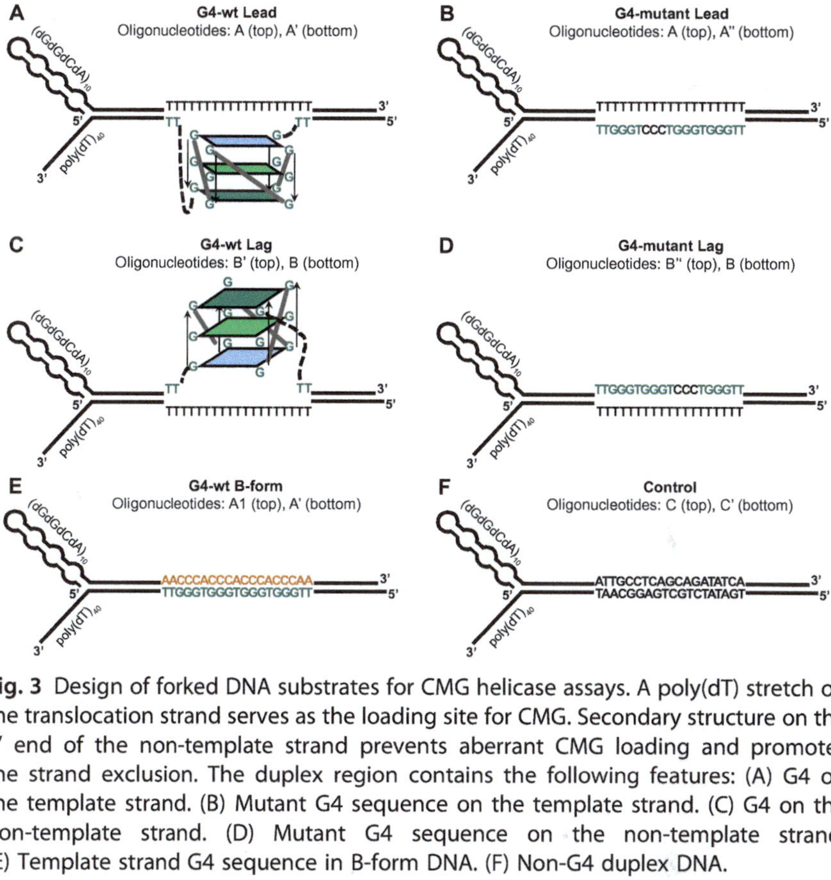

Fig. 3 Design of forked DNA substrates for CMG helicase assays. A poly(dT) stretch on the translocation strand serves as the loading site for CMG. Secondary structure on the 5′ end of the non-template strand prevents aberrant CMG loading and promotes the strand exclusion. The duplex region contains the following features: (A) G4 on the template strand. (B) Mutant G4 sequence on the template strand. (C) G4 on the non-template strand. (D) Mutant G4 sequence on the non-template strand. (E) Template strand G4 sequence in B-form DNA. (F) Non-G4 duplex DNA.

duplex region. To stabilize the G4 structure, the corresponding region in the complementary strand features a poly(dT) stretch that prevents reannealing of the G4 sequence (Fig. 3A + C). In control substrates, one of the G-stretches is mutated to CCC to interfere with the folding of the G4 structure (Fig. 3B + D). In additional control substrates, the G4 sequence is not pre-formed into a G4 structure and remains in B-form DNA (Fig. 3E).

To prepare the DNA substrates, oligonucleotides are annealed by boiling and subsequent slow cooling at 1 °C per minute using a thermocycler. The oligonucleotides are commercially synthesized as standard desalted oligonucleotides (Table 1). Oligonucleotides produced in this way generally have significant impurities and truncations that should be eliminated prior to substrate assembly. We, therefore, purify oligonucleotides by denaturing PAGE. Bands corresponding to full-length oligonucleotide are excised from the gel and the DNA is extracted and further purified by ethanol

Table 1 List of oligonucleotides.

Name	Sequence (5′–3′)
A	GGCAGGCAGGCAGGCAGGCAGGCAGGCAGGCAGGCAGGCA GGTTGGCGAATTCCCTTTTTTTTTTTTTTTTTTTCCTCAGC ACG ACGTTGTAAAACGAG
A'	GGCTCGTTTTACAACGTCGTGCTGAGGTTGGGTGGGTGGGT GGGTTGGGAATTCGCCAACCTTTTTTTTTTTTTTTTTTTT TTTTTTTTTTTTTTTTTTTT
A"	GGCTCGTTTTACAACGTCGTGCTGAGGTTGGGTGGGTCCCT GGGTTGGGAATTCGCCAACCTTTTTTTTTTTTTTTTTTTTT TTTT TTTTTTTTTTTTTTT
A1	GGCAGGCAGGCAGGCAGGCAGGCAGGCAGGCAGGCAGGCA GGTTGGCGAATTCCCAACCCACCCACCCACCCAACCTCAGCA CGACGTTGTAAAACGAG
B	GGCTCGTTTTACAACGTCGTGCTGAGGTTTTTTTTTTTTT TTTTTTGGGAATTCGCCAACCTTTTTTTTTTTTTTTTTTT TTTTTTTTT TTTTTTTTTTT
B'	GGCAGGCAGGCAGGCAGGCAGGCAGGCAGGCAGGCAGGCA GGTTGGCGAATTCCCTTGGGTGGGTGGGTGGGTTCCTCAGC ACGACGTTGTAAAACGAG
B"	GGCAGGCAGGCAGGCAGGCAGGCAGGCAGGCAGGCAGGCA GGTTGGCGAATTCCCTTGGGTGGGTCCCTGGGTTCCTCAGC ACGACGTTGTAAAACGAG
C	GGCAGGCAGGCAGGCAGGCAGGCAGGCAGGCAGGCAGGCA GGTTGGCGAATTCCCATTGCCTCAGCAGATATCACCTCAGCA CGACGTTGTAAAACGAG
C'	GGCTCGTTTTACAACGTCGTGCTGAGGTGATATCTGCTGAGG CAATGGGAATTCGCCAACCTTTTTTTTTTTTTTTTTTTTTT TTTTTTTTTTTTTTTTT

precipitation. PAGE-purified oligonucleotides are also available directly from the manufacturers. However, considering the yield, quality and turn-around times, we prefer in-house oligonucleotide purification. One of the oligonucleotides will be radiolabeled with γ-^{32}P-ATP and T4 polynucleotide kinase (T4 PNK) prior to annealing to the unlabeled complementary oligonucleotide. Alternatively, annealed DNA substrates may also be labeled following annealing. For this, the translocation strand features a 5' overhang of two basepairs to promote phosphorylation by T4 PNK. Annealed products are gel-purified after native PAGE to eliminate un-annealed oligonucleotides or higher-order inter- strand G4 structures.

3.1 Equipment

- Bio-Rad Mini-PROTEAN Tetra Cell—casting and electrophoresis systems
- Dark Reader (Clare Chemical Research)
- Eppendorf ThermoMixer F1.5
- Eppendorf 5424 centrifuge with Rotor FA-45-24-11
- Econospin mini spin columns for gel filtration resin (Epoch)
- Eppendorf 5417R centrifuge with Rotor F45-30-11
- Nanodrop 2000c (Thermo Scientific)
- Bio-Rad Universal hood II GelDoc.
- Bio-Rad C1000 thermocycler
- Konica SRX-101A Film processor

3.2 Buffers and reagents

- 1x TBE: 90 mM Tris-borate/2 mM EDTA
- 2x denaturing buffer: 95% (v/v) Formamide/5 mM EDTA/0.025% (w/v) Bromophenol Blue/0.025% (w/v) Xylene cyanol
- 1x TE: 10 mM Tris–HCl pH 8.0/1 mM EDTA
- Acrylamide elution buffer: 500 mM ammonium acetate/10 mM magnesium acetate/1 mM EDTA
- 3 M sodium acetate pH 5.2
- 1 M potassium acetate pH 7.6
- Sequagel urea gel system (National Diagnostics)
- Ethidium bromide solution (Thermo Scientific)
- γ-^{32}P-ATP (Perkin Elmer)
- T4 Polynucleotide kinase (NEB)
- 6x purple gel loading dye (NEB)

3.3 Oligonucleotide purification

Gel-isolation of oligonucleotides

1. Prepare a 12 mL solution of desired strength, using the Sequagel urea gel system. Typically, we cast gels with 8% monomer strength, which we find suitable for the purification of 70–120 nt long oligonucleotides.

2. Assemble the Bio-Rad mini-PROTEAN gel casting apparatus with 1.5 mm spacer plates as per the manufacturer's instructions; pour the monomer solution, insert five-well comb and set aside for polymerization.
3. Pre-run the gels in 1x TBE at 200 V for 30 min. After the pre-run, rinse the wells with reservoir buffer to flush out excess urea.
4. Mix 50 μL of each 100 μM oligonucleotide stock solution with an equal volume of 2x denaturing buffer, heat at 95 °C for 5 min and snap-chill on ice for 2 min.
5. Load 100 μL sample per well and electrophorese at 120 V. The length of the run will vary according to the length of the oligonucleotides and should be estimated based on the migration of tracking dyes. Typically, we run our gels for 85–100 min.
6. Transfer the gel to a staining tray containing 1 mg/mL ethidium bromide in 1x TBE buffer and stain for 10 min.
7. Visualize DNA bands on a blue-light transilluminator. Each sample should feature a prominent band corresponding to the full-length oligonucleotide. Depending on the quality of oligonucleotide synthesis, a smear of faster migrating truncated products and some higher molecular weight non-specific products may also be observed (Fig. 4A). The gel slices corresponding to full-length product are excised using razor blades.

Extraction of oligonucleotides

1. Transfer the excised slices to 1.5 mL centrifuge tubes and crush them with micro-pipette tips that have been melted close at the tip using an open flame.
2. Add 100 μL of 1x TE to the crushed gel, vortex and freeze at −80 °C for 30 min. Subsequently, heat the samples for 5 min at 95 °C.
3. Add 400 μL of acrylamide elution buffer to each tube and shake in a thermo-mixer at 37 °C and 1,400 rpm overnight.
4. Next day, transfer the gel suspension to an empty spin column and spin at 13,000 rpm for 1 min in an Eppendorf centrifuge to remove gel chunks. Collect the flow-through and transfer to a fresh 1.5 mL centrifuge tube.
5. Add 1 mL of ice-cold ethanol to each flow-through fraction, mix and incubate at −20°C for 30 min to allow precipitation of oligonucleotides.
6. Centrifuge tubes at 4 °C and 13,000 rpm for 10 min. Discard the supernatant, taking care not to disturb the pellet.
7. Dissolve the pellet in 200 μL of 1x TE and add 25 μL of 3 M sodium acetate. Precipitate oligonucleotides by adding 500 μL of ice-cold ethanol and incubate at −20 °C for 30 min.

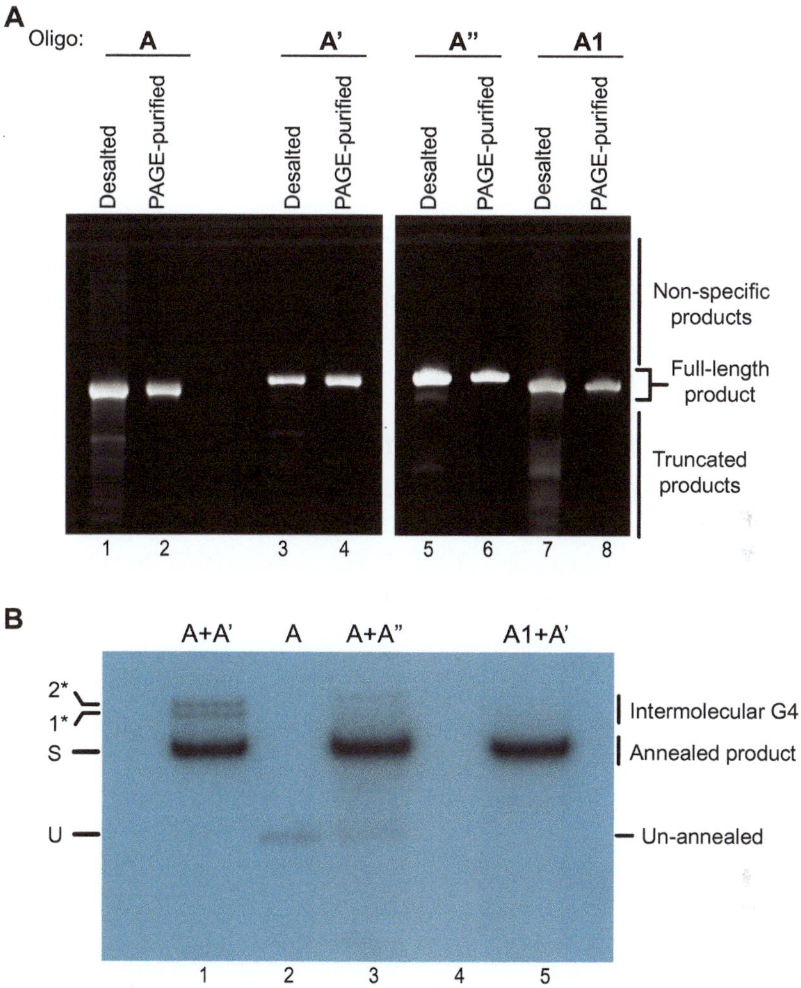

Fig. 4 DNA substrate preparation. (A) Denaturing polyacrylamide gel-analysis of oligonucleotides before (desalted) and after PAGE purification (PAGE-purified). The gels are stained with ethidium bromide. (B) Autoradiogram of annealing reaction products after native PAGE. U, unannealed oligonucleotide; S, substrate; 1* and 2*, Higher molecular weight products likely corresponding to intermolecular G4s.

8. Centrifuge samples at 4 °C and 13,000 rpm for 10 min, discard supernatant.
9. Wash pellets with 500 μL of 70% ethanol (v/v) and spin again at 4 °C and 13,000 rpm for 10 min.

10. Discard supernatant and air-dry pellets in tubes. The yield will depend on the length of the oligonucleotide, but typically we obtain 20–40 μg of purified oligonucleotide after extraction.

11. Resuspend pellets in 40 μL of 1x TE buffer. Determine the concentration of oligonucleotide using a Nanodrop and adjust to a final concentration of 10 μM.

12. Check a small aliquot on a denaturing urea-polyacrylamide gel. For this, we generally cast 10-well gels of 0.75 mm thickness which requires 6 mL of monomer solution per gel. The sample is electrophoresed and stained as above (Fig. 4A). The gel can be imaged using a gel documentation system.

3.4 Substrate formation and extraction

Radiolabeling of oligonucleotides

1. Perform a 5'-radiolabeling reaction with one of the oligonucleotides from a complementary pair (Table 2, Fig. 3) using PNK. Set up the reaction in a PCR tube as shown below:

Component	Volume (μL)
10x PNK buffer	3
ddH_2O	20.25
Oligonucleotide A/C (10 μM)	0.75
γ-P^{32}-ATP	4.5
PNK	1.5
Total	30

Table 2 List of DNA substrates.

Substrate	Oligonucleotide pair
G4-wt lead	**A** + A'
G4-mutant lead	**A** + A''
G4-wt lag	**B** + B'
G4-mutant lag	**B** + B''
G4-wt B-form	**A1** + A'
Control	**C** + C'

Bold denotes radiolabeled oligonucleotides.

2. Perform the reaction in a thermocycler which is programmed to maintain 37 °C for 1 h.

   Heat-inactivate the enzyme by raising the temperature to 80 °C for 20 min followed by 90 °C for 10 min.

Annealing of complementary oligonucleotides

1. Withdraw a 1 μL aliquot from the labeling reaction and keep aside. This will serve as a control during PAGE analysis alongside the annealed products.
2. Add 1.5 μL of 1 M KOAc to the remaining reaction to adjust the K^+ concentration to ~50 mM. Add 1 μL of complementary oligonucleotide (from 10 μM stock) to the reaction (Table 2).
3. Mix all the components thoroughly by pipetting and proceed with the annealing reaction in a thermocycler, programmed to heat the samples at 95 °C for 5 min and then slow-cool at a rate of 1 °C/min until the temperature reaches 10 °C.

Gel-isolation of annealed products

1. Prepare 6 mL acrylamide solution of 10% monomer strength in TBE to cast a native gel with acrylamide to bis-acrylamide ratio of 29:1.
2. Assemble the Bio-Rad mini-PROTEAN gel casting apparatus with 0.75 mm spacer plates as per the manufacturer's instructions; pour the above prepared monomer solution, insert a five-well comb and set aside for polymerization.
3. Mix each of the annealed reaction products with 6.5 μL of 6x purple gel loading dye (NEB) and electrophorese through 10% native gel in 0.5x TBE at 150 V for 45 min. Run 1 μL of the labeled unannealed oligonucleotide alongside the reaction products.
4. In the meantime, develop an unexposed autoradiography film. This film will be used as a transparent gel support for the excision of DNA bands containing the desired DNA substrates.
5. Place two phosphorescent stickers on the transparent support film and place the gel on the film. Wrap this setup in cling film to seal the gel and fix it in position.
6. Expose the gel to a second autoradiography film in a Hypercassette up to 1 min and develop the film (Fig. 4B). Place this film underneath the gel and align with the signals from the phosphorescent stickers to indicate the positions of the DNA bands in the gel.
7. An example autoradiogram depicting annealing products is shown in Fig. 4B. Lane 1 contains the annealed product of oligonucleotides A and A', which results in the formation of a fork substrate that has a

G4 on the translocation strand (Fig. 3A). Lane 3 contains the DNA substrate with the G4-mutant sequence, formed through annealing of oligonucleotides A and A" (Fig. 3B). Comparison of the gel-mobilities of the annealed products with that of un- annealed oligonucleotide A in Lane 2 reveals a prominent band shift of the annealed product. Lane 5 contains substrate with the G4 sequence in B-form DNA (Fig. 3E). We also observe higher molecular weight bands in Lane 1 (labeled 1* and 2*) that we attribute to intermolecular G4 species (Fig. 4B). Consistent with this interpretation, these species are not observed in substrates devoid of G4 structure.

8. Excise gel slices corresponding to the annealed substrates with a scalpel and transfer to a 1.5 mL centrifuge tube.
9. Crush the gel slices with a pipette tip that was sealed at the tip with a flame and add 100 μL of 1x TE supplemented with 50 mM potassium acetate to each tube.
10. Shake the slurry in a thermo-mixer at 22 °C and 1,400 rpm overnight to extract the DNA.
11. Next day, transfer the gel suspension to an empty spin column and spin at 13,000 rpm for 1 min to remove gel chunks. Collect the flow-through and transfer to a fresh 1.5 mL centrifuge tube. Stored at −20 °C.

3.5 Calculating substrate concentration

To determine the concentration of our final DNA substrate, we take the following considerations. We start with a known quantity of radiolabeled oligonucleotide which after annealing and gel-electrophoresis separates into distinct product bands: un-annealed oligonucleotide, fully annealed DNA substrate and non-specific annealing products. The total of all these species is equal to the starting material. Therefore, we quantify the total intensity of all bands in a lane and calculate the fraction contained in the band corresponding to the fully annealed DNA substrate. Furthermore, by measuring the radioactivity in the eluate and left behind in the separated gel chunks after the gel- extraction step we typically observe a gel extraction efficiency of 85%.

Given below is an example of how to calculate the concentration of a DNA substrate, using the substrate shown in Fig. 4B, lane 1. We radiolabeled 7.5 pmols of oligonucleotide A in a 30 μL reaction and removed 1 μL prior to the annealing reaction. Therefore, our starting material is:

$$7.5\,\text{pmols} \times \frac{29}{30} = 7.25\,\text{pmols}$$

After the annealing reaction, products are separated by native PAGE. We observe multiple bands (Fig. 4B, lane 1): (1) Un-annealed oligonucleotide A (U), (2) Substrate (S) and (3) Inter-strand G4 structures (1*, 2*). The intensity of each band is determined using ImageJ:

Band	Intensity
U	*not detected*
S	19461.6
1*	2098.1
2*	1717.0
Total	23276.7

The fraction of radiolabeled oligo in the correctly annealed DNA substrate therefore is:

$$\frac{19461.6}{23276.7} \times 7.25\,\text{pmols} = 6.06\,\text{pmols}$$

The annealed DNA substrate is extracted into 100 μL of buffer. At an efficiency of extraction of 85%, the final concentration of the substrate is:

$$\frac{0.85 \times 6.06\ \text{pmols}}{100\ \mu\text{L}} = 0.0515\ \text{pmols}/\mu\text{L} = 51.5\ \text{nM}$$

4. Helicase assay

For CMG helicase assays, CMG is initially loaded onto the DNA substrate in the presence of non-hydrolyzable ATP analogue. Addition of an excess of ATP subsequently initiates CMG translocation and concomitant DNA unwinding. Aliquots are removed at designated time points and the reaction is quenched by adding an excess of EDTA to chelate Mg^{2+} ions required for ATP hydrolysis. The stopped reaction aliquots are analyzed by native PAGE to separate forked DNA substrate from product single strands.

4.1 Equipment

- Slab gel dryer GD2000 (Hoefer)
- Imaging plate BAS MS 2040 (FUJIFILM)

- Typhoon FLA 7000 (GE)
- Eppendorf ThermoMixer F1.5
- Bio-Rad C1000 thermocycler
- Bio-Rad Mini-PROTEAN Tetra Cell—casting and electrophoresis systems

4.2 Buffers and reagents

- 1x TAE: 40 mM Tris-acetate/1 mM EDTA
- 10x Helicase Buffer: 250 mM HEPES-KOH pH 7.6/1 M potassium acetate/100 mM magnesium acetate
- Adenosine 5'-O-(3-thiotriphosphate) (ATPγS) [Roche]
- ATP (100 mM) (Thermo Scientific)

4.3 DNA unwinding assay

1. Cast 10-well native polyacrylamide gels of 10% monomer strength in 1x TAE/5% glycerol.
2. Prepare the following reaction mix:

Component	Volume
10x Helicase buffer	3.25 μL
DTT (100 mM)	0.81 μL
BSA (1 mg/mL)	3.25 μL
ATPγS (5 mM)	0.65 μL
Substrate	*x* μL for 4 nM
ddH_2O	ad 32.5 μL
Total	32.5 μL

3. Remove an aliquot of 3 μL from the above reaction mix and keep aside as no-protein control.
4. Add 20 nM of CMG to the remaining mix.
5. Mix the reactions thoroughly by pipetting and incubate in thermomixer at 30 °C for 1 h without shaking.
6. In the meantime, prepare the following solutions:

   2x ATP mix

Component	Volume
10x Helicase buffer	3.5 μL
DTT (100 mM)	0.9 μL
BSA (1 mg/mL)	3.5 μL
ATP (100 mM)	3.5 μL
ddH_2O	23.6 μL
Total	35 mL

No-ATP mix

As 2x ATP mix, but without ATP. Aliquot 3 μL into a 1.5 mL tube for each substrate.

2x Stop buffer

Dilute 6x Purple gel loading dye (NEB) two-fold with water and supplement with 40 mM EDTA and 0.24% SDS (w/v).

Component	Volume (μL)
6x Purple gel loading dye (NEB)	100
SDS (10% w/v)	2.4
EDTA (0.5 M)	12
ddH_2O	185.6
Total	300

7. Add 6 μL aliquots each of 2x Stop buffer to seven tubes that will receive reaction aliquots at the following time points [minutes]: 0.5, 1, 2, 3, 5, 15 and 60.
8. After the pre-incubation step, withdraw a 3 μL aliquot from the pre-incubation mix and add to the tube with 3 μL of no-ATP mix. This is a control reaction to ensure that DNA unwinding is ATP-dependent.
9. Add 3 μL of 2x ATP mix to the no-protein control tube. Place both the ATP- and no-protein control tubes back in the thermomixer for 1 h.
10. Mix the remaining ~27 μL pre-incubation reaction mix with 2x ATP mix and place in thermomixer to initiate DNA unwinding.

11. Remove aliquots of 6 μL at designated times and add to prepared tube containing 6 μL of 2x Stop buffer to stop the reaction. Store tubes on ice until they are ready to be loaded on the gel.
12. Dilute an equivalent amount of substrate—as used in the reaction—in water (leaving out the volume required for 10x Helicase buffer) and denature in thermocycler at 95 °C for 5 min, snap-cool on ice for 2 min. Add 10x Helicase buffer and 2x Stop buffer to the denatured substrate. This boiled control is a marker for unwound product.
13. At the 1 h mark, remove final reaction aliquot along with no-protein and no-ATP control reactions.
14. Assemble native polyacrylamide gel in the Bio-Rad mini-PROTEAN electrophoretic system and pre-run in 0.5x TAE at 110 V for 30 min. Rinse the wells with reservoir buffer and load 6 μL from each of the stopped reaction mixes. Perform gel-electrophoresis at 75 V for 2 h.
15. After electrophoresis, transfer the gels onto Blotting paper (VWR) and cover with cling film. Dry the gel on a gel dryer at 80 °C for 5 min.
16. Expose the dried gel to storage phosphor screen. Suitable exposure times and scanning parameters that avoid over-exposure and pixel saturation will depend on the specific activity of the DNA substrate and need to be determined empirically.
17. Scan exposed screen on Typhoon FLA 7000 laser scanner.
18. Analyze the scanned images by ImageJ to quantify band intensities: Use the rectangular selection tool to draw rectangle around individual lanes, measure the intensity profile across each lane and plot profiles using the plot lanes tool. Each lane will feature two distinct peaks, corresponding to the substrate and unwound product bands (Fig. 5). The intensity profiles are baseline corrected and the area under each of the bell-shaped intensity curves is determined. The values are then used to calculate the fraction of unwound product relative to the sum of substrate unwound product for each lane. Plot the values using GraphPad Prism software.

4.3.1 *Interpretation of results*

Here, we have performed the unwinding assays on three types of substrates: (1) Wild-type or mutant G4 on the translocating strand (Fig. 5A), (2) Wild-type or mutant G4 on the non-template strand (Fig. 5B) and (3) B-form duplex DNA with or without the G4 sequence (Fig. 5C). The 'boiled' control (lane 1 in respective gels) indicates the position of the unwound product, while the 'no CMG' controls in lane 2 of the

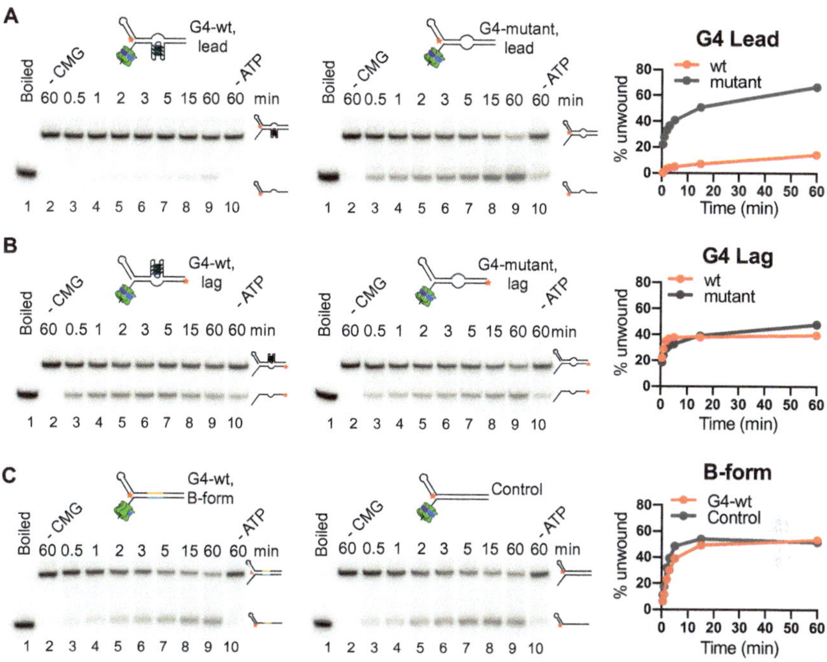

Fig. 5 CMG helicase assay. (A) Wildtype (left) or mutant (right) G4 on translocation strand; (B) Wildtype (left) or mutant (right) G4 on non-template strand. (C) Wildtype G4 sequence (left) or random DNA sequence (right) in B-form duplex DNA. Red star indicates the position of radiolabel. Reaction products were separated by native PAGE and imaged on a phosphor screen. Product formation is quantified using ImageJ and plotted using Prism 9 software.

respective gels indicate the position of the DNA substrate. When CMG is incubated with duplex DNA fork substrate (Fig. 3E and F and Fig. 5C), the unwound product accumulates over time, indicating substrate unwinding by CMG. The DNA unwinding activity of CMG is dependent on ATP, as can be inferred from "no ATP" control in lane 10 (compare to lane 9). The DNA unwinding activity of CMG is unaltered when the G4 sequence is incorporated in the duplex DNA. However, if the G4 structure is present on the translocation strand (Fig. 3A and B), unwinding is strongly inhibited (Fig. 5A). Disruption of the G4 structure by mutation of the G4 sequence relieves the inhibition of DNA unwinding by CMG, demonstrating that the inhibition is specific for the G4 structure. In contrast, the same G4 structure does not pose an impediment to CMG if present on the non-template strand (Fig. 5B).

4.3.2 Limitations

Bulk assays are limited in their ability to determine rates of helicase progression, as the observed DNA unwinding rates are a cumulative product of helicase progression rates and intermittent helicase pause or stall events. In addition, helicase loading onto the DNA substrate may be rate-limiting for product formation in this assay. To bypass the latter, we perform a CMG pre-loading step in the presence of the non-hydrolyzable ATP analog, ATPγS. It has recently been noted that adenylyl-imidodiphosphate (AMP-PNP) may be more suitable for the preloading step as CMG has been observed to exhibit limited DNA unwinding activity also in the presence of ATPγS (Yao, Zhang, Yurieva, & O'Donnell, 2022).

In the absence of competitor oligo-nucleotides that may trap free CMG after the initiation of DNA unwinding it is unclear if the extent of DNA unwinding observed over the course of the experiment is due to a single round of CMG loading onto the DNA substrate or whether it is the result of repeated initiation events. Under the conditions described here we have not noted a reduction in product formation in the presence of trapping oligo-nucleotides, indicating that repeated DNA unwinding events do not significantly contribute to the DNA unwinding activity observed here.

However, depending on the exact reaction conditions (e.g. CMG:DNA ratio) trapping oligo- nucleotides may be desirable to eliminate the potential for repeated CMG loading onto the DNA substrate for a more quantitative read-out of activity.

5. Summary and conclusions

In this chapter we have presented a procedure to assess the DNA helicase activity of budding yeast CMG and to determine the impact of G-quadruplexes on this activity. For this, we have tested the activity of CMG on DNA substrates harboring a G4 on either the template or non-template strand (Fig. 5). We observe that a G4 on the excluded non-template strand does not affect the CMG progression, which concurs with previous observations that bulky groups linked to the lagging strand can be readily bypassed by CMG (Fu et al., 2011; Kose et al., 2019). In contrast, a G4 on the template strand severely impedes DNA unwinding by CMG, suggesting that CMG is unable to resolve or bypass the G4 structure. Mutating a G-strand in the G4 forming sequence or pairing it with complementary sequence in B-form DNA relievies the inhibition of DNA unwinding by CMG, demonstrating that the inhibition of CMG activity is specific to the G4 structure and not the G4 sequence *per se*. The central channel of CMG has a variable width but can be ~25 Å wide (Yuan et al., 2016). This width is similar to the width of the G4 structure formed in the DNA substrates here (Do et al., 2011). The specific G4 structure formed here is symmetrical and can be approximated to a square cuboid with the plane parallel to the quartets forming a square (Fig. 1). A quartet which is 25 Å wide will thus have a diagonal of $\sqrt{2} \times 25\,\text{Å} = \sim 35\,\text{Å}$. Therefore, depending on the orientation in which the G4 enters the inner channel and depending on its conformational flexibility, even a minimal G4 may overwhelm the CMG, thus warranting the need for other helicases to resolve this obstruction (Sato et al., 2021). Future studies involving assays as described here but with distinct G4 structures should help elucidate the structural determinants that govern the impact of G4s on CMG activity.

Acknowledgment

This work was supported by NIGMS grants R01-GM127428 and R01-GM107239 (DR) and NIH/NCI Cancer Center Support Grant P30 CA008748.

References

Attali, I., Botchan, M. R., & Berger, J. M. (2021). Structural mechanisms for replicating DNA in eukaryotes. *Annual Review of Biochemistry*, *90*, 77–106.

Bando, M., Katou, Y., Komata, M., Tanaka, H., Itoh, T., Sutani, T., et al. (2009). Csm3, Tof1, and Mrc1 form a heterotrimeric mediator complex that associates with DNA replication forks. *Journal of Biological Chemistry*, *284*, 34355–34365.

Baretic, D., Jenkyn-Bedford, M., Aria, V., Cannone, G., Skehel, M., & Yeeles, J. T. P. (2020). Cryo-EM Structure of the Fork Protection Complex Bound to CMG at a Replication Fork. *Molecular Cell*, *78*(926-40), e13.

Bedrat, A., Lacroix, L., & Mergny, J. L. (2016). Re-evaluation of G-quadruplex propensity with G4Hunter. *Nucleic Acids Research*, *44*, 1746–1759.

Bhattacharyya, D., Arachchilage, G. M., & Basu, S. (2016). 'Metal Cations in G-Quadruplex Folding and Stability', *Frontiers. Chemistry*, *4*.

Biffi, G., Tannahill, D., McCafferty, J., & Balasubramanian, S. (2013). Quantitative visualization of DNA G-quadruplex structures in human cells. *Nature Chemistry*, *5*, 182–186.

Bochman, M. L., Paeschke, K., & Zakian, V. A. (2012). DNA secondary structures: stability and function of G-quadruplex structures. *Nature Reviews. Genetics*, *13*, 770–780.

Bosch, P. C., Segura-Bayona, S., Koole, W., van Heteren, J. T., Dewar, J. M., Tijsterman, M., et al. (2014). FANCJ promotes DNA synthesis through G-quadruplex structures. *EMBO Journal*, *33*, 2521–2533.

Byrd, A. K., Bell, M. R., & Raney, K. D. (2018). Pif1 helicase unfolding of G-quadruplex DNA is highly dependent on sequence and reaction conditions. *Journal of Biological Chemistry*, *293*, 17792–17802.

Cali, F., Bharti, S. K., Di Perna, R., Brosh, R. M., Jr., & Pisani, F. M. (2016). Tim/Timeless, a member of the replication fork protection complex, operates with the Warsaw breakage syndrome DNA helicase DDX11 in the same fork recovery pathway. *Nucleic Acids Research*, *44*, 705–717.

Chambers, V. S., Marsico, G., Boutell, J. M., Di Antonio, M., Smith, G. P., & Balasubramanian, S. (2015). High-throughput sequencing of DNA G-quadruplex structures in the human genome. *Nature Biotechnology, 33*, 877-+.

Chen, M. C., Tippana, R., Demeshkina, N. A., Murat, P., Balasubramanian, S., Myong, S., et al. (2018). Structural basis of G-quadruplex unfolding by the DEAH/RHA helicase DHX36. *Nature, 558*, 465-+.

Dahan, D., Tsirkas, I., Dovrat, D., Sparks, M. A., Singh, S. P., Galletto, R., et al. (2018). Pif1 is essential for efficient replisome progression through lagging strand G-quadruplex DNA secondary structures. *Nucleic Acids Research, 46*, 11847–11857.

De, S., & Michor, F. (2011). DNA secondary structures and epigenetic determinants of cancer genome evolution. *Nature Structural & Molecular Biology, 18*, 950–U115.

Do, N. Q., Lim, K. W., Teo, M. H., Heddi, B., & Phan, A. T. (2011). Stacking of G-quadruplexes: NMR structure of a G-rich oligonucleotide with potential anti-HIV and anticancer activity. *Nucleic Acids Research, 39*, 9448–9457.

Douglas, M. E., Ali, F. A., Costa, A., & Diffley, J. F. X. (2018). The mechanism of eukaryotic CMG helicase activation. *Nature, 555*, 265-+.

Frigola, J., Remus, D., Mehanna, A., & Diffley, J. F. (2013). ATPase-dependent quality control of DNA replication origin licensing. *Nature, 495*, 339–343.

Fu, Y. V., Yardimci, H., Long, D. T., Ho, T. V., Guainazzi, A., Bermudez, V. P., et al. (2011). Selective Bypass of a Lagging Strand Roadblock by the Eukaryotic Replicative DNA Helicase. *Cell, 146*, 930–940.

Georgescu, R. E., Langston, L., Yao, N. Y., Yurieva, O., Zhang, D., Finkelstein, J., et al. (2014). Mechanism of asymmetric polymerase assembly at the eukaryotic replication fork. *Nature Structural & Molecular Biology, 21*, 664–670.

Georgescu, R., Yuan, Z. N., Bai, L., Santos, R. D. A., Sun, J. C., Zhang, D., et al. (2017). Structure of eukaryotic CMG helicase at a replication fork and implications to replisome architecture and origin initiation. *Proceedings of the National Academy of Sciences of the United States of America, 114*, E697–E706.

Ghaemmaghami, S., Huh, W. K., Bower, K., Howson, R. W., Belle, A., Dephoure, N., et al. (2003). Global analysis of protein expression in yeast. *Nature, 425*, 737–741.

Heddi, B., Cheong, V. V., Martadinata, H., & Phan, A. T. (2015). Insights into G-quadruplex specific recognition by the DEAH-box helicase RHAU: Solution structure of a peptide-quadruplex complex. *Proceedings of the National Academy of Sciences of the United States of America, 112*, 9608–9613.

Huppert, J. L., & Balasubramanian, S. (2005). Prevalence of quadruplexes in the human genome. *Nucleic Acids Research, 33*, 2908–2916.

Ilves, I., Petojevic, T., Pesavento, J. J., & Botchan, M. R. (2010). Activation of the MCM2-7 helicase by association with Cdc45 and GINS proteins. *Molecular Cell, 37*, 247–258.

Kose, H. B., Larsen, N. B., Duxin, J. P., & Yardimci, H. (2019). Dynamics of the eukaryotic replicative helicase at lagging-strand protein barriers support the steric exclusion model. *Cell Reports, 26*, 2113.

Kumar, C., Batra, S., Griffith, J. D., & Remus, D. (2021). The interplay of RNA:DNA hybrid structure and G-quadruplexes determines the outcome of R-loop-replisome collisions. *eLife, 10*.

Langston, L. D., Mayle, R., Schauer, G. D., Yurieva, O., Zhang, D., Yao, N. Y., et al. (2017). Mcm10 promotes rapid isomerization of CMG-DNA for replisome bypass of lagging strand DNA blocks. *eLife, 6*.

Langston, L., & O'Donnell, M. (2017). Action of CMG with strand-specific DNA blocks supports an internal unwinding mode for the eukaryotic replicative helicase. *eLife, 6*.

Lerner, L. K., Holzer, S., Kilkenny, M. L., Svikovic, S., Murat, P., Schiavone, D., et al. (2020). Timeless couples G-quadruplex detection with processing by DDX11 helicase during DNA replication. *EMBO Journal, 39*, e104185.

Lerner, L. K., & Sale, J. E. (2019). Replication of G quadruplex DNA. *Genes, 10.*
Marsico, G., Chambers, V. S., Sahakyan, A. B., McCauley, P., Boutell, J. M., Di Antonio, M., et al. (2019). Whole genome experimental maps of DNA G-quadruplexes in multiple species. *Nucleic Acids Research, 47*, 3862–3874.
Mukundan, V. T., & Phan, A. T. (2013). Bulges in G-quadruplexes: Broadening the definition of G-quadruplex-forming sequences. *Journal of the American Chemical Society, 135*, 5017–5028.
Paeschke, K., Bochman, M. L., Garcia, P. D., Cejka, P., Friedman, K. L., Kowalczykowski, S. C., et al. (2013). Pif1 family helicases suppress genome instability at G-quadruplex motifs. *Nature, 497*, 458.
Parker, M. W., Botchan, M. R., & Berger, J. M. (2017). Mechanisms and regulation of DNA replication initiation in eukaryotes. *Critical Reviews in Biochemistry and Molecular Biology, 52*, 107–144.
Parkinson, G. N., Lee, M. P. H., & Neidle, S. (2002). Crystal structure of parallel quadruplexes from human telomeric DNA. *Nature, 417*, 876–880.
Petojevic, T., Pesavento, J. J., Costa, A., Liang, J. D., Wang, Z. J., Berger, J. M., et al. (2015). Cdc45 (cell division cycle protein 45) guards the gate of the Eukaryote Replisome helicase stabilizing leading strand engagement. *Proceedings of the National Academy of Sciences of the United States of America, 112*, E249–E258.
Piazza, A., Adrian, M., Samazan, F., Heddi, B., Hamon, F., Serero, A., et al. (2015). Short loop length and high thermal stability determine genomic instability induced by G-quadruplex-forming minisatellites. *EMBO Journal, 34*, 1718–1734.
Posse, V., Johansson, E., & Diffley, J. F. X. (2021). Eukaryotic DNA replication with purified budding yeast proteins. *Methods in Enzymology, 661*, 1–33.
Ribeyre, C., Lopes, J., Boule, J. B., Piazza, A., Guedin, A., Zakian, V. A., et al. (2009). The yeast Pif1 helicase prevents genomic instability caused by G-quadruplex-forming CEB1 sequences in vivo. *PLoS Genetics, 5*, e1000475.
Sarkies, P., Reams, C., Simpson, L. J., & Sale, J. E. (2010). Epigenetic instability due to defective replication of structured DNA. *Molecular Cell, 40*, 703–713.
Sato, K., Martin-Pintado, N., Post, H., Altelaar, M., & Knipscheer, P. (2021). Multistep mechanism of G-quadruplex resolution during DNA replication. *Science Advances, 7.*
Singleton, M. R., Dillingham, M. S., & Wigley, D. B. (2007). Structure and mechanism of helicases and nucleic acid translocases. *Annual Review of Biochemistry, 76*, 23–50.
Spiegel, J., Adhikari, S., & Balasubramanian, S. (2020). The structure and function of DNA G-quadruplexes. *Trends in Chemistry, 2*, 123–136.
Szyjka, S. J., Viggiani, C. J., & Aparicio, O. M. (2005). Mrc1 is required for normal progression of replication forks throughout chromatin in S. cerevisiae. *Molecular Cell, 19*, 691–697.
Todd, A. K., Johnston, M., & Neidle, S. (2005). Highly prevalent putative quadruplex sequence motifs in human DNA. *Nucleic Acids Research, 33*, 2901–2907.
Tourriere, H., Versini, G., Cordon-Preciado, V., Alabert, C., & Pasero, P. (2005). Mrc1 and Tof1 promote replication fork progression and recovery independently of Rad53. *Molecular Cell, 19*, 699–706.
Vannier, J. B., Sandhu, S., Petalcorin, M. I. R., Wu, X. L., Nabi, Z., Ding, H., et al. (2013). RTEL1 is a replisome-associated helicase that promotes telomere and genome-wide replication. *Science, 342*, 239–242.
Voter, A. F., Qiu, Y. P., Tippana, R., Myong, S., & Keck, J. L. (2018). A guanine-flipping and sequestration mechanism for G-quadruplex unwinding by RecQ helicases. *Nature Communications, 9.*
Wanrooij, P. H., Uhler, J. P., Shi, Y., Westerlund, F., Falkenberg, M., & Gustafsson, C. M. (2012). A hybrid G-quadruplex structure formed between RNA and DNA explains the extraordinary stability of the mitochondrial R-loop. *Nucleic Acids Research, 40*, 10334–10344.

Yao, N. Y., Zhang, D., Yurieva, O., & O'Donnell, M. E. (2022). CMG helicase can use ATPgammaS to unwind DNA: Implications for the rate-limiting step in the reaction mechanism. *Proceedings of the National Academy of Sciences of the United States of America, 119.*

Yardimci, H., Loveland, A. B., Habuchi, S., van Oijen, A. M., & Walter, J. C. (2010). Uncoupling of sister replisomes during eukaryotic DNA replication. *Molecular Cell, 40*, 834–840.

Yeeles, J. T. P., Janska, A., Early, A., & Diffley, J. F. X. (2017). How the eukaryotic replisome achieves rapid and efficient DNA replication. *Molecular Cell, 65*, 105–116.

Yuan, Z. N., Bai, L., Sun, J. C., Georgescu, R., Liu, J., O'Donnell, M. E., et al. (2016). Structure of the eukaryotic replicative CMG helicase suggests a pumpjack motion for translocation. *Nature Structural & Molecular Biology, 23*, 217–224.

Zhou, J. C., Janska, A., Goswami, P., Renault, L., Abid Ali, F., Kotecha, A., et al. (2017). CMG-Pol epsilon dynamics suggests a mechanism for the establishment of leading-strand synthesis in the eukaryotic replisome. *Proceedings of the National Academy of Sciences of the United States of America, 114*, 4141–4146.

CHAPTER TWELVE

Detecting G4 unwinding

Stefan Juranek and Katrin Paeschke*
Department of Oncology, Hematology and Rheumatology, University Hospital Bonn, Bonn, Germany
*Corresponding author: e-mail address: katrin.paeschke@ukbonn.de

Contents

Abstract

DNA can, in addition to the B-DNA conformation, fold into a variety of additional conformations. Among them are G-quadruplex structures that have gained a lot of attention in recent years. G-quadruplex structures (G4s) are highly stable nucleic acid structures that can fold within DNA and RNA molecules. They form in guanine-rich regions that harbor a specific G4 motif. The three-dimensional structure forms via Hoogsteen hydrogen bonding, where the guanines form hydrogen bonds to each other in order to generate G quartets, which stack in order to become G4 structures.

The existence and relevance of G4s was controversial as discussed in the past. However, accumulating data was published that supported the model that G4s form in living cells and importantly support biological processes. G4 formation and unfolding is tightly regulated *in vivo*. If G4s persist in the cell, they can lead to cellular defects such as genome instability. To avoid G4 accumulation in cells, and by this prevent cellular defect, cells has evolved a variety of proteins, mostly helicases, that efficiently unfold G4 DNA and RNA structures. Here, we describe a detailed protocol to monitor G4 structure unfolding by helicases.

Methods in Enzymology, Volume 672
ISSN 0076-6879
https://doi.org/10.1016/bs.mie.2022.03.034

1. Introduction

The central dogma of molecular biology describes the propagation of genetic information from DNA to RNA to protein. Only in the last several years has it become obvious that this process is not linear. DNA and RNA molecules can directly influence this process and each other by the formation of alternative structures (Shapiro, 2009). In recent years, a variety of alternative DNA and RNA structures have been described in the literature (Wu & D'Souza, 2020). The formation of alternative structures expands the function and relevance of given nucleic acid regions and extends their biological function. These alternative nucleic acid structures can also pose a threat to the cell if their formation is not controlled and they form at "wrong" regions or too many at a specific time in the cell (Kosiol, Juranek, Brossart, Heine, & Paeschke, 2021; Richard & Manley, 2017; Shao & Zhang, 2020; Spiegel, Adhikari, & Balasubramanian, 2020; Svoboda & Di Cara, 2006). To prevent negative downstream consequences of "misregulated" alternative nucleic acid structures, their formation and unfolding is controlled by proteins. The most prominent alternative nucleic acids structures are hairpins, cruciform DNA, Z-DNA, and G-quadruplexes (Bochman, Paeschke, & Zakian, 2012). So far, only a few have been shown to be relevant *in vivo*. Among the best described alternative DNA and RNA conformation is the G-quadruplex structure (G4). G4 structures can form in guanine-rich sequences containing a specific sequence motif ($G{\geq}_2N_xG{\geq}_2N_yG{\geq}_2N_zG{\geq}_2$), called G4 motif. Because of this motif, the guanines self-assemble into (at least two) $\pi-\pi$ stacking subplanar guanine tetrads, which are stabilized by Hoogsteen hydrogen bonding and monovalent cations (Bochman et al., 2012; Burge, Parkinson, Hazel, Todd, & Neidle, 2006; Spiegel et al., 2020).

G4 motifs have been identified in all surveyed organisms (Huppert & Balasubramanian, 2005; Lavezzo et al., 2018; Marsico et al., 2019; Vannutelli, Belhamiti, Garant, Ouangraoua, & Perreault, 2020; Wu et al., 2021; Yadav et al., 2021; Yadav, Hemansi, Kim, Tuteja, & Yadav, 2017; Zyner et al., 2022). The genomic locations of G4 motifs are conserved and enriched at key genomic regions such as telomeres, promoters, splicing sites, origins of replication, untranslated regions of mRNAs and immunoglobulins switch regions, where they support cellular functions (Bohalova, Dobrovolna, Brazda, & Bidula, 2021; Capra, Paeschke, Singh, & Zakian, 2010; Huppert, 2010; Marsico et al., 2019; Wu et al., 2021). Therefore, the current model is that G4 structures are nucleic acid-based tools that support cellular functions.

In this model, specific G4s form based on exogenous or endogenous stimuli and support cellular pathways such as changes in gene expression (Bochman et al., 2012; Linke, Limmer, Juranek, Heine, & Paeschke, 2021; Lipps & Rhodes, 2009; Spiegel et al., 2020). In contrast to this regulatory function of G4s is the fact that G4 structures can, under certain conditions, also impair cellular functions and even lead to genome instability (De Magis et al., 2019; Gong et al., 2021; Hansel-Hertsch et al., 2020; Hui, Simeone, Zyner, Tannahill, & Balasubramanian, 2021, 2022; Paeschke et al., 2013; Paeschke, Capra, & Zakian, 2011; Wang et al., 2019). G4s can block the processivity of polymerases or form at incorrect times and locations on the DNA (RNA) as a protein loading hub to alter cellular changes (e.g., gene expression, epigenetic changes). It has been shown that persisting G4 structures can alter gene expression, splicing, mRNA abundance, stress response, telomerase activation, DNA damage accumulation and the efficiency of the DNA repair and replication machineries (Bryan, 2020; Kim, 2019; Kosiol et al., 2021; Robinson, Raguseo, Nuccio, Liano, & Di Antonio, 2021). Different proteins have been identified that impact G4 formation. These proteins can be grouped into two classes: G4 forming/stabilizing proteins and G4 unwinding proteins including helicases. So far, 25 eukaryotic helicases have been described that can unwind G4 structures. These helicases differ in directionality (5'-3' vs 3'-5'), their nucleic acid substrate (DNA vs RNA), and their mode in selection their targets. Furthermore, they differ in their G4-binding modes, as some helicases bind directly to the G4 structures, whereas others need a single-stranded tail in front of the G4s. Detailed structural analysis among helicases have shown that not only is their binding mode different, but the mechanism of G4 structures unwinding also differs among helicases (reviewed in (Caterino & Paeschke, 2021)).

Many processes in the cell depend on helicases. More than 31 DNA and 64 RNA helicases exist in humans (Umate, Tuteja, & Tuteja, 2011). Of these, 16 can bind to G4s. Most helicases can unwind G4s *in vitro,* and it is believed that more helicases will be identified that can unwind G4 structures *in vitro and vivo*. This raises two questions: (i) if nearly all helicases can unwind G4s, why is it worthwhile to characterize this function? (ii) If all helicases can unwind G4s *in vitro* does this still point to a special function or relevance *in vivo*. In our opinion, both questions can be answered with a clear "yes." Current experiments show that G4s have a function *in vivo* and that their formation and unfolding changes dynamically and depends on the cell type and cell state. This dynamic regulation is primarily dependent on helicases. As described above, without G4-unwinding helicases, G4s accumulate in cells and lead to cellular defects (e.g., genome instability, stress, slow growth, apoptosis (Beauvarlet et al., 2019; Byrd et al., 2016; Lee et al., 2021; Qi et al., 2006; Sauer et al., 2019; Teng et al., 2021)). As most helicases can unwind G4s, this points to a significant role of G4s *in vivo*. By understanding which helicases can unwind G4s with which processivity and target preference, the function and relevance of G4 structures *in vivo* can be determined. The importance of helicase function at G4s is underlined by the current model that helicases support each other at G4s and can compensate for each other's loss at G4s. How helicase compensation at G4s is mediated and how they select their specific G4 targets is not currently clear. Detailed biochemical studies are required to understand if and how helicases unwind specific G4s. This knowledge is not only interesting for fundamental research to understand helicase function at G4s, but it is also of great interest for diagnostic and therapeutical aspects. Most human cancer cells have higher abundance of G4s, which may be connected to the mutational burden, cancer subtype generation and tumorigenesis (Biffi, Di Antonio, Tannahill, & Balasubramanian, 2014; Hansel-Hertsch et al., 2020; Hui et al., 2021, 2022; Kosiol et al., 2021). In some cancers, helicases that regulate G4s are mutated causing elevated G4s (Mohaghegh & Hickson, 2001; Suhasini & Brosh, 2013). A greater understanding on G4—helicase interaction will be essential to understand how G4s and helicases contribute to different cancer states and whether this can be used in the future as a diagnostic tool. In addition, the connection of G4s and helicases may also be relevant for novel therapeutic ideas. In recent years, chemical G4 ligands (e.g., PDS, BRACO-19, Telomestatin, CX5461) have been designed with the goal to stabilize G4s as a targeted approach for anti-cancer treatment (reviewed in (Kosiol et al., 2021; Sanchez-Martin, Soriano, & Garcia-Salcedo, 2021)). Multiple studies have shown that stabilized G4s cause alterations in the cancer cells that lead to reduced tumor growth and increased survival (Kosiol et al., 2021). So far, these ligands are very promising, but they can also cause many additional changes in the cells from undirected G4 stabilization. One interesting alternative approach could be to target specific G4s by modulating/blocking helicases (and their associated proteins) that contribute to G4 unwinding. This modulation of helicase activity will also increase G4 levels (of specific, selected G4s) and may also be useable as a therapeutic approach.

Multiple methods have been developed to study helicase function at G4s (reviewed in (Caterino & Paeschke, 2021)). We have limited the scope of this review on simple methods to study G4 unwinding by helicases using a gel-based assay. We will explain in detail G4 target selection, G4 structure folding, controlling G4 structure formation and helicase activity assays.

2. Before you begin

Key to most *in vitro* assays is a detailed planning and small adjustments to experimental settings will be required in dependency on the selected helicase. Therefore, buffer preparation (pH, salt, source of NTP (most likely ATP)) and target selection can take quite some time, but once that is accomplished, data can be generated quickly. The following points needs to be considered and discussed before the start of the experiment:

- Selection of a helicase.

  Depending on your experimental setting, you must select a helicase to test on a given G4 substrate. Helicases are classified according to conserved sequence motifs into superfamilies (SF1-SF6). SF1 and SF2 are grouped based on similarities of two helicase domains (Fairman-Williams, Guenther, & Jankowsky, 2010). The SF1 helicase family shares a GDxxQ motif. SF2 helicases share structural similarities in the catalytic core. Based on the superfamily, you can deduce whether the helicase is a DNA or RNA helicase, which is the preferred NTP utilized for hydrolysis (e.g., ATP), and which unwinding polarity/directionality (5'-3, 3'-5') it employs. This information is essential to select and design targets for helicase activity assays.
- Purification of the helicase

  G4 unwinding by a helicase is a biochemical assay which requires recombinant expressed and purified helicase. The purification strategy is always dependent on the helicase. The helicase can be expressed and purified from *E. coli*, insect cells, yeasts or even in small amounts from cell culture cell that overexpress the helicase. Protocols how to express and purify different helicases can be found in the literature and adopted to the helicase of interest. For example, purification protocols for Pif1 or RecQ helicase members (Boule & Zakian, 2007; Gu, Masuda, & Kamiya, 2008; Karow, Chakraverty, & Hickson, 1997; Nickens & Bochman, 2021; Orren et al., 1999) or DHX36 and DDX5 (Chen et al., 2018; Choi, Dutta, Fielding, & Tan, 2010; Xing, Wang, & Tran, 2017; Yangyuoru, Bradburn, Liu, Xiao, & Russell, 2018) can be found in the literature. In most cases, it is important to understand whether an active unwinding process takes place or if the activity is passive (Benhalevy et al., 2017). This can either be monitored by the amount of ATP hydrolyzed or by purifying mutated proteins that lack catalytic activity (e.g., mutations in the helicase core domains walker A or walker B box). For most cases described in the literature, the catalytic inactive helicases are expressed and purified with similar protocols to their active counterpart (Nickens & Bochman, 2021; Paeschke et al., 2013).

- Selection of DNA or RNA G4 targets

  It is also essential to select the target G4 structure to test your helicase. For this, it is advised to select either a standard G4-forming DNA or RNA sequence from the literature (for some examples that work in our hands see Table 1 or (Byrd & Raney, 2015)) or select a G4 motif from a genomic region or transcript, ideally of the same species of the helicase. A canonical G4 motif has four guanine tracts consisting of at least two guanines (see Fig. 1A for a cartoon of a G4 structure). These guanine tracts are separated by short intervening loop regions, which vary in their nucleotide composition. Different algorithms have successfully used this G4 motif to identify regions *in silico* that have a strong potential to fold into G4 structures (Brazda et al., 2019; Huppert & Balasubramanian, 2005; Lombardi & Londono-Vallejo, 2020; Miskiewicz, Sarzynska, & Szachniuk, 2021; Todd, Johnston, & Neidle, 2005). Different experimental findings also report G4 formation in guanine-rich regions that do not follow this canonical G4 motif (Jana, Mohr, Vianney, & Weisz, 2021). To map these G4 regions, a motif-independent screening tool has been developed that also identifies RNA G4 (Garant, Perreault, & Scott, 2017). Furthermore, polymerase stop assays and target-specific next generation sequencing approaches (e.g., chromatin immunoprecipitation) using G4-specific antibodies and probes have been performed in different organism that also experimentally determine G4 formation at specific DNA and RNA loci (Jamroskovic et al., 2019; Kwok, Marsico, & Balasubramanian, 2018; Marsico et al., 2019; Tu et al., 2021; Yang et al., 2018).

- Selection of control targets

  To estimate and compare helicase activity on G4 structures, a selection of control substrates is recommended. The choice of the control DNA or RNA regions depends on the scientific question. Most frequently, G-rich forked nucleic acid constructs were used as well as other secondary DNA or RNA structures such as hairpins or cruciform (Bochman et al., 2012; Ribeyre et al., 2009).

- Synthesizing DNA or RNA targets

  Selected regions can be ordered as standard unmodified oligonucleotides from commercial sources. Lyophilized oligonucleotides are reconstituted at

Table 1 Table provides examples of G-rich sequences that can fold G4 structures and have been used in the past for helicase unwinding experiments.

Sequence of G4 motif	Name of G4	Description	Topology
GGGGTTTTGGGGTTTTGGGGTTTTGGGG	Oxy Telo	Telomeric G4 sequence from *Stylonychia lemnea*	Antiparallel
GGGTTGGGTTGGGTTGGG	telo G4	Reduced loop size	Antiparallel
ACTGTCGTACTTGATATTTTGGGGTTTTGG GG	bi-Oxy G4	Bi-molecular telomeric G4 sequence from *Stylonychia lemnea*	Antiparallel
GGGGTGGGTAGGGTGGGTAA	c-MYC G4	G4 within c-MYC	Parallel
TTTCGGGCGGGCGCGAGGGAGGG	c-KIT G4	G4 within c-KIT	Parallel
TTTGGGAGGGAGAGGGGGDGGG	HIF-1alpha G4	G4 within HIF-1alpha	Parallel
TTTGGTTGGTGTGGTTGG	TBA G4	Thrombin binding G4 aptamer	Antiparallel
TGGACCAGACCTAGCAGCTATGGGGGAGCT GGGGAAGGTGGGAATGTGA	TP G4	Murine immunoglobulin switch sequence	Parallel
CCCAACCCAACCCTTCCC	rev_telo	Sequence use in helicase stop assay	No G4

Table provides sequences of G4 motif, name of the region used in the literature, description where this sequence origins from and the experimentally determined G4 topology.

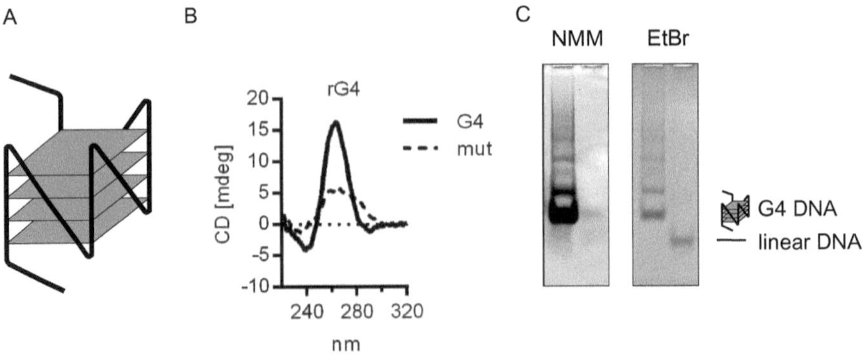

Fig. 1 Control of G4 formation. (A) Cartoon of a parallel G4 structure (B) CD spectra of a G4- and non-G4-forming oligodeoxynucleotide. (C) NMM- and ethidium bromide-(EtBr) stain of G4-forming oligodeoxynucleotide. EtBr gels detects all loaded nucleic acids. NMM gels visualizes only those nucleic acids that have a folded G4 structures. First lane G4, second lane control DNA (no G4 formation). Cartoon on the right indicates topology of the nucleic acid within the gel.

100 μM with ultrapure water. Oligonucleotides can be stored at −20 °C. RNA targets can be synthesized by *in vitro* transcription. For this. the target sequence will be in front of a T7 promoter and the RNA will be generated by reverse transcription. Due to the tight conformation of most G4s, we always add three adenines to the 5' end of each oligo. This renders the 5'-end free and available for end labeling instead of being buried within the G4. Also, we note that some helicases bind directly to the G4 (e.g., Pif1). Other helicases load in front of the G4 (e.g., DHX36 or BLM). These helicases require also an extension tail next to the G4. While designing these constructs that favor helicase binding and unwinding, please keep the directionality of your helicase in mind.

- Labeling of DNA or RNA target

  For later detection, DNA or RNA molecules need to be labeled. This protocol utilizes a radioactivity to label the DNA or RNA, however, alternative 5' or 3'-end labeling including biotinylation or addition of a fluorophore can be used. Labeling position needs to be carefully considered as to not hinder the formation of the G4 or affect helicase function.

3. Materials and equipment

3.1 General equipment

- Vertical polyacrylamide gel electrophoresis equipment (if using acrylamide gels)

- Horizontal gel electrophoresis equipment (if using agarose gels)
- Centrifuges (table-top) capable of cooling to 4 °C
- Incubators for 1.5 mL tubes (21–95 °C)
- Phosphoimager and phosphoimager screen (e.g. Typhoon FLA 7000) (alternative: X-ray films and developer). pH meter
- Circular dichroism (CD) instrument (e.g., Jasco-1500 spectropolarimeter)
- Quartz cell cuvette
- UV shadowing lamp
- Gel dryer (Biorad)
- Shaking incubators capable or heating (≥37 °C) and cooling (≤18 °C)

3.2 Reagents

- NaCl
- KCL
- Glycerol
- ß-Mercaptoethanol
- HEPES
- EDTA
- Tris
- HCl
- NaOH
- NaOAc
- $MgCl_2$
- Boric acid
- Agarose (if using agarose gels)
- Ethidium bromide
- 19:1 acrylamide:bis-acrylamide solution (PAA)
- Ammonium persulfate (APS) (if using acrylamide gels)
- *N,N,N',N'*-Tetramethylethylenediamine (TEMED) (if using acrylamide gels)
- Sodium dodecyl sulphate (SDS)
- Ammonium acetate
- Triton-X100
- Dithiothreitol (DTT)
- T4 polynucleotide kinase (PNK)
- G25 columns
- [sγ-^{32}P]ATP
- *N*-Methylmesoporphyrin IX (NMM)
- 10x TBE (1.3 M Tris (pH 7.6), 450 mM boric acid, 25 mM EDTA)
- Fresh 10% APS solution in water (if using acrylamide gels)

3.3 Oligonucleotides

Oligonucleotides can either be designed, taken from literature (Byrd & Raney, 2015) or be mined using biocomputational means (Brazda et al., 2019; Huppert & Balasubramanian, 2005; Lombardi & Londono-Vallejo, 2020; Miskiewicz et al., 2021; Todd et al., 2005). Examples are shown in Table 1.

4. Step-by-step method details

4.1 G4 formation

Different factors influence the formation of G4 structures including the type of salt, temperature, and the concentration of the oligonucleotide (Kim, Evans, Dubins, & Chalikian, 2015; Lane, Chaires, Gray, & Trent, 2008; You et al., 2017; Yuan et al., 2020). In general, use of K^+ leads to more stable G4s than Na^+, whereas Li^+ leads to less stable G4s (Bardin & Leroy, 2008; Kim et al., 2015). Of those ions, K^+ is discussed to be the more physiological relevant salt condition (Ghosh, Largy, & Gabelica, 2021; Zhang, Xu, Kumar, Zhang, & Wu, 2019).

As for many other protocols, there are different protocols described in the literature how to fold an G4 motif into a G4 structure *in vitro*. Here, we describe the two most common methods that were successfully used in our lab.

4.1.1 DNA G4 formation

DNA G4 formation

1. Selected DNA oligonucleotides (see above) were dissolved at 100 µM in G4 folding buffer (10 mM Tris (pH 7.5), 100 mM KCl, 1 mM EDTA). Depending on the experimental question KCl can be exchanged to NaCl or LiCl (see above). Once dissolved, we keep this as a stock, which is handy to freeze in aliquots at −20 °C. It is advised not to freeze/thaw the folded G4 oligonucleotides too often, which would prevent proper G4 formation.
2. Before the start of the experiment, the oligonucleotides will be diluted to 10 µM in G4 folding buffer.
3. Samples will be heat denatured for 10 min at 95 °C using a heating block.
4. Following denaturation, we slowly cool down samples to room temperature. We have achieved excellent results by decreasing the temperature

by 2 °C min^{-1} using either a PCR machine or temperature controlled incubators (Eppendorf thermomixer C).

5. Once the temperature reaches 21 °C, the G4 structures are folded properly.
6. Optional: To control for the efficiency of G4 formation, it is advised to separate the unfolded, partly folded and folded G4 structures from each other. This can be done by using vertical native gels (12% native acrylamide (19,1 acrylamide:bisacrylamide, 1x TBE, 1% APS, 0.1% TEMED). Gels will be run slowly using 10 V cm^{-1}. After ultraviolet shadowing, the correct size can be excised from the gel. Gel pieces can be cut into tiny pieces. Folded G4s are eluted overnight at 4 °C in TE (10 mM Tris, 1 mM EDTA pH 8.0) buffer. G4s are then ethanol-precipitated, dried and dissolved in water.

4.1.2 RNA G4 formation

1. Selected RNA G4 motifs were either synthesized or experimentally produced.
2. Selected RNA G4 oligonucleotides were dissolved at 2 μM in 10 mM Tris (pH 7.5) and 100 mM KCl (or NaCl/LiCl).
3. Samples will be heat denatured for 10 min at 95 °C using a heating block.
4. Following denaturation, slowly cool down the samples to room temperature. We have achieved excellent results by decreasing the temperature by 2 °C min^{-1} using either a PCR machine or temperature controlled incubators (e.g. Eppendorf Thermomixer C).
5. Once the temperature reaches 21 °C, the G4 structures are folded properly.

   Optional: To control for the efficiency of G4 formation, it is advised to separate unfolded, partly folded and fully folded G4 structures from each other. This can be done by using vertical native gels. After ultraviolet shadowing, the correct size can be excised from the gel. Gel pieces can be cut into tiny pieces. RNA fragment will be eluted overnight 4 °C in elution buffer (1 mM EDTA, 0.1% SDS and 0.5 M ammonium acetate). RNA G4s are then ethanol-precipitated, dried, and dissolved in water.

4.2 Control G4 structure formation

Different G4 structures conformations described in the literature can form parallel, anti-parallel, or as hybrid (Burge et al., 2006; Popenda, Miskiewicz, Sarzynska, Zok, & Szachniuk, 2020). It is known that protein binding and helicase function may differ dependent on the G4 topology. Different methods have been developed to monitor the formation of G4 structures *in vitro*. Among the most commonly used are circular dichroism (CD) measurements (Del Villar-Guerra, Trent, & Chaires, 2018; Kejnovska, Renciuk, Palacky, & Vorlickova, 2019; Randazzo, Spada, & da Silva, 2013), *N*-methylmesoporphyrin (NMM) gels (Smith & Johnson, 2010), gel mobility changes in native gels (Moon & Jarstfer, 2010; Paeschke et al., 2013; Sun & Hurley, 2010), NMR (Lin, Dickerhoff, & Yang, 2019; Sket & Plavec, 2010; Webba da Silva, 2007), thermal melting experiments (Gray & Chaires, 2011), FRET (especially single molecule FRET (Juskowiak & Takenaka, 2006; Luo, Granzhan, Verga, & Mergny, 2021; Maleki, Budhathoki, Roy, & Balci, 2017)), and TDS spectra (UV basically (Dao, Haselsberger, Michel-Beyerle, & Phan, 2011; Mergny, Li, Lacroix, Amrane, & Chaires, 2005)). Here, we will describe CD and NMM gels in more detail.

4.2.1 CD

1. Either G4 folding buffer or water will be used as the blank in the spectropolarimeter (CD instrument). The spectral input of buffers, salts, and if required proteins were subtracted by using the software supplied with the instrument (Spectra Manager II Spectroscopy Software Suite).
2. Folded G4 structures (at least 2 μM, see above) will be pipetted into a 1 mL quartz cell cuvette with a path length of 1 mm.
3. CD spectra are recorded at 20 °C using a wavelength range of 220–320 nm.
4. Scanning speed should be 100 nm min^{-1} with a 1 s response time, 1 nm data pitch, and 1 nm bandwidth.
5. The measurement is repeated at least three times. Data will be presented as the average of three wavelength scans.
6. To analyze and interpret the data obtained by CD spectral analysis, distinct and characteristic maxima and minima peaks of G4 structures are essential. CD spectral features associated with specific G4 topologies, namely parallel G4 conformation have a distinct positive band at 264 nm and a relative shallow negative band at 245 nm; antiparallel G4 structures have a positive band at 295 nm and a negative band around 265 nm; hybrid G4 structures have two positive bands around 295 nm and 264 nm and a negative band around 245 nm. See Fig. 1A for an example of a parallel G4 structure.

4.2.2 N-mesoporphyrin IX (NMM) native gel electrophoresis

Different chemical probes (G4 ligands) have been generated that bind specifically to G4 structures. These G4 ligands bind and, in most cases, stabilize the G4 structure. These G4 ligands vary in their specificity toward G4s, and some exist that are either tagged or are fluorescent (Santos, Salgado, Cabrita, & Cruz, 2021; Umar, Ji, Chan, & Kwok, 2019). In this gel-based system. We use the G4-specific ligand NMM (K_D ~0.1–1 μM (Yett, Lin, Beseiso, Miao, & Yatsunyk, 2019)) to specifically stain parallel G4 structures in native TBE gels. thioflavin (ThT) also binds specifically to both parallel and antiparallel G4s and can be also used in this assay.

1. Prepare a native polyacrylamide TBE gel (1x TBE, 15% PAA (19,1), 1% APS, 0.1% TEMED). 15% is the optimal concentration to separate 10–100 bp.
2. After polymerization (30 min), assemble gel and add 1x TBE as running buffer to the chamber.
3. Wash wells thoroughly with 1x TBE to remove residual acrylamide.
4. Add to folded G4 structures (0.5 μg) 1x native loading dye (60 mM KCl, 10 mM Tris (pH 7.5), 50% (w/v) glycerol, 0.025% (w/v) xylene cyanol, 0.025% (w/v) bromophenol blue) and load samples in the wells.
5. Run electrophoresis for 90 min at 12 V cm^{-1}.
6. Prepare *N*-mesoporphyrin IX (NMM) stock solution. 4 mg NMM will be dissolved in 0.2 N HCL. Use dark tubes or wrap with aluminum foil and store the stock solution in the dark at room temperature.
7. After the electrophoresis is completed, stain the gel with 10 μg mL^{-1} NMM or 0.5 μM Thioflavin (ThT) in 1x G4 folding buffer for 15 min under agitation. Because NMM requires KCl, only KCl-containing staining buffers work for this reaction. If G4 structures are formed using either NaCl or LiCl, switch to a KCl-containing buffer. We keep the gels in the dark to avoid losing fluorescent signal.
8. G4 structures can be visualized using a gel-documentation system that can utilize excitation wavelengths of 399 nm (NMM) or 420 nm (ThT) (emission max: 610 and 487 nm, respectively). See Fig. 1B, left, for an example.
9. As a last step, we monitor all nucleic acids by staining with ethidium bromide. For this, gels will be incubated with 0.5 μg mL^{-1} ethidium bromide in 1x TBE for 15 min at room temperature. During this step we usually agitate the gels slowly.

10. Nucleic acids will be detected using a standard gel documentation system (e.g., BioRad, ChemiDoc) with an UV lamp (254 nm). See Fig. 1B, right, for an example.

4.3 Helicase activity assay

G4 structures can be unfolded by many helicases. Depending on the stability of the G4 structure and the length and position of an overhang (tail), helicase processivity differs among different helicase families and family members. Once the G4 structure is folded and controlled, established helicase assay for given helicase can be used. Different assays that measure helicase activity are possible with the generated substrates (e.g., FRET, ATP hydrolysis analysis, radioactive, and fluorescence gel-based assays). The buffers and timepoints described here are for Pif1 family members but work well for other helicases. In our lab, we have very good experience using radioactive helicase assays that will be analyzed on native gels. This has the advantage that small amounts of G4 structures and helicases are required for this assay.

1. Folded G4 DNA structures (e.g. telo G4, see Table 1) were 5' end labeled with ^{32}P using T4 PNK (NEB) according to the manufactures protocol.
2. After labeling free nucleotides were separated using a G25 column or similar to remove unincorporated radionucleotides.
3. Prepare 5x reaction buffer (100 mM Tris (pH 7.5), 200 mM NaCl, 500 μg mL^{-1} BSA, 10 mM DTT).
4. Prepare helicase STOP buffer (17% Ficoll, 50 mM EDTA, 1 μM unlabeled single stranded DNA oligonucleotide (see Table 1 for example sequence matching telo G4) that is reverse complement to the G4 motif, 0.05% bromophenol blue and 0.05% xylene cyanol).
5. Prepare a 12% native polyacrylamide (19:1, 1x TBE, 1% APS, 0.1% TEMED) vertical gel.
6. To characterize the processivity of a given helicase on a G4 substrate the parameters in the helicase reaction can be altered. First, the concentration of the helicase and second the length/timing of the reaction. We will first (a) describe the reaction with altered helicase concentrations and second (b) the reaction with different incubation times of the helicases unwinding reaction.
7. Prepare seven tubes that contain: 1x reaction buffer, labeled DNA substrate (10 nM), 5 mM magnesium ions. The final reaction volume (after starting the reaction with ATP (see step 9) is 5 μL.

4.3.1 Effect of helicase concentration on G4 unwinding

1. Add to these seven tubes different concentrations of purified helicase. If you do not know how processive your helicase is, it is advised to choose a wide concentration range in the first experiments (e.g. 1 nM–100 mM) and narrow down the concentration in subsequent experiments to calculate unwinding parameters.
2. 1 out of these 7 samples will be collected without ATP as a reference to control for no helicase activity. Add 5 μL of STOP buffer and samples will be placed at 4 °C.
3. A second tube will be set to the site that will be used as a reference for full unwinding. This tube will be incubated for 10 min at 95 °C. The boiling step will linearize the G4. After boiling, 5 μL of STOP buffer must be added to prevent re-formation of the G4. Sample will be placed at 4 °C.
4. After setting the control samples, the reaction can be started using 4 mM ATP. Helicase unwinding will be performed at 30 °C in a thermal block or water bath
   a. Note, 30 °C is optimal for yeast Pif1. Other helicases from other organisms will perform better at higher temperatures, or, if unstable, need to be incubated at 4 °C for longer.
5. After 10 min (for less processive helicases this time can be extended to 30 min), the reaction will be stopped using 5 μL of STOP buffer.
6. Pre-run for 10 min a native gel (10 V cm^{-1}) and wash wells after this thoroughly to remove soluble acrylamide and debits from wells.
7. Load total helicase reactions and run the gel for 30 min at 10 V cm^{-1} in 1x TBE buffer.
8. Dry the gels using a vacuum gel drier and expose the dried gel to a phosphoimager plate for up to 24 h. Scan the phosphoimager screen using a phosphoimager. The gels can also be frozen at −20 °C for phosphoimager or film exposure. Note: to prevent gel cracking during drying or freezing, minimal amounts of TEMED (1%) should be used. Also, degassing of TBE is advised, as we observe better results.
9. Quantify helicase activity by measuring the densitometry for bands of the folded G4 and unfolded/linear substrate using ImageQuant, ImageJ, or similar software. Subtract the background from both bands. The controls, without ATP and the boiled sample, are set to 100% folded or unfolded, respectively, and the percentage of unwinding will be calculated based on these parameters (equation: (unwinding-background)/(control-background) * 100)

4.3.2 Time course for G4 unwinding

1. Add to all seven tubes 10nM purified helicase.
2. As a 100% G4 folding control, take time-point 0. Here all oligos are 100% folded into a G4. The second control is the 100% unfolding control, which is prepared by boiling the tube for 10min at 95°C. STOP buffer will be added (5μL) to these controls and kept at 4°C.
3. After setting the control samples, the reaction can be started using 4mM ATP. Helicase unwinding will be performed using a thermal block or water bath. Unwinding activity will be monitored over time.
4. If the processivity of your helicase on G4s is unknown, first monitor G4 unfolding over a longer range over time. We usually start with time points: 0 (not ATP),1, 2, 5, 10, 15, 30, and 60min. Once the time frame of G4 unwinding is narrowed down, shorter and more detailed time slots can be selected (see Fig. 2A). All reaction will be stopped immediately using 5μL of STOP buffer.
5. Pre-run for 10min a native gel (10 V cm^{-1}) and afterwards wash wells thoroughly to remove soluble acrylamide and debits.
6. Load total helicase reaction and run reaction for 30min at 10 V cm^{-1} in 1x TBE buffer.
7. Dry the gels using a vacuum gel drier and expose the dried gel to a phosphoimager plate as described in 3.1.1. (see Fig. 2A for an example).
8. Quantify helicase activity by measuring the densitometry for bands as described in 3.1.1. (see Fig. 2B for an example of the quantification of Fig. 2A).

Here, we describe a standard helicase assay for G4 DNA, but similar protocols work for G4 RNA. Since G4 can also form at the RNA level, it is noted

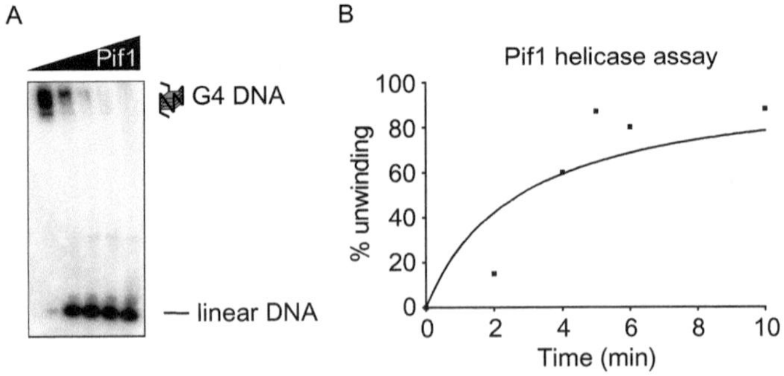

Fig. 2 G4 unwinding by Pif1 helicase (A), G4 unwinding over time was monitored by gel. Pif1 helicase action was monitored on Telo G4. (B) Quantification of unwinding reaction.

that to study G4 RNA unwinding by helicases, similar gel-based helicase unwinding protocols can be used. However, the amount and type of salt should be taken into consideration. Also, general good working practice for handling RNA should be observed, including using RNase-free reagents and plastic ware, reducing the heating times, and avoiding divalent cations.

Acknowledgments

We thank Philipp Schult for careful reading of the manuscript. Research in the Paeschke laboratory is fund by an ERC Stg Grant (638988-G4DSB) and by the Deutsche Forschungsgemeinschaft (DFG, German Research Foundation)—Project-ID 369799452—TRR237" as well as by the Deutsche Forschungsgemeinschaft (DFG, German Research Foundation) under Germany's Excellence Strategy—EXC2151—390873048.

References

Bardin, C., & Leroy, J. L. (2008). The formation pathway of tetramolecular G-quadruplexes. *Nucleic Acids Research*, *36*(2), 477–488.

Beauvarlet, J., Bensadoun, P., Darbo, E., Labrunie, G., Rousseau, B., Richard, E., et al. (2019). Modulation of the ATM/autophagy pathway by a G-quadruplex ligand tips the balance between senescence and apoptosis in cancer cells. *Nucleic Acids Research*, *47*(6), 2739–2756.

Benhalevy, D., Gupta, S. K., Danan, C. H., Ghosal, S., Sun, H. W., Kazemier, H. G., et al. (2017). The human CCHC-type zinc finger nucleic acid-binding protein binds G-rich elements in target mRNA coding sequences and promotes translation. *Cell Reports*, *18*(12), 2979–2990.

Biffi, G., Di Antonio, M., Tannahill, D., & Balasubramanian, S. (2014). Visualization and selective chemical targeting of RNA G-quadruplex structures in the cytoplasm of human cells. *Nature Chemistry*, *6*(1), 75–80.

Bochman, M. L., Paeschke, K., & Zakian, V. A. (2012). DNA secondary structures: Stability and function of G-quadruplex structures. *Nature Reviews. Genetics*, *13*(11), 770–780.

Bohalova, N., Dobrovolna, M., Brazda, V., & Bidula, S. (2021). Conservation and over-representation of G-quadruplex sequences in regulatory regions of mitochondrial DNA across distinct taxonomic sub-groups. *Biochimie*, *194*, 28–34.

Boule, J. B., & Zakian, V. A. (2007). The yeast Pif1p DNA helicase preferentially unwinds RNA DNA substrates. *Nucleic Acids Research*, *35*(17), 5809–5818.

Brazda, V., Kolomaznik, J., Lysek, J., Bartas, M., Fojta, M., Stastny, J., et al. (2019). G4Hunter web application: A web server for G-quadruplex prediction. *Bioinformatics*, *35*(18), 3493–3495.

Bryan, T. M. (2020). G-quadruplexes at telomeres: Friend or foe? *Molecules*, *25*(16).

Burge, S., Parkinson, G. N., Hazel, P., Todd, A. K., & Neidle, S. (2006). Quadruplex DNA: Sequence, topology and structure. *Nucleic Acids Research*, *34*(19), 5402–5415.

Byrd, A. K., & Raney, K. D. (2015). A parallel quadruplex DNA is bound tightly but unfolded slowly by pif1 helicase. *The Journal of Biological Chemistry*, *290*(10), 6482–6494.

Byrd, A. K., Zybailov, B. L., Maddukuri, L., Gao, J., Marecki, J. C., Jaiswal, M., et al. (2016). Evidence that G-quadruplex DNA accumulates in the cytoplasm and participates in stress granule assembly in response to oxidative stress. *The Journal of Biological Chemistry*, *291*(34), 18041–18057.

Capra, J. A., Paeschke, K., Singh, M., & Zakian, V. A. (2010). G-quadruplex DNA sequences are evolutionarily conserved and associated with distinct genomic features in Saccharomyces cerevisiae. *PLoS Computational Biology*, *6*(7), e1000861.

Caterino, M., & Paeschke, K. (2021). Action and function of helicases on RNA G-quadruplexes. *Methods*.

Chen, M. C., Tippana, R., Demeshkina, N. A., Murat, P., Balasubramanian, S., Myong, S., et al. (2018). Structural basis of G-quadruplex unfolding by the DEAH/RHA helicase DHX36. *Nature*, *558*(7710), 465–469.

Choi, Y. W., Dutta, S., Fielding, B. C., & Tan, Y. J. (2010). Expression, purification and preliminary crystallographic analysis of recombinant human DEAD-box polypeptide 5. *Acta Crystallographica. Section F, Structural Biology and Crystallization Communications*, *66*(Pt. 2), 192–194.

Dao, N. T., Haselsberger, R., Michel-Beyerle, M. E., & Phan, A. T. (2011). Following G-quadruplex formation by its intrinsic fluorescence. *FEBS Letters*, *585*(24), 3969–3977.

De Magis, A., Manzo, S. G., Russo, M., Marinello, J., Morigi, R., Sordet, O., et al. (2019). DNA damage and genome instability by G-quadruplex ligands are mediated by R loops in human cancer cells. *Proceedings of the National Academy of Sciences of the United States of America*, *116*(3), 816–825.

Del Villar-Guerra, R., Trent, J. O., & Chaires, J. B. (2018). G-quadruplex secondary structure obtained from circular dichroism spectroscopy. *Angewandte Chemie (International Edition in English)*, *57*(24), 7171–7175.

Fairman-Williams, M. E., Guenther, U. P., & Jankowsky, E. (2010). SF1 and SF2 helicases: Family matters. *Current Opinion in Structural Biology*, *20*.

Garant, J. M., Perreault, J. P., & Scott, M. S. (2017). Motif independent identification of potential RNA G-quadruplexes by G4RNA screener. *Bioinformatics*, *33*(22), 3532–3537.

Ghosh, A., Largy, E., & Gabelica, V. (2021). DNA G-quadruplexes for native mass spectrometry in potassium: A database of validated structures in electrospray-compatible conditions. *Nucleic Acids Research*, *49*(4), 2333–2345.

Gong, J. Y., Wen, C. J., Tang, M. L., Duan, R. F., Chen, J. N., Zhang, J. Y., et al. (2021). G-quadruplex structural variations in human genome associated with single-nucleotide variations and their impact on gene activity. *Proceedings of the National Academy of Sciences of the United States of America*, *118*(21).

Gray, R. D., & Chaires, J. B. (2011). Analysis of multidimensional G-quadruplex melting curves. *Current Protocols in Nucleic Acid Chemistry*, *17*. Unit 17.4.

Gu, Y., Masuda, Y., & Kamiya, K. (2008). Biochemical analysis of human PIF1 helicase and functions of its N-terminal domain. *Nucleic Acids Research*, *36*(19), 6295–6308.

Hansel-Hertsch, R., Simeone, A., Shea, A., Hui, W. W. I., Zyner, K. G., Marsico, G., et al. (2020). Landscape of G-quadruplex DNA structural regions in breast cancer. *Nature Genetics*, *52*(9), 878–883.

Hui, W. W. I., Simeone, A., Zyner, K. G., Tannahill, D., & Balasubramanian, S. (2021). Single-cell mapping of DNA G-quadruplex structures in human cancer cells. *Scientific Reports*, *11*(1), 23641.

Hui, W. W. I., Simeone, A., Zyner, K. G., Tannahill, D., & Balasubramanian, S. (2022). Publisher correction: Single-cell mapping of DNA G-quadruplex structures in human cancer cells. *Scientific Reports*, *12*(1), 908.

Huppert, J. L. (2010). Structure, location and interactions of G-quadruplexes. *The FEBS Journal*, *277*(17), 3452–3458.

Huppert, J. L., & Balasubramanian, S. (2005). Prevalence of quadruplexes in the human genome. *Nucleic Acids Research*, *33*(9), 2908–2916.

Jamroskovic, J., Obi, I., Movahedi, A., Chand, K., Chorell, E., & Sabouri, N. (2019). Identification of putative G-quadruplex DNA structures in S. pombe genome by quantitative PCR stop assay. *DNA Repair (Amst)*, *82*, 102678.

Jana, J., Mohr, S., Vianney, Y. M., & Weisz, K. (2021). Structural motifs and intramolecular interactions in non-canonical G-quadruplexes. *RSC Chemical Biology*, *2*(2), 338–353.
Juskowiak, B., & Takenaka, S. (2006). Fluorescence resonance energy transfer in the studies of guanine quadruplexes. *Methods in Molecular Biology*, *335*, 311–341.
Karow, J. K., Chakraverty, R. K., & Hickson, I. D. (1997). The Bloom's syndrome gene product is a 3'-5' DNA helicase. *The Journal of Biological Chemistry*, *272*(49), 30611–30614.
Kejnovska, I., Renciuk, D., Palacky, J., & Vorlickova, M. (2019). CD study of the G-quadruplex conformation. *Methods in Molecular Biology*, *2035*, 25–44.
Kim, N. (2019). The interplay between G-quadruplex and transcription. *Current Medicinal Chemistry*, *26*(16), 2898–2917.
Kim, B. G., Evans, H. M., Dubins, D. N., & Chalikian, T. V. (2015). Effects of salt on the stability of a G-quadruplex from the human c-MYC promoter. *Biochemistry*, *54*(22), 3420–3430.
Kosiol, N., Juranek, S., Brossart, P., Heine, A., & Paeschke, K. (2021). G-quadruplexes: A promising target for cancer therapy. *Molecular Cancer*, *20*(1), 40.
Kwok, C. K., Marsico, G., & Balasubramanian, S. (2018). Detecting RNA G-quadruplexes (rG4s) in the transcriptome. *Cold Spring Harbor Perspectives in Biology*, *10*(7).
Lane, A. N., Chaires, J. B., Gray, R. D., & Trent, J. O. (2008). Stability and kinetics of G-quadruplex structures. *Nucleic Acids Research*, *36*(17), 5482–5515.
Lavezzo, E., Berselli, M., Frasson, I., Perrone, R., Palu, G., Brazzale, A. R., et al. (2018). G-quadruplex forming sequences in the genome of all known human viruses: A comprehensive guide. *PLoS Computational Biology*, *14*(12), e1006675.
Lee, W. T. C., Yin, Y., Morten, M. J., Tonzi, P., Gwo, P. P., Odermatt, D. C., et al. (2021). Single-molecule imaging reveals replication fork coupled formation of G-quadruplex structures hinders local replication stress signaling. *Nature Communications*, *12*(1), 2525.
Lin, C., Dickerhoff, J., & Yang, D. (2019). NMR studies of G-quadruplex structures and G-quadruplex-interactive compounds. *Methods in Molecular Biology*, *2035*, 157–176.
Linke, R., Limmer, M., Juranek, S. A., Heine, A., & Paeschke, K. (2021). The relevance of G-quadruplexes for DNA repair. *International Journal of Molecular Sciences*, *22*(22).
Lipps, H. J., & Rhodes, D. (2009). G-quadruplex structures: In vivo evidence and function. *Trends in Cell Biology*, *19*(8), 414–422.
Lombardi, E. P., & Londono-Vallejo, A. (2020). A guide to computational methods for G-quadruplex prediction. *Nucleic Acids Research*, *48*(3), 1603.
Luo, Y., Granzhan, A., Verga, D., & Mergny, J. L. (2021). FRET-MC: A fluorescence melting competition assay for studying G4 structures in vitro. *Biopolymers*, *112*(4), e23415.
Maleki, P., Budhathoki, J. B., Roy, W. A., & Balci, H. (2017). A practical guide to studying G-quadruplex structures using single-molecule FRET. *Molecular Genetics and Genomics*, *292*(3), 483–498.
Marsico, G., Chambers, V. S., Sahakyan, A. B., McCauley, P., Boutell, J. M., Antonio, M. D., et al. (2019). Whole genome experimental maps of DNA G-quadruplexes in multiple species. *Nucleic Acids Research*, *47*(8), 3862–3874.
Mergny, J. L., Li, J., Lacroix, L., Amrane, S., & Chaires, J. B. (2005). Thermal difference spectra: A specific signature for nucleic acid structures. *Nucleic Acids Research*, *33*(16), e138.
Miskiewicz, J., Sarzynska, J., & Szachniuk, M. (2021). How bioinformatics resources work with G4 RNAs. *Briefings in Bioinformatics*, *22*(3).
Mohaghegh, P., & Hickson, I. D. (2001). DNA helicase deficiencies associated with cancer predisposition and premature ageing disorders. *Human Molecular Genetics*, *10*(7), 741–746.
Moon, I. K., & Jarstfer, M. B. (2010). Preparation of G-quartet structures and detection by native gel electrophoresis. *Methods in Molecular Biology*, *608*, 51–63.
Nickens, D. G., & Bochman, M. L. (2021). Characterization of the telomerase modulating activities of yeast DNA helicases. *Methods in Enzymology*, *661*, 327–342.

Orren, D. K., Brosh, R. M., Jr., Nehlin, J. O., Machwe, A., Gray, M. D., & Bohr, V. A. (1999). Enzymatic and DNA binding properties of purified WRN protein: High affinity binding to single-stranded DNA but not to DNA damage induced by 4NQO. *Nucleic Acids Research*, *27*(17), 3557–3566.

Paeschke, K., Bochman, M. L., Garcia, P. D., Cejka, P., Friedman, K. L., Kowalczykowski, S. C., et al. (2013). Pif1 family helicases suppress genome instability at G-quadruplex motifs. *Nature*, *497*(7450), 458–462.

Paeschke, K., Capra, J. A., & Zakian, V. A. (2011). DNA replication through G-quadruplex motifs is promoted by the Saccharomyces cerevisiae Pif1 DNA helicase. *Cell*, *145*(5), 678–691.

Popenda, M., Miskiewicz, J., Sarzynska, J., Zok, T., & Szachniuk, M. (2020). Topology-based classification of tetrads and quadruplex structures. *Bioinformatics*, *36*(4), 1129–1134.

Qi, H., Lin, C. P., Fu, X., Wood, L. M., Liu, A. A., Tsai, Y. C., et al. (2006). G-quadruplexes induce apoptosis in tumor cells. *Cancer Research*, *66*(24), 11808–11816.

Randazzo, A., Spada, G. P., & da Silva, M. W. (2013). Circular dichroism of quadruplex structures. *Topics in Current Chemistry*, *330*, 67–86.

Ribeyre, C., Lopes, J., Boule, J. B., Piazza, A., Guedin, A., Zakian, V. A., et al. (2009). The yeast Pif1 helicase prevents genomic instability caused by G-quadruplex-forming CEB1 sequences in vivo. *PLoS Genetics*, *5*(5), e1000475.

Richard, P., & Manley, J. L. (2017). R loops and links to human disease. *Journal of Molecular Biology*, *429*(21), 3168–3180.

Robinson, J., Raguseo, F., Nuccio, S. P., Liano, D., & Di Antonio, M. (2021). DNA G-quadruplex structures: More than simple roadblocks to transcription? *Nucleic Acids Research*, *49*(15), 8419–8431.

Sanchez-Martin, V., Soriano, M., & Garcia-Salcedo, J. A. (2021). Quadruplex ligands in cancer therapy. *Cancers (Basel)*, *13*(13).

Santos, T., Salgado, G. F., Cabrita, E. J., & Cruz, C. (2021). G-quadruplexes and their ligands: Biophysical methods to unravel G-quadruplex/ligand interactions. *Pharmaceuticals (Basel)*, *14*(8).

Sauer, M., Juranek, S. A., Marks, J., De Magis, A., Kazemier, H. G., Hilbig, D., et al. (2019). DHX36 prevents the accumulation of translationally inactive mRNAs with G4-structures in untranslated regions. *Nature Communications*, *10*(1), 2421.

Shao, Y., & Zhang, Q. C. (2020). Targeting RNA structures in diseases with small molecules. *Essays in Biochemistry*, *64*(6), 955–966.

Shapiro, J. A. (2009). Revisiting the central dogma in the 21st century. *Annals of the New York Academy of Sciences*, *1178*, 6–28.

Sket, P., & Plavec, J. (2010). Tetramolecular DNA quadruplexes in solution: Insights into structural diversity and cation movement. *Journal of the American Chemical Society*, *132*(36), 12724–12732.

Smith, J. S., & Johnson, F. B. (2010). Isolation of G-quadruplex DNA using NMM-sepharose affinity chromatography. *Methods in Molecular Biology*, *608*, 207–221.

Spiegel, J., Adhikari, S., & Balasubramanian, S. (2020). The structure and function of DNA G-quadruplexes. *Trends Chemistry*, *2*(2), 123–136.

Suhasini, A. N., & Brosh, R. M., Jr. (2013). DNA helicases associated with genetic instability, cancer, and aging. *Advances in Experimental Medicine and Biology*, *767*, 123–144.

Sun, D., & Hurley, L. H. (2010). Biochemical techniques for the characterization of G-quadruplex structures: EMSA, DMS footprinting, and DNA polymerase stop assay. *Methods in Molecular Biology*, *608*, 65–79.

Svoboda, P., & Di Cara, A. (2006). Hairpin RNA: A secondary structure of primary importance. *Cellular and Molecular Life Sciences*, *63*(7-8), 901–908.

Teng, Y. C., Sundaresan, A., O'Hara, R., Gant, V. U., Li, M., Martire, S., et al. (2021). ATRX promotes heterochromatin formation to protect cells from G-quadruplex DNA-mediated stress. *Nature Communications*, *12*(1), 3887.

Todd, A. K., Johnston, M., & Neidle, S. (2005). Highly prevalent putative quadruplex sequence motifs in human DNA. *Nucleic Acids Research*, *33*(9), 2901–2907.

Tu, J., Duan, M., Liu, W., Lu, N., Zhou, Y., Sun, X., et al. (2021). Direct genome-wide identification of G-quadruplex structures by whole-genome resequencing. *Nature Communications*, *12*(1), 6014.

Umar, M. I., Ji, D., Chan, C. Y., & Kwok, C. K. (2019). G-quadruplex-based fluorescent turn-on ligands and aptamers: From development to applications. *Molecules*, *24*(13).

Umate, P., Tuteja, N., & Tuteja, R. (2011). Genome-wide comprehensive analysis of human helicases. *Communicative & Integrative Biology*, *4*(1), 118–137.

Vannutelli, A., Belhamiti, S., Garant, J. M., Ouangraoua, A., & Perreault, J. P. (2020). Where are G-quadruplexes located in the human transcriptome? *NAR Genomics and Bioinformatics*, *2*(2), lqaa035.

Wang, Y., Yang, J., Wild, A. T., Wu, W. H., Shah, R., Danussi, C., et al. (2019). G-quadruplex DNA drives genomic instability and represents a targetable molecular abnormality in ATRX-deficient malignant glioma. *Nature Communications*, *10*(1), 943.

Webba da Silva, M. (2007). NMR methods for studying quadruplex nucleic acids. *Methods*, *43*(4), 264–277.

Wu, M. T., & D'Souza, V. (2020). Alternate RNA structures. *Cold Spring Harbor Perspectives in Biology*, *12*(1).

Wu, F., Niu, K., Cui, Y., Li, C., Lyu, M., Ren, Y., et al. (2021). Genome-wide analysis of DNA G-quadruplex motifs across 37 species provides insights into G4 evolution. *Communications Biology*, *4*(1), 98.

Xing, Z., Wang, S., & Tran, E. J. (2017). Characterization of the mammalian DEAD-box protein DDX5 reveals functional conservation with S. cerevisiae ortholog Dbp2 in transcriptional control and glucose metabolism. *RNA*, *23*(7), 1125–1138.

Yadav, V., Hemansi, Kim, N., Tuteja, N., & Yadav, P. (2017). Quadruplex in plants: A ubiquitous regulatory element and its biological relevance. *Frontiers in Plant Science*, *8*, 1163.

Yadav, P., Kim, N., Kumari, M., Verma, S., Sharma, T. K., Yadav, V., et al. (2021). G-quadruplex structures in bacteria: Biological relevance and potential as an antimicrobial target. *Journal of Bacteriology*, *203*(13), e0057720.

Yang, S. Y., Lejault, P., Chevrier, S., Boidot, R., Robertson, A. G., Wong, J. M. Y., et al. (2018). Transcriptome-wide identification of transient RNA G-quadruplexes in human cells. *Nature Communications*, *9*(1), 4730.

Yangyuoru, P. M., Bradburn, D. A., Liu, Z., Xiao, T. S., & Russell, R. (2018). The G-quadruplex (G4) resolvase DHX36 efficiently and specifically disrupts DNA G4s via a translocation-based helicase mechanism. *The Journal of Biological Chemistry*, *293*(6), 1924–1932.

Yett, A., Lin, L. Y., Beseiso, D., Miao, J., & Yatsunyk, L. A. (2019). N-methyl mesoporphyrin IX as a highly selective light-up probe for G-quadruplex DNA. *Journal of Porphyrins and Phthalocyanines*, *23*(11n12), 1195–1215.

You, J., Li, H., Lu, X. M., Li, W., Wang, P. Y., Dou, S. X., et al. (2017). Effects of monovalent cations on folding kinetics of G-quadruplexes. *Bioscience Reports*, *37*(4).

Yuan, W. F., Wan, L. Y., Peng, H., Zhong, Y. M., Cai, W. L., Zhang, Y. Q., et al. (2020). The influencing factors and functions of DNA G-quadruplexes. *Cell Biochemistry and Function*, *38*(5), 524–532.

Zhang, M. L., Xu, Y. P., Kumar, A., Zhang, Y., & Wu, W. Q. (2019). Studying the potassium-induced G-quadruplex DNA folding process using microscale thermophoresis. *Biochemistry*, *58*(38), 3955–3959.

Zyner, K. G., Simeone, A., Flynn, S. M., Doyle, C., Marsico, G., Adhikari, S., et al. (2022). G-quadruplex DNA structures in human stem cells and differentiation. *Nature Communications*, *13*(1), 142.

CHAPTER THIRTEEN

Helicase mediated vectorial folding of telomere G-quadruplex

Tapas Paul[a] and Sua Myong[a,b,*]
[a]Department of Biophysics, Johns Hopkins University, Baltimore, MD, United States
[b]Physics Frontier Center (Center for Physics of Living Cells), University of Illinois, Urbana, IL, United States
*Corresponding author: e-mail address: smyong@jhu.edu

Contents

Abstract

The G-rich single-stranded telomere overhang can self-fold into G-quadruplex (G4) structure both *in vivo* and *in vitro*. In somatic cells, telomeres shorten progressively due to the end-replication. In stem cells, however, telomeres are replenished by a special enzyme, telomerase which synthesizes single-stranded telomere overhang. The active extension by the telomerase releases G-rich overhang segmentally in 5′ to 3′ direction as the overhang folds into G4 structure after successive elongation. To replicate such vectorial G4 folding process, we employed a superhelicase, Rep-X

Methods in Enzymology, Volume 672
ISSN 0076-6879
https://doi.org/10.1016/bs.mie.2022.03.065

to release the G-rich sequence gradually. Using single-molecule assay we demonstrated that the folded conformation achieved by the vectorial folding is inherently different from the post-folding where the entire overhang is allowed to fold at once. In addition, the vectorially folded overhangs are less stable and more accessible to a complementary C-rich strand and the telomere binding protein, POT1 compared to the post-folded state. The higher accessibility may have implications for the facile loading of shelterin proteins after DNA replication.

1. Introduction

1.1 Telomeric DNA

The linear eukaryotic chromosomes are terminated by telomeres which bears tandem repeats of guanine-rich DNA sequence (Neidle & Parkinson, 2003). The human telomeric DNA consisting of TTAGGG repeats ranges in size from 2 to 20 kilobases of duplex followed by a terminal 3' single-stranded (ss) overhang (~50 to 300 nt) (Cimino-Reale et al., 2001). The TTAGGG in ssDNA self-assembles into a non-canonical DNA structure, G-quadruplex (G4) formed by Hoogsteen hydrogen bonding between guanines in the presence of monovalent cations (Tang et al., 2008). The G4 structure can limit the accessibility to telomere binding proteins (Wang et al., 2011) termed shelterin which serve to prevent inappropriate activation of DNA damage repair pathways (Baumann & Cech, 2001; Palm & de Lange, 2008).

1.2 Telomerase action

In most somatic cells, telomeres are gradually shortened due to the end-replication problem (Blasco, 2005). Highly proliferative cells such as stem cells and cancer cells can replenish telomere length by a special reverse transcriptase enzyme called telomerase (Zhao et al., 2009). Telomerase is a ribonucleoprotein complex that specifically binds the single-stranded telomere overhang and synthesizes TTAGGG repeats (Collins & Gandhi, 1998). Templated by the telomerase RNA (TER), telomerase elongates TTAGGG repeat overhang in 5' to 3' direction while the complementary strand is synthesized by a DNA polymerase to maintain the telomere length homeostasis (Srinivas, Rachakonda, & Kumar, 2020). POT1 is a member of the shelterin complex which binds exclusively to the telomere overhang with high affinity and sequence specificity (Paul, Liou, Cai, Opresko, & Myong, 2021). POT1 binding helps regulating telomerase access and activity (Gu et al., 2021).

1.3 Vectorial folding

Telomere overhang is synthesized directionally (5' to 3') and released segmentally (Masutomi et al., 2003). Therefore, the successive rounds of [TTAGGG]n overhang synthesis leads to G4 folding immediately following the synthesis (Jansson et al., 2019). To depict such a vectorial folding process we employed a highly processive superhelicase, Rep-X which can unwind duplex DNA to release the TTAGGG repeats in 5' to 3' direction (Arslan, Khafizov, Thomas, Chemla, & Ha, 2015) thereby mimicking the vectorial nature of the telomerase-catalyzed overhang synthesis. A recent study reported a Rep-X mediated vectorial folding of a telomere overhang using an optical tweezers method (Mitra & Ha, 2019).

2. Key resources table

Reagent or resource	Source	Identifier
Chemicals peptide, and recombinant proteins		
Alconox™ Powdered Precision Cleaner	Fisher Scientific	Cat#16000104
Biotin	Sigma-Aldrich	Cat#2003993
Biotin-PEG-SVA, MW 5,000	Laysan Bio Inc.	Cat#Biotin-PEG-SVA-5000-100 mg
Catalase from bovine liver	Sigma-Aldrich	Cat#C3155
Glucose oxidase from *Aspergillus niger*	Sigma-Aldrich	Cat#G2133
mPEG-Succinimidyl Valerate, MW 5,000	Laysan Bio Inc.	Cat#MPEG-SVA-5000-1g
N-(2-Aminoethyl)-3-aminopropyltrimethoxysilane	United Chemical Technologies	Cat#1760-24-3
NeutrAvidin Protein	Thermo Scientific	Cat#31000
Trolox(R), 97%, ACROS Organics™	Fisher Scientific	Cat#AC218940010
Oligonucleotides		
G4-Cy3: 5'-TGG CGA CGG CAG CGA GGC TTA GGG TTA GGG TTA GGG TTA GGG/Cy3/-3'	Integrated DNA Technologies	N/A

Continued

—cont'd

Reagent or resource	Source	Identifier
C4-T15: 5'- CCC TAA CCC TAA CCC TAA CCC TAA TTT TTT TTT TTT TTT-3'	Integrated DNA Technologies	N/A
C2: 5'-CCC TAA CCC TAA-3'	Integrated DNA Technologies	N/A
Cy5-18mer-Bio: 5'-GCC T/Cy5/CG CTG CCG TCG CCA/Bio/-3'	Integrated DNA Technologies	N/A
Software and algorithms		
IDL	Harris Geospatial	https://www.l3harrisgeospatial.com/Software-Technology/IDL
MATLAB	Mathworks	https://www.mathworks.com/products/matlab.html
smCamera	Roy, Hohng, and Ha (2008)	https://cplc.illinois.edu/research/tools
Other		
5 Minute® Rapid-Curing, General Purpose Adhesive Epoxy, 25 mL Tube	All-Spec	Cat#14250
Chemyx Inc fusion 200 Touch Syringe Pump	Fisher Scientific	Cat#NC0670590
ETT PTFE Tubing Size 26, 100 ft, Natural Color	Weico Wire	Cat#ETT-26
Quartz Microscope Slides	G Finkenbeiner Inc.	N/A
Scotch® 665 Permanent Double-Sided Tape, 1/2" x 250", Clear, Pack Of 3 Rolls	Office Depot	Cat#391775
TIRF Microscope	Olympus	N/A
532 and 634-nm diode laser	Compass 315M; Coherent, Santa Clara, CA	N/A

3. Experimental preparation for vectorial folding

3.1 DNA sample preparation

The synthetic DNA oligonucleotides containing both Cy3 and Cy5 for FRET measurements and biotin for surface immobilization were purchased from Integrated DNA Technology, USA. The DNA constructs containing telomere overhang (TTAGGG repeats) were designed in two different formats. For post folding experiments (where the telomere overhang is free to fold at once): each appropriate pair of ssDNA was mixed together at molar ratio of 1:1.2 (biotinylated:nonbiotinylated) and annealed in a buffer containing 10 mM Tris–HCl (pH 7.5) and 100 mM KCl. For vectorial folding experiments (where the telomere overhang was annealed to the complementary strand containing 3' poly-T tail for Rep-X loading): each appropriate pair of ssDNA was mixed together and annealed in a buffer containing 10 mM Tris–HCl (pH 7.5) and 5 mM $MgCl_2$. All DNA constructs were annealed in a thermal cycler with the following program as reported earlier: 95 °C for 2 min; gradual cooling to 40 °C at the rate of 2 °C/min; further cooling by 5 °C/min until 4 °C (Paul, Bera, Agnihotri, & Mishra, 2016; Paul & Mishra, 2016).

3.2 Imaging surface preparation

We follow a well-established surface passivation protocol for smFRET measurements (Paul, Ha, & Myong, 2021; Paul, Mohapatra, & Mishra, 2021). Quartz slides and coverslips were cleaned thoroughly by washing steps involving soap, Milli-Q water, methanol, acetone and etched by 1 M potassium hydroxide. Then the slides are burned for one minute and coverslips quickly sterilized by passing through a flame for 4–5 times to remove any source of fluorescence. Afterwards, both slides and coverslips are treated with aminosilane for 30–45 min and coated with a mixture of 2% biotinylated PEG (biotin-PEG-5000, Laysan Bio, Inc) and 98% methoxy-polyethylene glycol (mPEG-5000, Laysan Bio, Inc.) which form a passivation layer that prevents nonspecific binding of molecules to the surface. The slides and coverslips are then washed and dried using nitrogen gas. The slides are then vacuum-packed in 50-mL corning tube using a vacuum sealed plastic bag and stored at −20 °C for future experiments (Paul & Myong, 2022).

3.3 Slide assembly

On the day of the experiment, a set of slide and coverslip pair is taken out from −20 °C and thawed at room temperature for 15 min. Using a double-sided tape, microfluidic sample chamber is created by sandwiching the plasma cleaned slide and the coverslip, both coated with PEG and biotin-PEG. The coverslip edges are sealed with a quick-drying epoxy. For real-time smFRET experiment, a reservoir is built by using a cut piece of plastic tip attached at the top of one hole for sample loading and the epoxy mold to the corresponding hole for syringe attachment (Paul & Myong, 2022; Yang & Ha, 2018).

3.4 Imaging buffer preparation

All single-molecule (sm) FRET measurements are carried out in an imaging buffer containing 10 mM Tris–HCl (pH 7.5), 100 mM KCl, 1 mM $MgCl_2$, and 10% glycerol. To minimize blinking and photobleaching of fluorescent dyes, the imaging buffer is supplemented with the oxygen scavenging system composed of 10 mM trolox, 0.5% glucose, 1 mg/mL glucose oxidase and 4 μg/mL catalase (Roy et al., 2008).

3.5 Single-molecule FRET assay

The smFRET data are acquired using a home-built prism-type total internal reflection (PTIR) inverted fluorescence microscope (Olympus IX 71) equipped with an electron-multiplying CCD camera (EMCCD). Freshly annealed partial duplex DNA labeled with Cy3, Cy5 and biotin is diluted to 20–25 pM and applied to the NeutrAvidin (50 μg/mL) coated slide surface. Free (unannealed) DNA is removed by a buffer wash and imaging buffer applied for smFRET measurements. We observe ~300–500 fluorescent spots in one field of view of 25 x 75 μm area (Joo, Balci, Ishitsuka, Buranachai, & Ha, 2008). If the number of spots is too high (above 600) or too low (below 100), PEG-biotin or applied DNA concentration can be adjusted accordingly. All smFRET experiments are carried out at room temperature (~23 °C ± 2 °C).

3.6 Data acquisition and analysis

A solid-state diode laser of either 532 or 634 nm (Compass 315M, Coherent) was used to generate an evanescent field of illumination through PTIR to excite the fluorophores (Cy3 or Cy5) in the sample chamber. The fluorescent emission from Cy3 (donor) and Cy5 (acceptor) are simultaneously collected using a water immersion Olympus objective (60X, NA 1.2). A dichroic mirror (cutoff: 630 nm) separates the fluorescence signal from the sample into two wavelengths which are detected by an EMCCD camera (Andor). All data were recorded with 100 ms frame integration time (Bera, Paul, Iyengar, & Mishra, 2018; Paul, Bera, & Mishra, 2017).

The recorded data were processed by IDL script (http://www.exelisvis.co.uk/ProductsServices/IDL.aspx) and analysis was carried out by scripts written in MATLAB (https://www.mathworks.com/). The FRET histograms were generated from ~4000 molecules collected from 10 to 20 imaging spots. To avoid collecting donor-only molecules, red laser (634 nm) is used first to collect all the acceptor containing molecules, followed by a green laser (532 nm). Donor leakage is corrected based on the FRET values of the donor-only molecules. The smFRET histograms are normalized and fitted to Gaussian distributions using Origin 2018 (https://www.originlab.com/) with unconstrained peak position. The peak center positions are then used to restrain the peak position in the C2 or POT1 bound histograms.

3.7 Real-time folding

The smFRET real-time vectorial folding assays using Rep-X is carried out in a micro-fluidic imaging flow chamber (Paul & Myong, 2022; Yang & Ha, 2018). Flow chamber is prepared by attaching a small cut pipet tip placed above one hole at one end of the chamber and a syringe pump with silicone tube connected to the other hole (Harvard Apparatus, Holliston, MA). FRET labeled DNA is immobilized in the imaging chamber and 10 nM Rep-X with 1 mM or lower concentrations of ATP in imaging buffer is loaded into the reservoir. Real-time FRET image of vectorial folding was collected while the solution is syringed through the silicone tubing at a rate of 20 μL/s. Similarly, the real-time C-strand and POT1 binding is carried out in the same flow chamber as above where either C2 or POT1 with Rep-X and ATP is applied together. Real-time FRET images are collected during and after the sample injection with a frame rate of 100 ms and recorded data are analyzed using MATLAB scripts. The flow heatmaps are generated by combining more than 100 individual single-molecule real-time flow traces by synchronizing the low to high FRET transition using a MATLAB script. The C2 or POT1 binding kinetics are analyzed by collecting the time interval between the folded state (high FRET) and the subsequent FRET decline from individual FRET traces. For slow reactions that did not have FRET transition in 3 min, short movies are collected for longer time duration to generate FRET histograms and binding kinetics calculated based on the bound fraction.

4. Helicase induced vectorial folding (vf)

Helicases unwind nucleic acid by moving directionally along the nucleic acid phosphodiester backbone and thereby separating a duplex into single strands by using the energy from ATP hydrolysis (Paul et al., 2020). Here, to mimic the pathway of telomerase induced vectorial G4 folding process, we used a superhelicase, Rep-X (an enhanced version of Rep helicase) which can unwind the telomeric duplex DNA to release the single stranded TTAGGG repeats in 5' to 3' direction (Fig. 1). Rep-X unwinds in 3'–5' direction, similar to other superfamily I helicases (Arslan et al., 2015; Paul et al., 2020). Our smFRET construct contained biotin for surface immobilization and, donor (Cy3) and acceptor (Cy5) fluorophores positioned such that the G4 folding induced by the Rep-X unwinding will result in an increase in FRET. For all experiments, we used a physiological potassium concentration (100 mM KCl).

4.1 Telomeric G4 folding at high ATP

Telomere overhang contains tandem TTAGGG repeats which readily fold into G4 structure in the presence of a monovalent cation such as sodium or potassium. The conventional experimental setting in which the full sequence is allowed to fold at once is termed post-folding (pf). In a buffer containing 100 mM KCl and 1 mM $MgCl_2$, the pf of G4 showed two FRET peaks at ~0.7 (major) and ~0.85 (minor), indicating two different conformations, in agreement with a previous finding (Fig. 1A) (Lee et al., 2020). We asked if pf vs vf will lead to different outcomes. The vf construct contains a partial telomeric duplex with 3' poly-T15 tail in the C-rich strand for helicase loading. Rep-X binds 3' poly-T15 tail and unwind the duplex to release the G-rich DNA in the 5' to 3' direction, which mimics the telomere extension (Fig. 1B). Upon addition of Rep-X (10 nM) with 1 mM ATP to the vf construct immobilized on single molecule surface, the FRET peak at ~0.25 corresponding to duplex state shifted to ~0.7 (minor) and

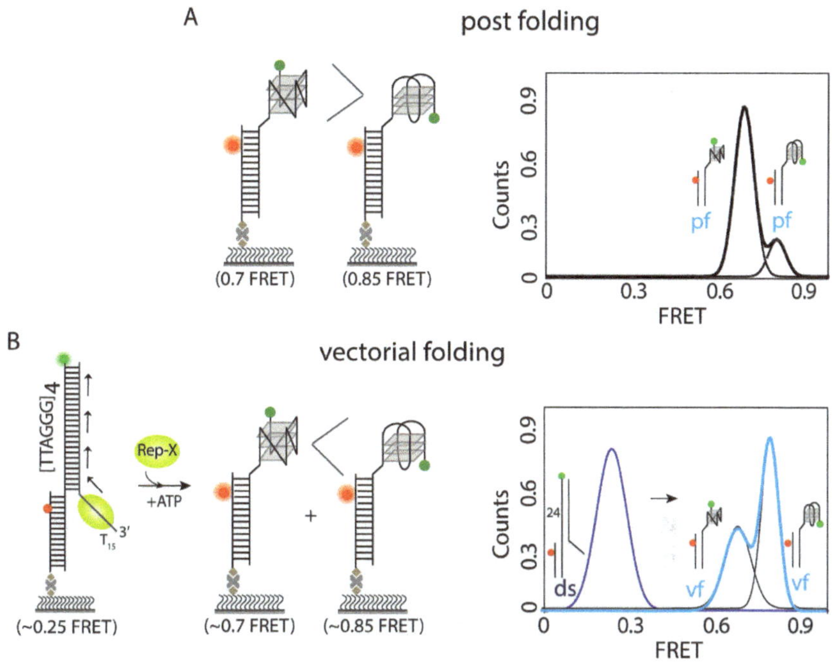

Fig. 1 Post folding and vectorial folding of telomere G4 overhang. (A) Schematic diagram of smFRET constructs depicting post folding (pf) of G4 ($(TTAGGG)_4$) telomere overhang and beside contain the model histogram where the FRET ~0.7 is the major and FRET ~0.85 is the minor population. (B) Schematic diagram of smFRET construct depicting vectorial folding (vf) of G4 overhang where the unfolding of telomere-duplex via Rep-X (10 nM and 1 mM ATP) leads to subsequent folding of telomere overhang. Beside contain the model histogram where the FRET ~0.7 is the minor and FRET ~0.85 is the major population.

~0.85 (major), which represent the G4 folded conformations (Fig. 1B). Interestingly, both pf and vf of G4 showed two similar FRET peak position at ~0.7 and ~0.85, yet with strikingly different distribution.

4.2 Folding pathway

Substrate immobilization enables monitoring the helicase induced vf of telomeric G4 folding. Real-time measurement is done as described in Section 3.7 (Fig. 2A). Rep-X (10 nM) with 1 mM ATP is applied to the vf DNA immobilized surface. Single-molecule trajectories show instantaneous FRET change from ~0.25 (duplex) to 0.7 or 0.85 (G4 folded)

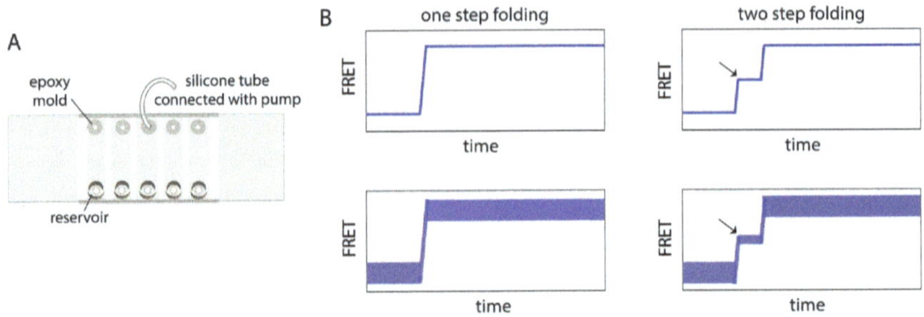

Fig. 2 Real-time flow experiment. (A) Schematic of PEG passivated slide assembly with reservoir in one end for sample loading and other end with epoxy mold which is connected to a silicone tube. (B) Model smFRET real-time G4 folding traces shows one and two steps of folding pathway. Below, a model heatmap of the respective folding traces are displayed.

(Fig. 2B). While some molecules exhibit abrupt one-step transition from low to high FRET, others showed an intermediate FRET state before reaching the high FRET (Fig. 2B). Many (>100) single-molecule time traces are combined to generate a heatmap where one step and two steps are clearly distinguished (Fig. 2B).

Since, Rep-X releases TTAGGG strand in 5' to 3' direction, the one-step folding suggests that the folding starts upon the release of the 3' end or that the folding starts from the 3' end, which is consistent with a previous finding (Mitra & Ha, 2019). In contrast, the two-step folding may arise from the 5' end or stepwise folding from 3' end or folding intermediates such as G-hairpin and G-triplex may form before the final structure emerges. Rep-X may take pauses during unwinding, yet it is unlikely based on the high speed of translocation (Arslan et al., 2015).

4.3 Telomeric G4 folding at low ATP

Helicase mediated unwinding and translocation both depend on ATP hydrolysis (Paul et al., 2020). Therefore, helicase activity can be tuned by ATP concentration. Here, we used 50 μM ATP to slow down the helicase activity to match the speed of the telomerase extension of ~1 nt/s (Hwang, Opresko, & Myong, 2014). The low ATP yielded a gradual FRET increase followed by an abrupt increase to a high FRET corresponding to the G4 folded state (Fig. 3). The gradual FRET increase likely indicates slow translocation of Rep-X followed by a rapid release and folding of the G4 DNA (Fig. 3). The completely unwound state matched that of the vf-G4. Thus, even the gradual release of the overhang leads to the instantaneous G4

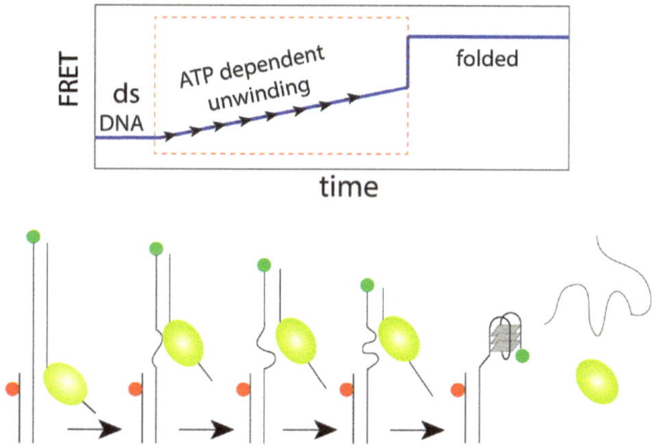

Fig. 3 Rep-X induced G4 vectorial folding at low ATP concentration. Diagram of a real-time smFRET trace of Rep-X induced gradual unwinding of dsDNA that leads to folding telomere G4 overhang. Schematic model of stepwise unwinding and folding process.

folding, suggesting a 3' end directed folding (Fig. 1B). This suggests that the folding mechanism is inherent to the vectorial process regardless of the rate of overhang synthesis or release.

5. Accessibility test

5.1 C2 and POT1 binding

The above result suggests that the G4 conformations induced by vf vs pf are different which raises a question about their thermal stability. Complementary C-strand ([CCCTAA]2 i.e. C2) and POT1 can be applied to both constructs to test if the different conformations yield different level of accessibility (Fig. 4A) (Lee et al., 2020). POT1 is a member of the shelterin protein complex which binds exclusively to the single-stranded telomere overhangs with high affinity and sequence specificity (Paul, Liou, et al., 2021). Real-time measurement of C2 and POT1 addition to vf and pf-G4 constructs are carried out as described in Section 3.7 (Fig. 2A). Rep-X with ATP and either C2 or POT1 is flowed in to the immobilized vf constructs. A resulting high FRET means that the folded structure is inaccessible while a low FRET indicates that the C2 anneals to or POT1 binds to the vf-G4. Interestingly, the vf-G4 show ~75% accessibility for C2 and ~90% accessibility for POT1 whereas pf-G4 leads to less than 10% accessibility for both C2 and POT1 (Fig. 4B–E). The low accessibility

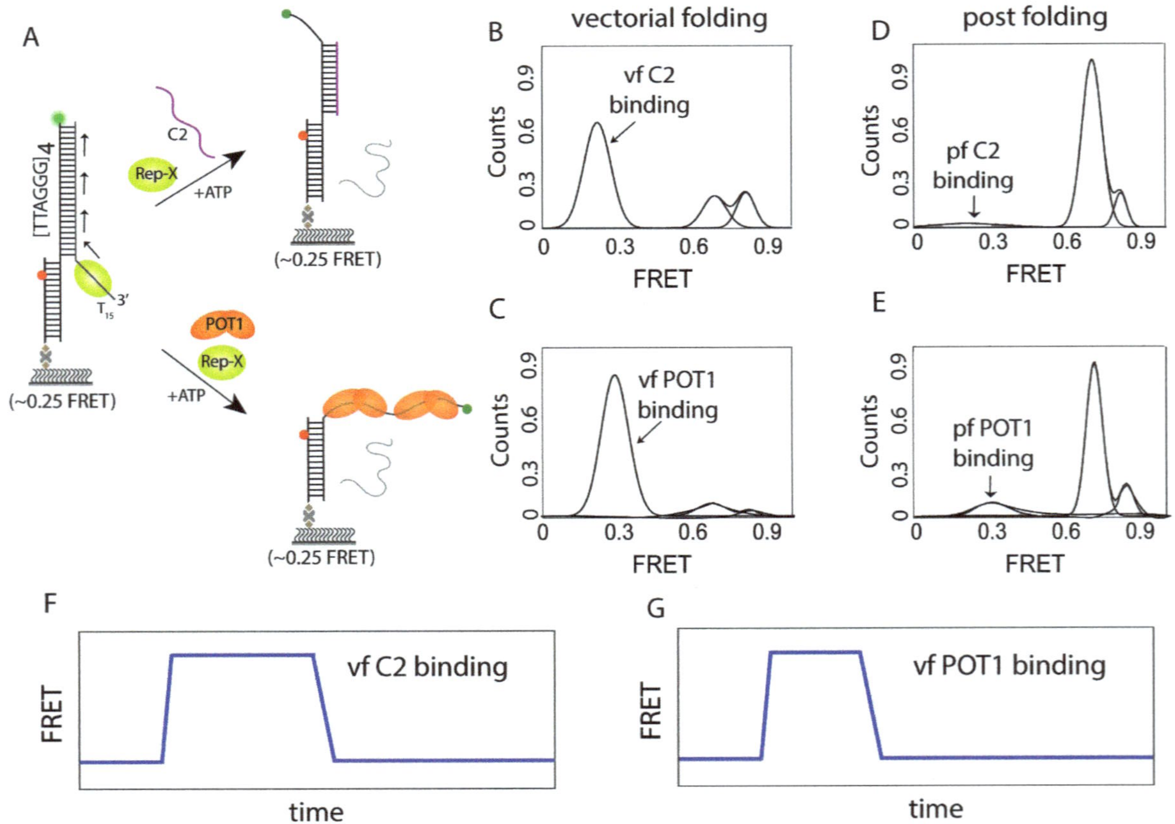

Fig. 4 Accessibility test by C2 and POT1 to vf G4. (A) Schematic model of vf and simultaneous annealing of C2 strand, (CCCTAA)2 or POT1 binding. (B, C) Model histogram of C2 and POT1 binding to vf G4 and pf G4 (D, E). (F, G) Expected real-time smFRET traces of C2 and POT1 binding during vf of G4.

of C2 and POT1 to pf G4 is consistent with our previous result (Lee et al., 2020). The smFRET trace shows a clear transition to high FRET due to the vf, followed by a shift to low FRET due to the C2 and POT1 binding (Fig. 4F and G).

6. Conclusion

Telomeres shorten every cell division and telomere length is maintained by upregulating the telomerase activity in certain cells. The Rep-X mediated release and subsequent folding of the G4 strand mimics the telomerase-based overhang synthesis where G4 folds vectorially. Using our single-molecule assay, we show that the G4 folded conformation achieved by the vectorial folding is significantly different from the conventional post folding where the entire length is allowed to fold at once. We can control the Rep-X unwinding speed by lowering ATP concentration to 50 μM ATP to match the speed of telomerase extension. The same conformation of vf-G4 achieved in different ATP concentrations reveal an inherent folding mechanism stimulated by the vectorial folding process (Paul et al., 2020). Importantly, such vf of G4 yields a higher accessibility to C2 and POT1 indicating a less stable structure. The vectorial folding platform provides a unique stage which offers an important implication in telomere maintenance.

Acknowledgments

We thank Dr. Momčilo Gavrilov and Dr. Patricia L. Opresko for providing the Rep-X and POT1 protein respectively. We thank Dr. Taekjip Ha for helpful comments and scientific discussion. This research was funded by National Institute of Health General Medicine [1R01GM115631-01A1 to T.P. and S.M.]; National Science Foundation Physics Frontiers Center Program [0822613] through the Center for the Physics of Living Cells; [1R01-CA 207342-01A1 and 1RF1 NS113636-01 to S.M.].

References

Arslan, S., Khafizov, R., Thomas, C. D., Chemla, Y. R., & Ha, T. (2015). Engineering of a superhelicase through conformational control. *Science*, *348*(6232), 344–347.

Baumann, P., & Cech, T. R. (2001). Pot1, the putative telomere end-binding protein in fission yeast and humans. *Science*, *292*(5519), 1171–1175.

Bera, S. C., Paul, T., Iyengar, A. S., & Mishra, P. P. (2018). Direct observation of the external force mediated conformational dynamics of an IHF bound Holliday junction. *Faraday Discussions*, *207*, 251–265.

Blasco, M. A. (2005). Telomeres and human disease: Ageing, cancer and beyond. *Nature Reviews Genetics*, *6*(8), 611–622.

Cimino-Reale, G., Pascale, E., Battiloro, E., Starace, G., Verna, R., & D'Ambrosio, E. (2001). The length of telomeric G-rich strand 3′-overhang measured by oligonucleotide ligation assay. *Nucleic Acids Research, 29*(7), e35.

Collins, K., & Gandhi, L. (1998). The reverse transcriptase component of the tetrahymena telomerase ribonucleoprotein complex. *Proceedings of the National Academy of Sciences of the United States of America, 95*(15), 8485–8490.

Gu, P., Jia, S., Takasugi, T., Tesmer, V. M., Nandakumar, J., Chen, Y., et al. (2021). Distinct functions of POT1 proteins contribute to the regulation of telomerase recruitment to telomeres. *Nature Communications, 12*(1), 1–16.

Hwang, H., Opresko, P., & Myong, S. (2014). Single-molecule real-time detection of telomerase extension activity. *Science Reports, 4*(1), 1–7.

Jansson, L. I., Hentschel, J., Parks, J. W., Chang, T. R., Lu, C., Baral, R., et al. (2019). Telomere DNA G-quadruplex folding within actively extending human telomerase. *Proceedings of the National Academy of Sciences of the United States of America, 116*(19), 9350–9359.

Joo, C., Balci, H., Ishitsuka, Y., Buranachai, C., & Ha, T. (2008). Advances in single-molecule fluorescence methods for molecular biology. *Annual Review of Biochemistry, 77*, 51–76.

Lee, H.-T., Sanford, S., Paul, T., Choe, J., Bose, A., Opresko, P. L., et al. (2020). Position-dependent effect of guanine base damage and mutations on telomeric G-quadruplex and telomerase extension. *Biochemistry, 59*(28), 2627–2639.

Masutomi, K., Evan, Y. Y., Khurts, S., Ben-Porath, I., Currier, J. L., Metz, G. B., et al. (2003). Telomerase maintains telomere structure in normal human cells. *Cell, 114*(2), 241–253.

Mitra, J., & Ha, T. (2019). Streamlining effects of extra telomeric repeat on telomeric DNA folding revealed by fluorescence-force spectroscopy. *Nucleic Acids Research, 47*(21), 11044–11056.

Neidle, S., & Parkinson, G. N. (2003). The structure of telomeric DNA. *Current Opinion in Structural Biology, 13*(3), 275–283.

Palm, W., & de Lange, T. (2008). How shelterin protects mammalian telomeres. *Annual Review of Genetics, 42*, 301–334.

Paul, T., Bera, S. C., Agnihotri, N., & Mishra, P. P. (2016). Single-molecule FRET studies of the hybridization mechanism during noncovalent adsorption and desorption of DNA on graphene oxide. *The Journal of Physical Chemistry B, 120*(45), 11628–11636.

Paul, T., Bera, S. C., & Mishra, P. P. (2017). Direct observation of breathing dynamics at the mismatch induced DNA bubble with nanometre accuracy: A smFRET study. *Nanoscale, 9*(18), 5835–5842.

Paul, T., Ha, T., & Myong, S. (2021). Regeneration of PEG slide for multiple rounds of single-molecule measurements. *Biophysical Journal, 120*(9), 1788–1799.

Paul, T., Liou, W., Cai, X., Opresko, P. L., & Myong, S. (2021). TRF2 promotes dynamic and stepwise looping of POT1 bound telomeric overhang. *Nucleic Acids Research, 49*(21), 12377–12393.

Paul, T., & Mishra, P. P. (2016). Direct observation of spatial configuration and structural stability of locked Y-shaped DNA structure. *RSC Advances, 6*(105), 103270–103274.

Paul, T., Mohapatra, P., & Mishra, P. P. (2021). Direct observation of adsorption and desorption of ds-DNA on graphene oxide and graphene oxide-gold nanoparticle hybrid material: A kineto-mechanistic investigation. *Applied Surface Science, 151696.*

Paul, T., & Myong, S. (2022). Protocol for generation and regeneration of PEG-passivated slides for single-molecule measurements. *STAR Protocols, 3*(1), 101152.

Paul, T., Voter, A. F., Cueny, R. R., Gavrilov, M., Ha, T., Keck, J. L., et al. (2020). E. coli Rep helicase and RecA recombinase unwind G4 DNA and are important for resistance to G4-stabilizing ligands. *Nucleic Acids Research, 48*, 6640–6653.

Roy, R., Hohng, S., & Ha, T. (2008). A practical guide to single-molecule FRET. *Nature Methods*, *5*(6), 507–516.

Srinivas, N., Rachakonda, S., & Kumar, R. (2020). Telomeres and telomere length: A general overview. *Cancers*, *12*(3), 558.

Tang, J., Kan, Z.-Y., Yao, Y., Wang, Q., Hao, Y.-H., & Tan, Z. (2008). G-quadruplex preferentially forms at the very 3′ end of vertebrate telomeric DNA. *Nucleic Acids Research*, *36*(4), 1200–1208.

Wang, Q., Liu, J.-Q., Chen, Z., Zheng, K.-W., Chen, C.-Y., Hao, Y.-H., et al. (2011). G-quadruplex formation at the 3′ end of telomere DNA inhibits its extension by telomerase, polymerase and unwinding by helicase. *Nucleic Acids Research*, *39*(14), 6229–6237.

Yang, O., & Ha, T. (2018). Single-molecule studies of ssDNA-binding proteins exchange. *Methods Enzymology*, *600*, 463–477.

Zhao, Y., Sfeir, A. J., Zou, Y., Buseman, C. M., Chow, T. T., Shay, J. W., et al. (2009). Telomere extension occurs at most chromosome ends and is uncoupled from fill-in in human cancer cells. *Cell*, *138*(3), 463–475.

CHAPTER FOURTEEN

Production of long linear DNA substrates with site-specific chemical lesions for single-molecule replisome studies

Gurleen Kaur[a,b], Lisanne M. Spenkelink[a,b], Jacob S. Lewis[a,b,†], Slobodan Jergic[a,b], Nicholas E. Dixon[a,b], and Antoine M. van Oijen[a,b,*]
[a]Molecular Horizons and School of Chemistry and Molecular Bioscience, University of Wollongong, Wollongong, NSW, Australia
[b]Illawarra Health and Medical Research Institute, Wollongong, NSW, Australia
*Corresponding author: e-mail address: vanoijen@uow.edu.au

Contents

† Present address: The Francis Crick Institute, 1 Midland Road, London, United Kingdom.

Methods in Enzymology, Volume 672
ISSN 0076-6879
https://doi.org/10.1016/bs.mie.2022.02.022

Abstract

Single-molecule imaging studies using long linear DNA substrates have revealed unanticipated insights into the dynamics of multi-protein systems. The use of long DNA substrates allows for the study of protein–DNA interactions with observation of the movement and behavior of proteins over distances accessible by fluorescence microscopy. Generalized methods can be exploited to generate and optimize a variety of linear DNA substrates with plasmid DNA as a simple starting point using standard biochemical techniques. Here, we present protocols to produce high-quality plasmid-based 36-kb linear DNA substrates that support DNA replication by the *Escherichia coli* replisome and that contain chemical lesions at well-defined positions. These substrates can be used to visualize replisome–lesion encounters at the single-molecule level, providing mechanistic details of replisome stalling and dynamics occurring during replication rescue and restart.

1. Introduction

Single-molecule characterization has emerged as an important tool to study protein–DNA interactions. In particular, fluorescence imaging has allowed for the study of dynamic interactions on DNA (Kaur, Lewis, & van Oijen, 2019). An important factor in the establishment of single-molecule assays is the development of generalized methods to produce high-quality linear DNA substrates that are compatible with the resolution of fluorescence microscopy and can be easily customized to accommodate various experimental variables (Kaur & Spenkelink, 2021). Plasmid DNA is a simple starting point to create sufficiently long linear DNA with relative ease using standard biochemical techniques. Similar to phage λ DNA, plasmid sequences can be constructed to allow for the attachment of DNA handles needed to tether the substrate to a surface or beads.

A generalized method has been outlined previously to create a linear DNA substrate for single-molecule studies from a single plasmid (Mueller, Spenkelink, van Oijen, & Lewis, 2020). The method was used to produce forked linear DNA substrates of homogeneous structure and length with the insertion of an internal Cy5 fluorophore. Similarly, the method can be used to specifically incorporate modifications including UV lesions and tertiary structures like G-quadruplexes. Such a substrate can be used to visualize helicase unwinding activity, DNA synthesis, and DNA replication–roadblock encounters, in a controlled system. In this case, the forked linear

DNA acted as an efficient substrate for *in vitro* DNA replication using reconstituted T7 replisome proteins. The replication rates measured (193 ± 44 bp/s) were similar to previous studies (Duderstadt et al., 2016; Geertsema, Duderstadt, & van Oijen, 2015). On average complete replication of a single double-tethered substrate required about 1.5 min, sufficient time to capture events and additional replication dynamics. However, the *E. coli* (*E. coli*) replisome is a highly efficient and much faster model replication system. An average event on this substrate is estimated to take only 30–40 s (~18,300 bp substrate at an average rate of replication of 500 bp/s) (Lewis et al., 2017; Tanner et al., 2009). Such a short time poses significant experimental challenges. To capture *E. coli* replication dynamics more efficiently, a longer substrate would be appropriate. While small (<20 kb) plasmid DNA-based substrates are easily constructed, the development of longer substrates remains challenging. However, this generalized method can be exploited to optimize a variety of substrate parameters including length, site-specific sequences, tertiary structures, or chemical modifications.

Here we report a method to generate high-quality plasmid-based long linear DNA substrates. These long substrates support the visualization of fast replisomes. We validate these substrates by visualizing DNA replication by the *E. coli* replisome at the single-molecule level. Finally, we illustrate that this method can be used to visualize replisome–lesion encounters at the single-molecule level.

2. Materials and methods

2.1 List of reagents and buffer components

See Table 1 for list of reagents and Table 2 for buffers components.

2.2 Custom oligonucleotides

Oligonucleotides (Table 3) were purchased from Integrated DNA Technologies (USA). Oligonucleotides were dissolved in sterilized TE buffer to a stock concentration of 100 μM and stored at −80 °C. The site-specific cyclobutane pyrimidine dimer (CPD) lesion oligonucleotide was a generous gift from Professor Michael M. Cox (Wisconsin, USA), originally purchased from Trilink BioTechnologies (USA).

2.3 Construction of pUBER plasmid

The plasmid pUBER (Addgene plasmid 80716) was generated as previously described (Mueller et al., 2020). Briefly, two synthetic double-stranded gene

Table 1 List of reagents.

Supplier	Chemical and reagents
Ajax Finechem	Acetic acid
	Acetone
	Ethanol
Astral Scientific	Dithiothreitol (DTT)
Alfa Aesar	3-aminopropyltriethoxysilane (APTES)
Bioline	100 mM 2′-deoxyadenosine-5′-triphosphate (dATP),
	100 mM 2′-deoxycytidine-5′-triphosphate (dCTP),
	100 mM 2′-deoxyguanosine-5′-triphosphate (dGTP),
	100 mM 2′-deoxythymidine-5′-triphosphate (dTTP),
	Agarose (molecular grade)
	100 mM adenosine 5′-triphosphate (ATP),
	100 mM cytidine 5′-triphosphate (CTP),
	100 mM guanosine 5′-triphosphate (GTP),
	100 mM uridine 5′-triphosphate (UTP)
Laysan Bio	Biotin PEG
	MPEG (MW 5000)
Life Technologies	10,000 x SYBR safe
	10,000 x SYBR orange
New England Biolabs (NEB)	6 × DNA Gel Loading Dye
	β-Agarase I (M0392L)
	*Bst*XI (R0113L)
	Low molecular weight DNA ladder (N3233S)
	T4 DNA ligase (M0202L)
Sigma Aldrich	2-amino-2-hydroxymethyl-propane-1,3-diol (Trizma/Tris)
	Adenosine 5′-triphosphate (ATP)
	Ethylenediaminetetraacetic acid disodium salt dihydrate (EDTA)
	Bovine serum albumin (BSA)

Table 1 List of reagents.—cont'd

Supplier	Chemical and reagents
	Hydrochloric acid (HCl)
	Low gelling temperature agarose
	Magnesium acetate
	PEG6000
	Potassium chloride (KCl)
	Potassium glutamate
	Potassium hydroxide
	Sodium acetate
	Sodium chloride (NaCl)
	Tween-20
Sylgard	Polydimethylsiloxane (PDMS)
Thermo Fisher Scientific	GeneRuler DNA ladder 100–10,000 bp
	λ/*Hind*III Marker
	Neutravidin (cat 31000)

Table 2 List of buffer components.

Buffer	Components
Blocking buffer	50 mM Tris–HCl, pH 7.6, 50 mM KCl, 2% Tween-20
CutSmart buffer	50 mM potassium acetate, 20 mM Tris-acetate, 10 mM magnesium acetate, 100 μg/mL BSA, pH 7.9 @ 25 °C
Elution buffer	10 mM Tris–HCl, pH 7.6, 1 mM EDTA, 300 mM NaCl
NEBuffer 3.1	100 mM NaCl, 50 mM Tris–HCl, 10 mM $MgCl_2$, 100 μg/mL BSA, pH 7.9 @ 25 °C
Replication buffer	25 mM Tris–HCl, pH 7.6, 10 mM magnesium acetate, 50 mM potassium glutamate, 40 μg/mL BSA, 0.1 mM EDTA, 5 mM DTT, 0.0025% (v/v) Tween20
Tris acetate EDTA (TAE) buffer	40 mM Tris, 20 mM acetic acid, 1 mM EDTA, pH 8.0
Tris EDTA (TE) buffer	10 mM Tris–HCl pH 7.6, 1 mM EDTA pH 8.0

Table 3 Custom oligonucleotides.

Oligonucleotide	Sequence
FORK	
E61Lg	5′-TTTTTTTTTTTTTTTTTTTTTTTTTTTTTTCCAATTCTCACT TCCTACCACATCGGTCGAT-3′
E61Lg-Cy3[a]	5′-/5Cy3/TTTTTTTTTTTTTTTTTTTTTTTTTTTTTTCCA ATTCTCACTTCCTACCACATCGGTCGAT-3′
E81Ld	5′-/5Phos/ACCGATGTGGTAGGAAGTGAGAATTGGCCGAGGAAAGAAATGT TGGTGAGGGTTGGGAAGTGGAAGGATGGGCTCGAGAGG-3′
EC57Primer	5′-/5Biosg/TTTTTTTTTTTTTTTTTTTTCCTCTCGAGCCCATCCTTCC ACTTCCCAACCCTCACC-3′
EC57Primer-Cy5[a]	5′-/5Biosg/TTTTTTTTT/iCy5/TTTTTTTTTTCCTCTCGAGCCCATCCTTC CACTTCCCAACCCTCACC-3′
LINKER	
LgL	5′-/5Phos/AGTCGCAGCTATAGGTGGCATTTCAGGCAG-3′
LdL	5′-/5Phos/CTGAAATGCCACCTATAGCTGCGACTCATG-3′
CPD LINKER	
LinkerLg	5′-/5Phos/GCGTAATGGAAGAATTCGAACCTACCCGCGAT-3′
LinkerCPDLd LinkerLd-Cy5[a]	5′-/5Phos/CGGGTAGGTTCGAA**TT**CTTCCATTACGCCTGC-3′ 5′-/5Phos/CGGGTAGGTTCGAAT/iCy5/TCTTCCATTACGCCTGC-3′
CAPPING	
BlockLg	5′-/5Phos/AGTCGCAGCTATAGGTGGCATTTCAG-3′
BlockLd	5′-/5Biosg/CTGAAATGCCACCTATAGCTGCGACTCATG-3′

[a]Cy3 or Cy5 dye is covalently attached to the ribose at 5′ terminus or internally of the oligonucleotide (Moreira, You, & Owczarzy, 2015).

blocks were inserted into pSuperCos1 plasmid (van Loenhout, De Grunt, & Dekker, 2012). The key gene block insertion used for this protocol was inserted between *Mlu*I and *Bst*XI restriction sites to reduce the distance between *Bst*XI sites to allow for simple purification of the digested linear plasmid (DNA). Inserted sequences were verified by nucleotide sequencing. Plasmid DNA (18,409 bp; 2 mg) was purified by Aldevron (USA). Although plasmid preparation is possible to do in-house using various purification methods, it is important to produce high-quality plasmid DNA as any nicks or damage on the DNA ultimately translates to the single-molecule substrate preparation.

2.4 Construction of an unmodified forked 18-kb linear substrate

To construct an unmodified forked linear DNA substrate (Fig. 1A), typically 50 μg of pUBER plasmid DNA (4.6 mg/mL) was treated with 1 unit of *Bst*XI per μg DNA for 16 h at 37 °C in 1 × NEBuffer 3.1 (New England Biolabs) that cuts the plasmid twice. The resulting long 18.3 kb linear fragment was resolved by gel filtration chromatography on a Sepharose 4B column (0.7 × 25 cm; Sigma-Aldrich) equilibrated with elution buffer (Note 1) flown at 0.2 mL/min. The fractions containing linear 18-kb fragment were pooled and dialyzed in sterile TE buffer (Note 1) overnight and then concentrated in a Savant SpeedVac (DNA 130) at room temperature to increase the DNA concentration to ~100 μg/mL. Next, the fork and capping oligonucleotides (Table 3) were annealed in the following molar ratios respectively: 1:6:60 (E61Lg:E81Ld:EC57Primer or labeled oligonucleotides where specified) and 1:1 (BlockLd:BlockLg). Annealed oligonucleotides were separated on a 3% agarose gel run in 1 × TAE buffer (Note 1) at 75 V and visualized by staining the gel with SYBR safe. Confirming the annealing, both the synthetic replication fork and capping oligonucleotides were ligated onto the linear 18-kb fragment using 1 unit of T4 DNA ligase per μg DNA in 1 × CutSmart buffer supplemented with 1 mM ATP and 10 mM dithiothreitol (DTT) (Note 2). Excess oligonucleotides and T4 ligase were removed by purification on a Sepharose 4B column in elution buffer as before. The resolution of the column efficiently separates 18-kb DNA from the smaller oligonucleotides, however separation of single-tethered and double-tethered 18-kb substrates is not possible. High efficiency in the annealing and ligation steps ensures only a small fraction of the final single-molecule DNA substrate is single-tethered. DNA concentrations of each collected fraction were measured using a NanoDrop 2000 spectrophotometer.

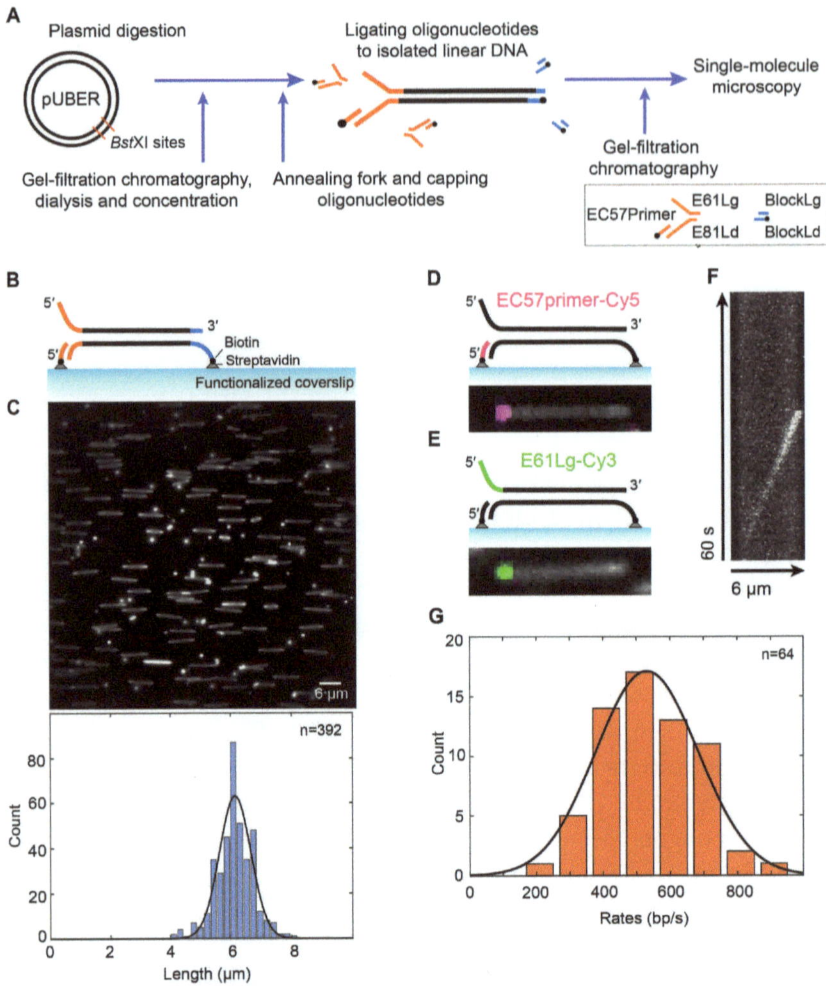

Fig. 1 Construction and characterization of unmodified 18-kb forked linear substrate. (A) Schematic outline of the synthesis of the unmodified 18-kb forked linear substrate from the pUBER plasmid. (B) Side view schematic of 18-kb DNA substrate on a functionalized coverslip double-tethered *via* streptavidin–biotin linkages. (C) Snapshot of a single-molecule TIRF microscopy field of view with double-tethered 18-kb DNA substrate stained with SxO. Histogram of 18-kb DNA-substrate lengths. The black line represents a Gaussian fit giving an average DNA length of 6.12 ± 0.04 μm (mean ± S.E.M.). (D) (top) Schematic of 18-kb substrate containing Cy5-labeled primer (EC57Primer-Cy5). (bottom) Snapshot of a single double-tethered 18-kb DNA substrate generated with a labeled primer oligonucleotide (magenta spot). (E) (top) Schematic of 18-kb substrate containing Cy3-labeled lagging-strand oligonucleotide (E61Lg-Cy3). (bottom) Snapshot of a single double-tethered 18-kb DNA substrate generated with a labeled lagging-strand oligonucleotide (green spot). (F) Representative kymograph of an *in vitro E. coli* DNA replication event. (G) Single-molecule histogram of *E. coli* replication rates. The black line represents a Gaussian fit with an average rate of 540 ± 40 bp/s (mean ± S.E.M.).

2.5 Construction of modified forked 36-kb linear substrate containing a site-specific CPD lesion

Modifications were made to the existing unmodified substrate protocol to incorporate a site-specific internal CPD lesion (Fig. 2A). The substrate was constructed in two separate blocks with the total length of the substrate being 36.8 kb. Block 1 was constructed by annealing the fork and linker oligonucleotides (Table 3) in the following molar ratios, respectively: 1:6:60 (E61Lg:E81Ld:EC57Primer) and 1:1 (LdL:LgL). Block 2 was constructed by annealing of the CPD linker (LinkerLg and LinkerCPDLd or labeled oligonucleotides where specified) and capping oligonucleotides (Table 3), each mixed in a 1:1 M ratio. The respective sets of annealed oligonucleotides were then ligated to the purified 18-kb linear DNA using 1 unit of T4 DNA ligase per μg DNA in 1 × CutSmart buffer supplemented with 1 mM ATP and 10 mM DTT. Excess oligonucleotides and T4 ligase were removed from each block using gel filtration chromatography on a Sepharose 4B column in elution buffer, as before. Fractions pooled for each block were dialyzed separately in sterile TE buffer overnight. Each block was then concentrated in the SpeedVac at room temperature to increase the DNA concentration to ~100 μg/mL. Next, block 1 and block 2 were ligated at a 1:1 M ratio using 1 unit of T4 DNA ligase per μg DNA in 1 × CutSmart buffer supplemented with 1 mM ATP and 10 mM DTT and 15% PEG6000 (Note 3). The ligation resulted in 80% of 36-kb product being formed (Fig. S1 in the online version at https://doi.org/10.1016/bs.mie.2022.02.022) with unligated block 1 and block 2 in solution. The ligation product was dialyzed in sterile TE buffer overnight at 4 °C to remove PEG and purified separation on a 0.5% low gelling temperature agarose gel until sufficient separation between the 36-kb and 18-kb bands was achieved. The 36-kb band was excised from the gel and melted at 50 °C in TE buffer supplemented with 150 mM sodium acetate (Note 4) until completely dissolved. The solution was cooled to 42 °C and the molten agarose was digested with 1 unit of β-agarase I at 42 °C for 1 h. The digested agarose can be removed by separation on a Sepharose 4B column in elution buffer or dialyzed in sterile TE buffer overnight.

2.6 *In vitro* single-molecule microscopy

Single-molecule microscopy experiments were carried out on an Eclipse Ti-E inverted microscope (Nikon) fitted with a CFI Apo TIRF 100 × oil immersion objective (NA 1.49, Nikon), as previously described (Geertsema et al., 2015). The temperature was maintained at 33 °C by using

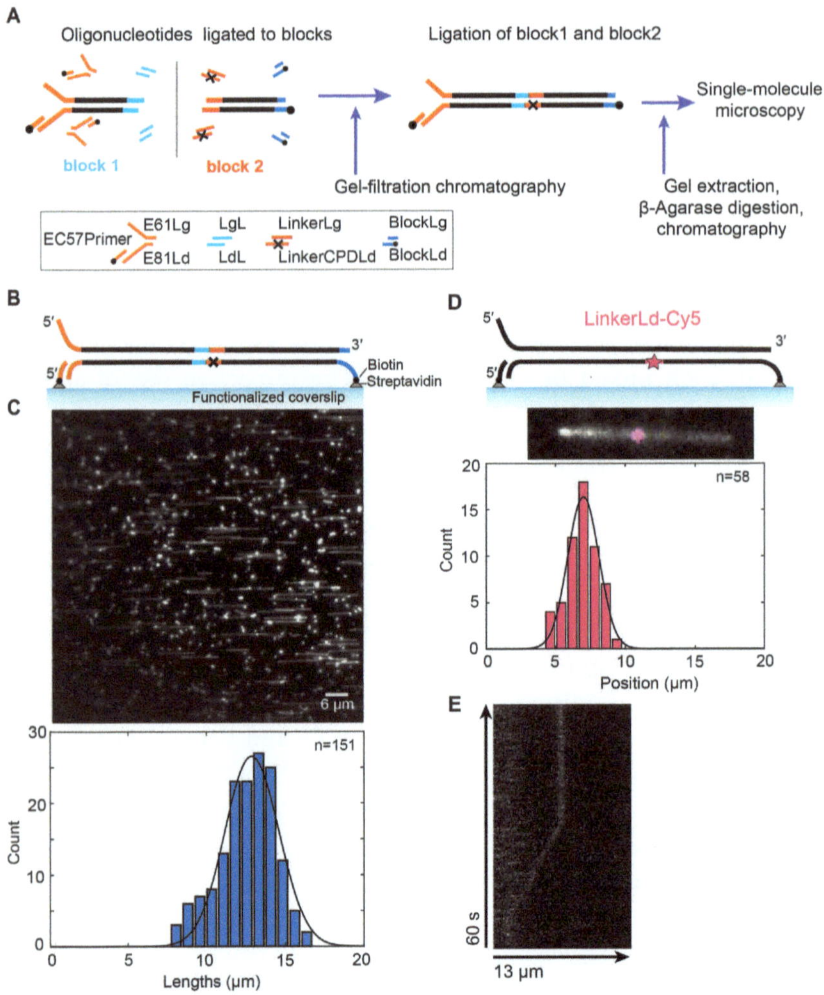

Fig. 2 Construction and characterization of modified 36-kb forked linear substrate. (A) Schematic outline of the synthesis of the modified 36-kb forked linear substrate containing a CPD lesion. (B) Side view schematic of 36-kb DNA substrate containing a CPD lesion (black cross) on a functionalized coverslip double-tethered *via* streptavidin- biotin linkages. (C) (top) Snapshot of a single-molecule TIRF field of view with double-tethered 36-kb DNA substrate stained with SYTOX Orange. (bottom) Histogram of 36-kb DNA substrate lengths. The black line represents a Gaussian fit giving an average DNA length of 12.6 ± 0.4 μm (mean ± S.E.M.). (D) (top) Schematic of 36-kb substrate containing an internal Cy5 modification (magenta star). (middle) Snapshot of a single double-tethered forked linear DNA molecule generated with an internal Cy5 modification (magenta spot). (bottom) Histogram of Cy5 position on the 36-kb substrate. The black line represents a Gaussian fit giving an average Cy5 position of 6.9 ± 0.9 μm (mean ± S.E.M.). (E) Kymograph of DNA replication stalling on modified 36-kb DNA substrate containing a CPD lesion.

an electrically heated chamber (Okolab). Images were captured with a Hamamatsu C9100-13 EM-CCD camera. Fluorescently stained DNA molecules were imaged with a 514-nm laser (Coherent, Sapphire 514–150 CW) at $0.23\,W/cm^2$. Cy5 and Cy3 labels were imaged with a 647-nm laser (Coherent, Obis 647–100 CW) at $0.14\,W/cm^2$ and 488 nm (Coherent, Sapphire 488–200 CW) at $3.17\,W/cm^2$, respectively.

2.7 Single-molecule *E. coli* DNA replication assay

Replication reactions were carried out in microfluidic flow cells constructed from a PDMS flow chamber placed on top of a PEG-biotin-functionalized microscope coverslip as previously described (Geertsema et al., 2015; Lewis et al., 2017; Tanner et al., 2008, 2009). Once assembled, all surfaces of the flow-cell including connecting tubing were blocked against non-specific binding by introduction of 100 μL of blocking buffer (Note 1). The linear DNA substrates were surface tethered inside the flow chamber by flowing for 20 min at 17 μL/min in the presence of 200 μM chloroquine (Note 5). The DNA was visualized by flowing in 150 nM SYTOX Orange (SxO) in replication buffer (Note 1). *E. coli* DNA replication proteins were produced as described previously: β_2 clamp (Oakley et al., 2003), SSB_4 (Mason et al., 2013), $DnaB_6(DnaC)_6$ (Spinks et al., 2021), DnaG (Stamford, Lilley, & Dixon, 1992), the $\tau_3\delta\delta'\psi\chi$ clamp loader (Tanner et al., 2008), and $\alpha\varepsilon\theta$ core (Jergic et al., 2013; Lewis et al., 2017). *E. coli* replication reactions were carried out as previously described (Lewis et al., 2017). Briefly, 30 nM $DnaB_6(DnaC)_6$, 1 mM ATP, 10 mM DTT and 150 nM SxO in replication buffer was loaded into the flow cell and incubated for 10 min. Replication was initiated by flowing in replication buffer containing 30 nM $\alpha\varepsilon\theta$ core, 10 nM $\tau_3\delta\delta'\psi\chi$ clamp loader, 40 nM β_2 clamp, 100 nM DnaG, 100 nM SSB_4 and 150 nM SxO.

2.8 Notes

1. All buffers made were filtered through MF-Millipore membrane filter, 0.22 μm pore size.
2. Manufacturer recommendations for T4 DNA ligase is to use with T4 DNA ligase buffer. To avoid precipitation and degradation of ATP and DTT in this buffer, 1 × CutSmart buffer supplemented with fresh DTT and ATP is used instead.
3. Polyethylene glycol (PEG) is included in the ligation reaction to improve the reaction rate and the overall yield. PEG acts as a volume

excluder which effectively increases the concentration of both the ligase and the DNA substrates, improving the overall ligation efficiency.

4. Sodium acetate increases the melting temperature of the oligonucleotides preventing any from melting off.
5. Chloroquine is added to stretch the DNA substrate to a length that is similar to the crystallographic length of double-stranded DNA (Yardimci, Loveland, Habuchi, van Oijen, & Walter, 2010).

2.9 Analysis of substrate length and DNA replication

All data were analyzed using ImageJ/Fiji (1.51j) and MATLAB 2016b, with in-house built plugins. For length measurements, 1-min movies were averaged and a 1D line profile was manually drawn across the molecule of interest; the start- and end-points of double-stranded DNA were then located by taking the first-order differential of the 1D line profile. Lengths were derived by calculating the distance between the minimum and maximum peaks. Rates of replication of single replisomes were obtained as previously described (Lewis et al., 2020). All results were derived from at least three technical replicates.

3. Validation of methods and controls

3.1 Unmodified forked linear substrate

After purification of the substrate, its purity and homogeneity were confirmed at the single-molecule level. DNA substrates were attached to the surface of a microfluidic flow cell through a biotin–streptavidin linkage (Fig. 1B). DNA molecules were visualized through staining with the fluorescent double-strand DNA-specific dye SxO and imaged in the absence of buffer flow (Fig. 1C). Under these conditions, 85.9% $\pm$ 0.4% (mean $\pm$ S.E.M.) of DNA substrates were double tethered to the surface, presenting with a low fraction of single-tethered substrates. The double-tethered 18-kb substrates were measured with a length of 6.12 $\pm$ 0.04 μm (mean $\pm$ S.E.M.); (Fig. 1C). This measurement agrees with the theoretical crystallographic length (6.24 μm) of the forked linear substrate (Mueller et al., 2020).

To validate the full assembly of the synthetic fork, linear substrates were constructed with fluorescently labeled oligonucleotides (Table 3). The presence of the biotinylated primer is expected in all double tethered substrates, but this can also be demonstrated by generating forked linear DNA substrates containing EC57Primer-Cy5. The fluorescent primer was visualized in 90.3 ± 0.6% of DNA substrates (Fig. 1D). The presence of the primer is only possible with the presence of E81Ld. Therefore, presence of the primer automatically confirms the presence of E81Ld. However, the E61Lg oligonucleotide can possibly be lost in the case of a failed ligation. To confirm the presence of E61Lg and thereby the quality of our protocol, linear DNA substrates were also generated containing E61Lg-Cy3, with 75.5% ± 0.9% of substrates visualized to have a fluorescent E61Lg oligonucleotide (Fig. 1E).

Finally, the full *E. coli* replisome was reconstituted to assess the efficiency of DNA replication on the linear DNA substrates. In the absence of flow, surface-tethered DNA molecules stained with SxO were imaged and the replication events were monitored through real-time single-molecule TIRF microscopy. Replication events were identified through the growth of the lagging-strand product, presented as a growing bright fluorescent spot that moves unidirectional along the length of DNA. Concurrently, the brightness of the leading-strand product remained unchanged indicating its persistent double-stranded nature (Fig. 1F). Under these conditions, the measured *E. coli* replication rates (540 ± 40 bp/s; mean ± S.E.M.) are similar to those measured in previous studies (Fig. 1G) (Lewis et al., 2017; Spenkelink et al., 2019; Spinks et al., 2021; Tanner et al., 2009), with 80.4% ± 0.8% of replisomes replicating the entire length of the DNA substrate. High processivity indicates that a large proportion of DNA substrates bound to the surface are free from nicks, as nicks in the leading- or lagging-strand upon collision with an elongating replisome would result in cleavage of the linear DNA substrate or loss of the product DNA (moving fluorescent spot), respectively.

3.2 Modified forked linear substrate containing a site-specific CPD lesion

To demonstrate the ability to introduce internal modifications such as a CPD lesion using this method, an internally modified forked DNA substrate was generated using an oligonucleotide with an internal site-specific CPD lesion (Fig. 2B). After purification of the substrate, the purity and homogeneity were visualized at the single-molecule level (Fig. 2C). Under these conditions, 65.7% ± 0.9% (mean ± S.E.M.) of DNA substrates were double tethered to the surface with a length of 12.6 ± 0.4 μm (mean ± S.E.M.) (Fig. 2C). To illustrate the incorporation of an internally modified oligonucleotide, the modified 36-kb substrate was generated by including LinkerLd-Cy5 instead of the LinkerCPDLd oligonucleotide (see Table 3). Under single-molecule imaging conditions, the position of the Cy5 label was measured from the nearest tethered end to be $6.9 \pm 0.9\,\mu m$ (mean ± S.E.M.; Fig. 2D).

As a proof of concept, we utilized *E. coli* replication proteins to assess the efficiency of the CPD lesion as a block to replication on the leading strand. In the absence of flow, surface-tethered DNA molecules stained with SxO were imaged and stalled replication events were monitored through real time single-molecule TIRF microscopy. Replication was confirmed through the unidirectional movement of the lagging-strand product, which stalled at the position of the CPD lesion (Fig. 2E). Under these conditions, 100% of events ($N = 42$) stalled at the CPD lesion.

It is generally accepted that DNA replication can be restarted after collision with a lagging-strand lesion through the normal Okazaki-fragment priming cycle. The advantage of this protocol design is that it could be easily customized to include a CPD lesion on the lagging strand to directly visualize the lagging-strand lesion collision dynamics in the future.

4. Summary and discussion

Here, we have adapted a generalized method to design and produce long modified DNA substrates derived from a single plasmid. Our characterization of the unmodified substrate using various fluorescent and biochemical techniques ensures the production of a high-quality substrate. The substrate can be used to visualize helicase unwinding activity, priming activity, and DNA synthesis at the single-molecule level. Here we demonstrate observation of *E. coli* DNA replication. This method produces reliable yields for single-molecule DNA replication studies, where a single preparation produces an amount of material sufficient for >200 experiments.

We use a site-specific CPD roadblock to demonstrate the suitability of this approach to study protein dynamics at the replication fork at sites of replisome stalling. Single-molecule visualization of DNA replication on linear templates has been observed using replisomes from bacteriophages T7 and T4 and also *Saccharomyces cerevisiae*. However, single-molecule studies of these replisomes encountering lesions are lacking. While we demonstrate the use of the lesion-containing template in DNA replication by the *E. coli* replisome, the template can easily be adapted for studies on other DNA replication systems. Similarly, the template can be adapted to study the influence of lesions on other DNA-based processes. For example, the pUBER plasmid contains promoter sites that allow transcription initiation and elongation.

The CPD lesion, although physiologically very relevant, comes with a high cost of production, limiting easy access to this specific substrate design. However, many other commercially available lesions, such as abasic lesions can be incorporated at a much lower cost. Furthermore, there are other approaches to customize this method to introduce a diverse range of road-blocks, such as secondary structures and dCas9 binding sites (Whinn et al., 2019). Furthermore, the protocol can be adapted to create chromatinized substrates to study the dynamic equilibrium of nucleosomes in the context of DNA damage. This substrate presents a uniquely customizable design that may prove useful in understanding fine mechanistic details of replisome stalling and dynamics during replication rescue and restart, as well as observing movement and behavior of other protein–DNA interactions over distances.

Conflicts of interest

The authors declare no competing interests.

Author contributions

Conceptualization, G.K., L.M.S., A.M.v.O., J.S.L.; Formal analysis, G.K; Funding acquisition, N.E.D, A.M.v.O.; Investigation, G.K.; Methodology, G.K., S.J.; Resources, G.K., S.J., Software, G.K., L.M.S.; Supervision, N.E.D, A.M.v.O.; Validation, G.K; Visualization, G.K.; Writing – original draft, G.K., L.M.S., A.M.v.O., reviewing and editing, G.K., L.M.S., A.M.v.O., S.J., N.E.D, J.S.L.

Acknowledgments

The authors thank Prof Michael M. Cox for the CPD oligonucleotide, and members of the van Oijen laboratory for helpful suggestions.

Funding

This work was supported by Australian Research Council Grants DP180100858 and DP210100167, and National Institutes of Health Grant 5RM1GM130450 (to A.M.v.O. and N.E.D.), National Health and Medical Research Council Investigator Grant 2007778 (to L.M.S.) and Australian Government Research Training Program Scholarship (to G.K).

Software availability

All home-built plugins used in this study are freely available on the Github repository for Single-molecule/Image analysis tools (https://github.com/SingleMolecule) or available upon request.

References

Duderstadt, K. E., Geertsema, H. J., Stratmann, S. A., Punter, C. M., Kulczyk, A. W., Richardson, C. C., et al. (2016). Simultaneous real-time imaging of leading and lagging strand synthesis reveals the coordination dynamics of single replisomes. *Molecular Cell, 64*(6), 1035–1047.

Geertsema, H. J., Duderstadt, K. E., & van Oijen, A. M. (2015). Single-molecule observation of prokaryotic DNA replication. *Methods in Molecular Biology, 1300*, 219–238.

Jergic, S., Horan, N. P., Elshenawy, M. M., Mason, C. E., Urathamakul, T., Ozawa, K., et al. (2013). A direct proofreader–clamp interaction stabilizes the Pol III replicase in the polymerization mode. *EMBO Journal, 32*(9), 1322–1333.

Kaur, G., Lewis, J. S., & van Oijen, A. M. (2019). Shining a spotlight on DNA: Single-molecule methods to visualize DNA. *Molecules, 24*(3), 491.

Kaur, G., & Spenkelink, L. M. (2021). Biology on track: Single-molecule visualization of protein dynamics on linear DNA substrates. *Essays in Biochemistry, 65*(1), 5–16.

Lewis, J. S., Spenkelink, L. M., Jergic, S., Wood, E. A., Monachino, E., Horan, N. P., et al. (2017). Single-molecule visualization of fast polymerase turnover in the bacterial replisome. *eLife, 6*, e23932.

Lewis, J. S., Spenkelink, L. M., Schauer, G. D., Yurieva, O., Mueller, S. H., Natarajan, V., et al. (2020). Tunability of DNA polymerase stability during eukaryotic DNA replication. *Molecular Cell*, 77(1), 17–25.

Mason, C. E., Jergic, S., Lo, A. T. Y., Wang, Y., Dixon, N. E., & Beck, J. L. (2013). Escherichia coli single-stranded DNA-binding protein: NanoESI-MS studies of salt-modulated subunit exchange and DNA binding transactions. *Journal of the American Society for Mass Spectrometry, 24*(2), 274–285.

Moreira, B. G., You, Y., & Owczarzy, R. (2015). Cy3 and Cy5 dyes attached to oligonucleotide terminus stabilize DNA duplexes: Predictive thermodynamic model. *Biophysical Chemistry, 198*, 36–44.

Mueller, S. H., Spenkelink, L. M., van Oijen, A. M., & Lewis, J. S. (2020). Design of customizable long linear DNA substrates with controlled end modifications for single-molecule studies. *Analytical Biochemistry, 592*, 113541.

Oakley, A. J., Prosselkov, P., Wijffels, G., Beck, J. L., Wilce, M. C. J., & Dixon, N. E. (2003). Flexibility revealed by the 1.85 Å crystal structure of the β sliding-clamp subunit of Escherichia coli DNA polymerase III. *Acta Crystallographica, Section D: Biological Crystallography, 59*(7), 1192–1199.

Spenkelink, L. M., Lewis, J. S., Jergic, S., Xu, Z.-Q., Robinson, A., Dixon, N. E., et al. (2019). Recycling of single-stranded DNA-binding protein by the bacterial replisome. *Nucleic Acids Research, 47*(8), 4111–4123.

Spinks, R. R., Spenkelink, L. M., Stratmann, S. A., Xu, Z.-Q., Stamford, N. P. J., Brown, S. E., et al. (2021). DnaB helicase dynamics in bacterial DNA replication resolved by single-molecule studies. *Nucleic Acids Research, 49*, 6804–6816.

Stamford, N. P. J., Lilley, P. E., & Dixon, N. E. (1992). Enriched sources of Escherichia coli replication proteins: The dnaG primase is a zinc metalloprotein. *Biochimica et Biophysica Acta - Gene Structure and Expression, 1132*(1), 17–25.

Tanner, N. A., Hamdan, S. M., Jergic, S., Loscha, K. V., Schaeffer, P. M., Dixon, N. E., et al. (2008). Single-molecule studies of fork dynamics in Escherichia coli DNA replication. *Nature Structural & Molecular Biology, 15*(2), 170–176.

Tanner, N. A., Loparo, J. J., Hamdan, S. M., Jergic, S., Dixon, N. E., & van Oijen, A. M. (2009). Real-time single-molecule observation of rolling-circle DNA replication. *Nucleic Acids Research, 37*(4), e27.

van Loenhout, M. T., De Grunt, M., & Dekker, C. (2012). Dynamics of DNA supercoils. *Science*, *338*(6103), 94–97.

Whinn, K. S., Kaur, G., Lewis, J. S., Schauer, G. D., Mueller, S. H., Jergic, S., et al. (2019). Nuclease dead Cas9 is a programmable roadblock for DNA replication. *Scientific Reports*, *9*(1), 1–9.

Yardimci, H., Loveland, A. B., Habuchi, S., van Oijen, A. M., & Walter, J. C. (2010). Uncoupling of sister replisomes during eukaryotic DNA replication. *Molecular Cell*, *40*(5), 834–840.

CHAPTER FIFTEEN

Replication fork collapse *in vitro* using *Xenopus* egg extracts

Sara C. Conwell[†], Matthew T. Cranford[†], Tamar Kavlashvili, and James M. Dewar*

Department of Biochemistry, Vanderbilt University School of Medicine, Nashville, TN, United States
*Corresponding author: e-mail address: james.dewar@vanderbilt.edu

Contents

[†] These authors contributed equally to this work.

Methods in Enzymology, Volume 672
ISSN 0076-6879
https://doi.org/10.1016/bs.mie.2022.03.017

Abstract

When a replication fork encounters a nick in the parental DNA, the replisome dissociates and the replication fork structure is lost. This outcome is referred to as replication fork "collapse." Collapsed forks can be highly cytotoxic and mutagenic if not appropriately repaired by the cell. However, the events that occur during and after replication fork collapse are unclear. Here, we describe an *in vitro* system to induce site specific, strand specific replication fork collapse using Xenopus egg extracts, which contain the full set of DNA replication and repair enzymes. We also describe simple assays to monitor the stability of DNA nicks and the different structures formed during replication fork collapse. This methodology permits detailed mechanistic analysis of collapsed forks *in vitro*.

1. Introduction

Complete and accurate DNA replication is crucial for genome stability and cell viability. DNA replication is frequently challenged by a variety of physiologically and clinically relevant genotoxins that interfere with DNA synthesis (Liu et al., 2016). One of the most harmful ways that genotoxins interfere with DNA synthesis is by creating a nick in the parental DNA. This can occur due to environmental agents or cancer therapeutics, for example, due to ionizing radiation, topoisomerase poisons, and PARP inhibitors (Dias, Moser, Ganesan, & Jonkers, 2021; Goodhead, 1994; Zagnoli-Vieira & Caldecott, 2020). When a replication fork encounters a discontinuity in the parental DNA, it causes dissociation of the core replication machinery and breakage of the replication fork structure (Cortez, 2019). We refer to this situation as "replication fork collapse," because the replication forks and their associated proteins are completely and prematurely disassembled. Collapsed forks lead to cytotoxicity and mutagenesis unless they are acted upon by DNA repair enzymes to restart DNA replication (Petermann, Orta, Issaeva, Schultz, & Helleday, 2010). However, despite the importance of this process, many questions remain about how collapse occurs and the mechanisms that subsequently act.

The DNA end formed following replication fork collapse is thought to be the substrate for break induced replication (BIR), which allows restart of DNA synthesis at a D-loop using the homologous recombination (HR) machinery (Wu & Malkova, 2021). BIR is well defined based on studies in yeast and is expected to also take place in metazoa, based on the roles of HR proteins in replication fork restart (Petermann et al., 2010), but has not been formally demonstrated. Restart of DNA synthesis from a DNA end has been demonstrated during mitosis in human cells, where it is called Mitotic DNA Synthesis (MiDAS) (Bhowmick, Minocherhomji, & Hickson, 2016; Sonneville et al., 2019). However, MiDAS is mechanistically distinct from BIR, suggesting there may be multiple mechanisms that deal with collapsed forks. It is thus imperative to understand the events that occur during and after replication fork collapse.

We recently adapted the *Xenopus* egg extract system to force replication fork collapse at a site specific and strand specific DNA nick *in vitro* (Vrtis et al., 2021). The exact *Xenopus* egg extract system we use (Walter & Newport, 2000) allows for replication of plasmid DNA *in vitro* using the full set of cellular proteins (Wühr et al., 2014, 2015). Using this system, site-specific DNA lesions can be engineered into the DNA to study how they impact DNA replication. This has allowed multiple detailed mechanisms of replication-coupled DNA repair to be elucidated (Duxin, Dewar, Yardimci, & Walter, 2014; Räschle et al., 2008; Semlow, Zhang, Budzowska, Drohat, & Walter, 2016). However, DNA nicks are rapidly religated, which previously precluded analysis of replication fork encounter with site specific DNA nicks. To overcome this limitation, we generated site-specific nicks flanked by *tet* operator (*tetO*) sequences. When bound by *tet* repressor (TetR), relegation of nicks is prevented (Vrtis et al., 2021). Importantly, TetR binding does not interfere with DNA replication and TetR is readily displaced by the replication machinery (Vrtis et al., 2021). Thus, TetR-protected DNA nicks can be replicated in *Xenopus* egg extracts to study replication fork collapse. These nicks can be engineered adjacent to a *lac* repressor (LacR) bound *lac* operator (*lacO*) array (Dewar, Budzowska, & Walter, 2015) to prevent convergence of the opposing fork and thus simplify analysis of a single collapsed fork (Vrtis et al., 2021). Importantly, nicks can be positioned on either the leading or lagging strand template to examine the different consequences (Vrtis et al., 2021). This has led to a working model for how leading and lagging strand collapse differ (Fig. 1).

In this chapter, we describe preparation of TetR, LacR, and nicked DNA, which are the key reagents required to study replication fork collapse. We also describe simple assays to measure stability of the DNA nicks and monitor the formation of collapsed replication forks (Figs. 2–3). We do not cover preparation of *Xenopus* egg extracts, as has been described elsewhere (Lebofsky, Takahashi, & Walter, 2009; Sparks & Walter, 2019). We also do not cover other *Xenopus* egg extract systems that can be used to study replication of DNA nicks (Hashimoto, Puddu, & Costanzo, 2011). These approaches do not permit analysis of site and strand specific

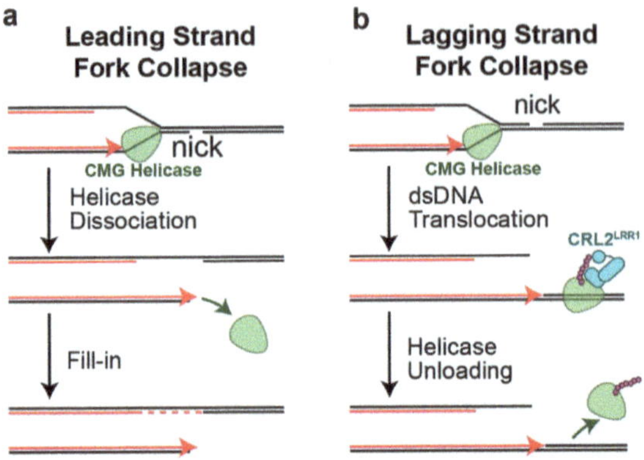

Fig. 1 Overview of replication fork collapse in *Xenopus* egg extracts. (A) When a replication fork encounters a DNA nick on the leading strand template the nick is in the translocating strand of the replicative CMG helicase. Once CMG encounters the nick, it translocates off the 5′ end of the parental DNA strand (black), leading to helicase dissociation. This results in a full-length DNA strand and a truncated DNA strand. Leading strand DNA synthesis occurs up to the very end of the truncated DNA strand, resulting in an essentially blunt end. The full-length strand contains a gap that is rapidly filled in (dashed red line) resulting in a continuous full length DNA molecule that is composed of both parental DNA (black-black) as well as lagging strands (red) and gap-filled nascent DNA (dashed red). (B) When a replication fork encounters a DNA nick on the lagging strand template the nick is in the non-translocating strand of the replicative CMG helicase. Once CMG reaches the nick, the lagging strand template is fully unwound and dissociates. This results in a full-length DNA strand and a truncated DNA strand. CMG then translocates over the downstream parental DNA because the lack of a 5′ flap prevents continued unwinding (Kang, Galal, Farina, Tappin, & Hurwitz, 2012; Vrtis et al., 2021). Nascent leading strands are rapidly extended up to the 5′ end of the parental lagging strand template, which is presumably facilitated by the loss of CMG from the underlying DNA. Once CMG translocates over double-stranded DNA it is removed from DNA by the same mechanism that normally operates during termination (Dewar & Walter, 2017). First, the ubiquitin ligase CRL2^{LRR1} binds to and ubiquitylates CMG due to the lack of a lagging strand flap (Deegan, Mukherjee, Fujisawa, Polo Rivera, & Labib, 2020; Dewar, Low, Mann, Räschle, & Walter, 2017; Jenkyn-Bedford et al., 2021; Low, Chistol, Zaher, Kochenova, & Walter, 2020; Sonneville et al., 2017). Second, p97 recognizes ubiquitylated CMG and extracts the complex from DNA (Maric, Maculins, De Piccoli, & Labib, 2014; Moreno, Bailey, Campion, Herron, & Gambus, 2014). Note that the truncated strand contains a 3′ single-stranded DNA overhang, which arises from the gap that normally exists on the lagging strand template.

nicks but have made important contributions to our understanding of replication fork collapse (Hashimoto et al., 2011). Overall, the techniques described in this chapter support a powerful new approach to study replication fork collapse.

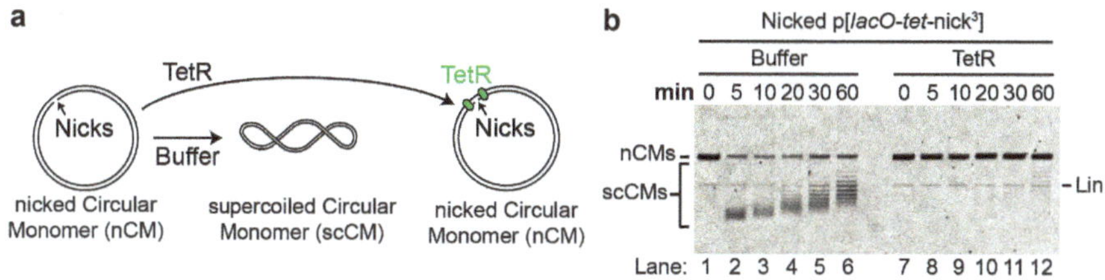

Fig. 2 Analysis of nick stability in *Xenopus* egg extracts. (A) Plasmid DNA containing three tandem DNA nicks was pre-incubated with buffer control or TetR before being added to High Speed Supernatant (HSS). In the buffer control the nicks are rapidly religated, resulting in covalently closed circular monomers that undergo compensatory supercoiling (supercoiled Circular Monomers, scCMs) once the reaction is stopped and the DNA is deproteinized. Nicks are stabilized by the addition of TetR, which binds to *tetO* sites adjacent to the nicks, resulting in persistence of nicked circular monomers (nCMs). (B) Plasmid DNA from (A) was separated on a chloroquine-containing agarose gel and detected by SYBR gold staining. Most nicks were rapidly religated in the buffer control resulting in conversion of nCMs to scCMs (lanes 1–2), which exist as different topoisomers due to differing amounts of chromatinization (lanes 2–6). Nicks were stable in the presence of TetR and did not appreciably change over the course of the experiment (lanes 7–12). Note that a small fraction of linear (Lin) molecules were present in both conditions due to double-strand breaks in the parental DNA.

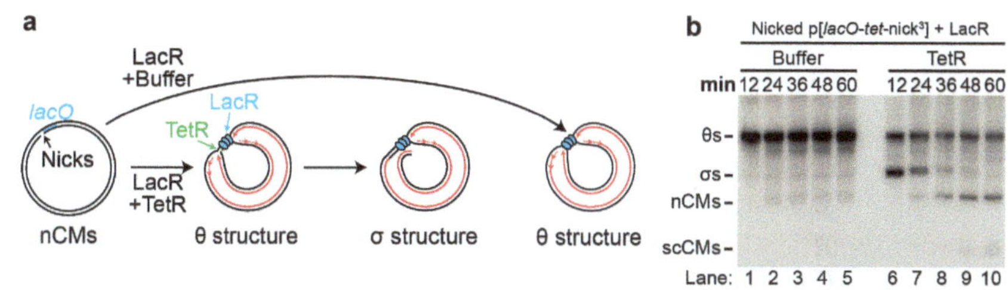

Fig. 3 Analysis of replication fork collapse in *Xenopus* egg extracts. (A) Plasmid DNA containing three tandem DNA nicks adjacent to a *lacO* array was pre-incubated with LacR and either buffer control or TetR before undergoing DNA replication in *Xenopus* egg extracts. In the presence of TetR the nicks are stabilized (as in Fig. 2) resulting in replication fork collapse when the nick is encountered (as in Fig. 1A). LacR binds to the *lacO* array and blocks convergence of the opposing fork, which allows the conversion of replication fork structures (θ structures) to collapsed structures (σ structures) to be detected. In the presence of buffer control the nicks are readily religated (as in Fig. 2) resulting in persistence of θ structures. (B) Replication was performed as indicated in (A) in the presence of radiolabeled dATP to allow detection of replicating DNA molecules. Reactions were stopped at different time points and DNA structures were separated on an agarose gel then detected by autoradiography. In the buffer control θ structures were the most abundant species (lanes 1–5), as expected. In the presence of TetR, σ structures initially accumulated due to fork collapse (lanes 6–7) but then gave rise to θ structures (lanes 8–10), as well as nicked and supercoiled circular monomers (nCMs and scCMs, respectively) that arose from complete replication.

2. Purification of *lac* repressor

The strong DNA-binding activity of *lac* repressor (LacR) allows it to function as a potent barrier to DNA replication (Dewar et al., 2015). The DNA constructs described in this chapter contain a *lacO* array adjacent to the nick sites. This allows convergence of the opposing fork to be prevented so that the fate of the collapsed fork can be unambiguously monitored. In this section, we describe how to express and purify biotinylated LacR from bacteria. However, we note that His-tagged LacR has also been used successfully for the same purpose (Duxin et al., 2014).

2.1 Materials and equipment

1. T7 Express cells (New England Biolabs)
2. pBirAcm (Avidity)
3. pET11a[lacI::avi] (pJD72 from (Dewar et al., 2015))
4. Lysogeny Broth (LB) medium: 10 g/L tryptone; 5 g/L yeast extract; 5 g/L NaCl
5. Ampicillin stock: 50 mg/mL in water
6. Chloramphenicol stock: 34 mg/mL in ethanol
7. LB/Amp/Cam medium: 1 × LB medium; 100 μg/mL ampicillin; 17 μg/mL chloramphenicol
8. LB/Amp/Cam Agar plates: 1 × LB medium; 100 μg/mL ampicillin; 17 μg/mL chloramphenicol; 7.5% (*w*/*v*) agar
9. Isopropyl β-D-1-thiogalactopyranoside (IPTG) stock: 1 M in water
10. Biotin solution 1: 5 mM in 10 mM Bicine-NaOH, pH 8.3
11. High speed floor standing centrifuge
12. 25 × complete: Dissolve 1 tablet of complete protease inhibitor cocktail (Roche) in 2 mL of water
13. Freezing Buffer: 10% (*w*/*v*) sucrose; 5 mM EDTA; 50 mM Tris-HCl pH 7.5; 1 × complete
14. SoftLink resin (Promega)
15. Gravity flow column (10 mL)
16. Buffer 1: 4 mM EDTA; 10% (*w*/*v*) sucrose (w/v); 1 × Complete; 1 × Buffer A
17. Lysozyme: 10 mg/mL in 20 mM Tris-HCL pH 7.5
18. Brij 58: 10% (*w*/*v*) in water
19. 2 mL microcentrifuge tubes

20. Cooled Microcentrifuge
21. 4 × Buffer A (no DTT): 200 mM Tris-HCl, pH 7.5; 4 mM EDTA; 400 mM NaCl
22. 4 × Buffer A: add 4 mM DTT to 4 × Buffer A (no DTT) immediately prior to use
23. Extraction Buffer: 4 mM EDTA; 30 mM IPTG; 900 mM NaCl; 1 × Complete; 1 × Buffer A
24. Probe sonicator
25. Glass beaker (20 mL)
26. Stir bar
27. Stir plate
28. Polymin P: 1% (*v*/v) in 50 mM Tris-HCl, pH 7.4
29. Wash Buffer: 2.4 M NaCl; 1 × Complete, 1 × Buffer A
30. Buffer 2: 1 × Buffer A
31. Elution Buffer: 1 × Buffer A; add 5 mM biotin fresh
32. Dialysis Buffer: 150 mM NaCl; glycerol 38% (*v*/v); 1 × Buffer A
33. Dialysis cassette (10,000 molecular weight cutoff)
34. Nanodrop spectrophotometer
35. Amicron ultra centrifugal filter unit (10,000 molecular weight cutoff)

2.2 Procedure

Express biotinylated *lac* repressor (LacR):

1. Co-transform NEB T7 express cells with pBirAcm and pET11a[lacI::avi]. Grow transformants for 16 h at 37 °C on LB/Amp/Cam Agar plates
2. Store agar plates at 4 °C for up to 2 days
3. Inoculate a single colony into 20 mL of LB/Amp/Cam medium. Grow for 16 h at 37 °C with aeration
4. Add 20 mL of overnight culture to 1 L of LB/Amp/Cam medium. Grow with aeration to OD_{600} 0.5–0.6
5. Add 1 mM IPTG and 50 μM biotin to induce expression and biotinylation of LacR. Incubate for 2 h at 37 °C with aeration
6. Chill the culture on ice water then centrifuge at 4000 × *g*, 4 °C to pellet cells
7. Pour off the LB media, then resuspend the pellet(s) in 10 mL of freezing buffer. Flash freeze cell suspension in liquid nitrogen. Store at −80 °C until ready to use

Purify biotinylated *lac* repressor (LacR):

1. Prepare a gravity flow column containing 1 mL of Softlink resin (Promega) according to the manufacturer's instructions. Store at 4 °C according to the manufacturer's instructions until ready to use
2. Thaw the cell suspensions on ice, then centrifuge at 4000 × *g*, 4 °C to pellet cells
3. Remove supernatant and resuspend the cell pellet to 10 mL in Buffer 1
4. Add 1/49th volume of Lysozyme (10 mg/mL).
5. Add 1/100th volume of Brij 58 (10% (*w*/*v*)).
6. Incubate at room temperature for 30 min with end over end rotation
7. Split the suspension between 5 × 2 mL microcentrifuge tubes and centrifuge at 14,000 × *g*, 4 °C for 1 h in a microcentrifuge
8. Collect supernatants (soluble fractions) from Step 6 in a single 10 mL tube then flash in liquid nitrogen and store at −80 °C. Steps 9–34 can be performed later with this soluble fraction to yield a lower quantity of biotinylated LacR
9. Resuspend each pellet (insoluble fraction) from step 7 in 2 mL in Extraction Buffer then combine in a new 10 mL tube
10. Disrupt the cell suspension by sonication on ice
11. Split the suspension between 5 × 2 mL microcentrifuge tubes and centrifuge at 14,000 × *g*, 4 °C for 1 h in a microcentrifuge
12. Pool the supernatants and transfer to a 20 mL glass beaker. Perform all subsequent steps in a cold room at 4 °C
13. To precipitate DNA, slowly add 0.03–0.06 volumes of Polymin P (1%(*v*/v)) while stirring constantly with a stir bar
14. Stir for an additional 30 min
15. Split the suspension between 2 mL microcentrifuge tubes and centrifuge at 14,000 × *g*, 4 °C for 1 h in a microcentrifuge
16. Collect and combine supernatants (typically approximately 6.5 mL) in a 20 mL glass beaker
17. Slowly add ammonium sulfate to 37–60% saturation while stirring constantly with a stir bar
18. Stir for an additional 1 h or overnight
19. Split the suspension between 5 × 2 mL microcentrifuge tubes and centrifuge at 14,000 × *g*, 4 °C for 1 h in a microcentrifuge
20. Discard the supernatants then resuspend each pellet in 0.4 mL Wash Buffer and combine into a single 2 mL suspension
21. Wash the SoftLink column with 10 mL of Buffer 2
22. Apply 1 mL of the suspension from step 20 to the column. Incubate for 30 min

23. Apply the remaining 1 mL of suspension to the column. Incubate for 30 min
24. Wash the column 3 times with 5 mL of Wash Buffer
25. Apply 1 mL of Elution Buffer to the column
26. Incubate for 30 min to displace LacR
27. Apply 1 mL of Elution Buffer to the column to elute the first fraction
28. Repeat steps 26–27 twice more to perform two additional elutions
29. Pool elutes from step 28 and perform 3 rounds of dialysis against 1 L of Dialysis Buffer. Make sure to flash freeze and save an aliquot of the final round of dialysis buffer for subsequent dilutions of the protein and to serve as a negative control
30. During dialysis, regenerate the Softlink column according to the manufacturer's instructions
31. Measure the concentration of the dialyzed LacR using a Nanodrop assuming an extinction coefficient for the tagged protein of 28,545 (M^{-1} cm^{-1}) at 280 nm
32. Concentrate the LacR to approximately 1 μg/μL using an Amicon Ultra Centrifugal Filter unit and a cooled centrifuge
33. Flash freeze LacR into liquid nitrogen and store at −80 °C until ready to use
34. Analyze concentrated LacR alongside a BSA standard curve using SDS-PAGE to determine protein concentration

2.3 Notes

- A high-speed floor standing centrifuge can be used for centrifugation steps instead of a microcentrifuge. The LacR purification protocol is designed for use with a cooled microcentrifuge because these are typically more accessible and so are recommended to speed up the protocol
- Most LacR is initially bound to the insoluble fraction, presumably due to its low nonspecific DNA-binding activity. The soluble and insoluble fractions are separated because this provides a simple single-step removal of all soluble proteins from the prep before the LacR is solubilized
- It is essential that the SoftLink resin does not dry out. When preparing the SoftLink resin, it is advantageous to place a filter disk on top of the resin as well as below. This will stop the flow once the meniscus reaches the filter and thus prevent the resin from drying out. Alternatively, the resin can be monitored, and the flow can be stopped once the meniscus reaches the resin

- The length of each dialysis step can be varied. We recommend 2–16 h for each step and it is typically fine for the first step to be only 1 h if the second step is overnight
- We typically find that SDS-PAGE analysis of purified proteins alongside a BSA standard provides a more reproducible measure of protein concentration than absorbance at 280 nm. However, because absorbance at 280 nm is usually quite close and is much faster, we use this to concentrate the protein to approximately 1 μg/μL (25 μM). The concentration of working stocks can be subsequently adjusted to 18 μM for experiments according to the concentration determined by SDS-PAGE

3. Purification of *tet* repressor

The relatively weak DNA-binding activity of *tet* repressor (TetR) allows it to inhibit ligation of DNA nicks without interfering with DNA replication (Vrtis et al., 2021). The DNA constructs described in this chapter contain *tetO* sites flanking the nick sites to stabilize them in the presence of TetR. This allows the nicks to be destabilized by TetR until it is displaced by the replication machinery. In this section we describe how to express and purify HIS-tagged TetR from bacteria.

3.1 Materials and equipment

1. T7 Express cells (New England Biolabs)
2. pET28b[TetR::HIS] (pJD13, available upon request)
3. Lysogeny Broth (LB) medium: 10 g/L tryptone; 5 g/L yeast extract; 5 g/L NaCl
4. LB/Kan Agar plates: 1 × LB medium; 50 μg/mL Kanamycin (Kan); 7.5% (*w*/*v*) agar
5. LB/Kan medium: 1 × LB medium; 50 μg/mL Kan
6. IPTG: 1 M in water
7. 25 × complete: Dissolve 1 tablet of complete protease inhibitor cocktail (Roche) in 2 mL of water
8. Lysis Buffer: 20 mM Tris-HCl, pH 8.0; 1 M NaCl; 1 × complete
9. Lysozyme: 10 mg/mL in Lysis Buffer
10. High speed floor standing centrifuge
11. Ni-NTA agarose (Qiagen): 50% slurry
12. Gravity flow columns (10 mL)
13. Wash Buffer: 20 mM imidazole in Lysis Buffer
14. Elution Buffer: 500 mM imidazole in Lysis Buffer
15. Dithiothreitol (DTT): 1 M in water

16. Dialysis cassette (10,000 molecular weight cutoff)
17. Dialysis Buffer: 20 mM Tris-HCl, pH 8.0; 200 mM NaCl; 5 mM DTT

3.2 Procedure

Express HIS-tagged *tet* repressor (TetR):

1. Transform NEB T7 express cells with pET28b[TetR::HIS]
2. Grow transformants for 16 h at 37 °C on LB/Kan agar plates
3. Store agar plates at 4 °C for up to 2 days
4. Inoculate a single colony into 30 mL of LB medium containing kanamycin (50 μg/mL). Grow for 16 h at 37 °C with aeration
5. Add 20 mL of overnight culture to 1 L of LB medium containing kanamycin (50 μg/mL). Grow with aeration to OD_{600} 0.5–0.6
6. Add 1 mM IPTG to induce expression of TetR. Incubate for 3 h at 37 °C with aeration
7. Chill the culture on ice water then centrifuge at 3000 × *g*, 4 °C for 5 min to pellet cells
8. Pour off the LB media then flash freeze the cell pellet in liquid nitrogen. Store at −80 °C until ready to use

Purify HIS-tagged *tet* repressor (TetR):

1. Thaw the frozen cell pellet on ice. Perform all subsequent steps in a cold room at 4 °C
2. Resuspend the cell pellet in 40 mL of Lysis buffer
3. Add 1/9th volume of lysozyme (10 mg/mL). Rotate end-over-end for 1 h
4. Centrifuge the lysate at 21,000 × *g*, 4 °C for 30 min
5. During Step 4: Add 3 mL of 50% Ni-NTA agarose (50% slurry) to three gravity flow columns to yield 1.5 mL Ni-NTA agarose per column. Wash each Ni-NTA agarose column three times with 10 mL of Lysis buffer then cap the bottom of the column
6. Collect the supernatant from step 4 then add equal volumes of supernatant to each of the three Ni-NTA columns. Do not uncap the bottom of the columns
7. Cap the top of each column and incubate at 4 °C for 5 min with end over end rotation to allow TetR to bind to the Ni-NTA agarose
8. Mount the columns upright and allow the Ni-NTA agarose to settle at the bottom of the column. Uncap the top and then the bottom of the column so that the supernatant can flow out. Recap the bottom of the column once the meniscus reaches the resin

9. Add 8 mL of Wash buffer to each column, then cap the top of each column and incubate at 4 °C for 5 min with end over end rotation to wash the Ni-NTA agarose
10. Repeat steps 8–9 twice to perform two additional washes
11. Repeat step 8 to allow wash buffer to drain from the column
12. Apply 2 mL of Elution buffer to each column to elute the first fraction. Add 10 μL of DTT (1 M) to the elute and mix gently by inversion
13. Repeat step 12 a further five times, to perform five additional elutions
14. Analyze the elutes from step 13 by SDS-PAGE to identify fractions containing purified TetR
15. Pool the purified fractions and inject into a dialysis cassette. Perform 3 rounds of dialysis against 1 L of Dialysis Buffer. Make sure to flash freeze and save an aliquot of the final round of dialysis buffer for subsequent dilutions of the protein and to serve as a negative control
16. Aliquot and flash freeze TetR into liquid nitrogen and store at −80 °C until ready to use
17. Analyze the dialyzed TetR alongside a BSA standard curve using SDS-PAGE to determine protein concentration

3.3 Notes

- We recommend using three smaller gravity flow columns rather than one large column to improve the effectiveness of end-over-end washes and incubations
- When incubating gravity flow columns end-over-end we recommend applying parafilm to the top cap to ensure it does not come loose
- There is no need to concentrate down TetR prior to storage because this procedure typically yields ~100 μM TetR but only ~50 μM is needed for experiments. The concentration of working stocks can be diluted subsequently, according to the concentration determined by SDS-PAGE

4. Purification of nicked DNA

To generate site specific DNA nicks, plasmid DNA is treated with site specific nicking enzymes to introduce nicks at the appropriate locations. Residual proteins and other components of the nicking reaction are then removed by phenol: chloroform extraction and ethanol precipitation. Although commercial DNA purification kits can also be used, we recommend the approach outlined here because the purity is greater and the yield is typically ~2-fold higher.

4.1 Materials and equipment

1. p[*lacO-tet*-nick]: Miniprepped plasmid DNA containing 1 × or 3 × Nb. *Bsm*I sites flanked on each side by *tetO* sequences, as in (Vrtis et al., 2021). Available upon request
2. Nb.BsmI (New England Biolabs)
3. Extraction Stop: 25 mM EDTA pH 8.0; 0.5% SDS; 50 mM Tris-HCl, pH 7.5
4. Refrigerated Microcentrifuge
5. Safe-Lock Microcentrifuge Tubes (Eppendorf)
6. Tris-HCl: 10 mM, pH 8.0
7. Phenol/chloroform/isoamyl alcohol (25:24:1)
8. Chloroform/isoamyl alcohol (24:1)
9. 3 M sodium acetate, pH 5.5
10. Ethanol: 100%, molecular biology grade
11. Ethanol: 70% (*v*/v) in water
12. GELoader Tips (Eppendorf)
13. Nanodrop spectrophotometer

4.2 Procedure

1. Subclone and purify p[*lacO-tet*-nick] using QIAprep spin miniprep kit (Qiagen) according to the manufacturer's instructions
2. In a Safe-Lock microcentrifuge tube, digest 10 μg of p[*lacO-tet*-nick] in 100 μL of NEB Buffer 3.1 (1 ×) with 25 units Nb.*Bsm*I for 16 h at 37 °C. Perform subsequent steps at room temperature unless otherwise stated
3. Chill Ethanol (100% and 70% (v/v)) on ice
4. Dilute the digestion mix from step 2 to 110 μL with 10 mM Tris-HCl, pH 8.0
5. Add 110 μL of Extraction Stop to bring total volume to 220 μL
6. Mix by inversion and centrifuge briefly at 13,000 × *g* to collect all liquid at the bottom of the tube
7. Add 220 μL Phenol/chloroform/isoamyl alcohol (25:24:1). Mix thoroughly by inverting the tube 10–15 times and then centrifuge at 13,000 × *g* for 4 min
8. Transfer 200 μL of the aqueous phase to a new microcentrifuge tube containing 200 μL chloroform/isoamyl alcohol (24:1). Mix thoroughly by inverting the tube 10–15 times and then centrifuge at 13,000 × *g* for 4 min

9. Transfer 160 μL of the aqueous phase from step 7 to a new microcentrifuge tube containing 16 μL sodium acetate (3 M). Pipette up and down three times to mix
10. Add 400 μL chilled ethanol (100%) from step 3 then mix by gently inverting the microcentrifuge tube 5–10 times. Incubate on ice for 15 min
11. Centrifuge at 14,000 RPM, 4 °C for at least 30 min
12. Use a vacuum line to aspirate all but approximately 50 μL of supernatant
13. Add 400 μL chilled ethanol (70%) from step 3
14. Centrifuge at 14,000 RPM, 4 °C for at least 20 min
15. Use a vacuum line to aspirate all but approximately 50 μL of supernatant
16. Centrifuge at 14,000 RPM, 4 °C for at least 1 min to collect all liquid at the bottom of the tube
17. Carefully use a GELoader tip to aspirate most of the remaining ethanol from DNA pellet. Leave no more than 1–2 μL of ethanol behind. Cap the microcentrifuge tube and store on ice until ready to resuspend the DNA
18. Uncap the microcentrifuge tube then transfer to a room temperature rack and monitor the residual ethanol and the pellet. Once the remaining ethanol has evaporated and the pellet has turned translucent (typically no more than 5 min) wait an additional 4 min then add 11 μL of Tris-HCl (10 mM, pH 8.0) to yield approximately 300 ng/μL of DNA
19. Incubate the DNA at room temperature for 30 min
20. Centrifuge for 1 min to collect all liquid at the bottom of the tube, then mix by flicking the tube. Repeat twice more, then centrifuge one last time to collect all liquid
21. Incubate the DNA at room temperature for a further 30 min or at 4 °C overnight
22. Measure the concentration of the purified DNA using a Nanodrop then add Tris-HCl (10 mM, pH 8.0) to adjust the concentration to 225 ng/μL for experiments
23. Aliquot the DNA and store at −20 °C or − 80 °C

4.3 Notes

- Propagate p[*lac*O-*tet*-nick] in the presence of 5 mM IPTG to prevent the endogenous LacR from interfering with replication of the *lac*O array within the plasmid

- Different versions of p[*lacO-tet*-nick] have been generated that include a single nick or three tandem nicks on either the leading or lagging strand template as in Vrtis et al. (2021). All reagents are available upon request
- The DNA extraction outlined here has been empirically determined to leave sufficient aqueous phase at each step that the interface between aqueous and organic phases is not disrupted. This ensures the DNA is not contaminated by phenol. We recommend Eppendorf safe lock tubes because changes in tube dimensions can alter the depth of the aqueous phase
- This protocol describes purification of a single DNA template, but it can be scaled up as needed and multiple purifications can be performed in parallel
- During aspiration steps, the tubes should be kept upright to ensure the DNA pellet is not displaced
- If multiple purifications are performed in parallel, the tubes can be stored upright on ice during the aspiration steps without the pellet being dislodged

5. Analysis of nick stability in *xenopus* egg extracts

When plasmid DNA is introduced to *Xenopus* egg extracts it is quickly chromatinized (Laskey, Mills, & Morris, 1977). When the DNA is deproteinized any covalently closed DNA undergoes compensatory supercoiling while any open DNA (e.g., nicked) will remain relaxed. Thus, the detection of supercoiled DNA acts as a simple assay for DNA that is fully ligated. In this section, we describe how to monitor nicked and supercoiled DNA by gel electrophoresis (as in Fig. 2).

5.1 Materials and equipment

1. p[*lacO-tet*-nick] (225 ng/μL) from Section 4
2. TetR (50 μM) from Section 3
3. High Speed Supernatant (HSS): prepared as described (Lebofsky et al., 2009)
4. Phosphocreatine disodium salt (PC): 1 M in 10 mM potassium phosphate, pH 7.0. Store in 50 μL aliquots at −20 °C
5. Adenosine 5′-triphosphate disodium salt hydrate (ATP): dissolve in approximately 4.85 mL sterile water, then adjust the pH to 7.0 with approximately 0.15 mL 10 M NaOH. Adjust final concentration to 0.2 M ATP with water, as necessary. Store 50 μL aliquots at −20 °C

6. Creatine phosphokinase (CPK): 5 mg/mL in 50 mM NaCl, 50% glycerol and 10 mM HEPES-KOH, pH 7.5. Store 50 L aliquots at −20 °C for up to 6 months
7. ARS (ATP Regenerating System): Combine 10 μL PC, 5 μL ATP, and 0.5 μL CPK. Store on ice for up to 2 h prior to use
8. Nocodazole: 0.5 mg/mL in in DMSO. Store 50 μL aliquots at −20 °C
9. Extraction Stop: 0.5% (*w/v*) SDS; 25 mM EDTA, pH 8.0; 50 mM Tris–HCl, pH 7.5
10. 2 mg/mL RNase
11. 20 mg/mL Proteinase K (Roche): Dissolve in water. Store 50 μL aliquots at −20 °C
12. GTG Agarose (Seakam)
13. 10× TBE buffer: 0.89 M Tris Base; 0.89 M Boric Acid; 25.5 mM EDTA, pH 8.0
14. DNA Loading dye (6×): 15% Ficoll; 66 mM EDTA; 0.09% (w/v) Bromophenol blue; 20 mM Tris-HCl, pH 8.0
15. High voltage power supply
16. SYBR Gold Staining solution: 1× SYBR Gold (Invitrogen; 1× TBE buffer)
17. Plastic container for staining (light-protected container or covered in foil)
18. Gel imager with UV light source for visualization of SYBR Gold stain

5.2 Procedure

1. Combine 1 μL of p[*lacO-tet*-nick] (225 ng/μL) 2 μL of TetR (100 μM) His-TetR (Binding Mix). Mix gently by pipetting up and down. Incubate at room temperature for at least 1 h to allow TetR to bind DNA. Perform all subsequent steps at room temperature
2. Thaw a 33 μL aliquot of HSS then add 1 μL ARS and 0.2 μL Nocodozole. Mix by pipetting up and down 20–30 times
3. Centrifuge HSS at 13,000 × *g* for 5 min to remove any insoluble debris and provide time for the HSS to activate
4. Transfer 27 μL of HSS into a new tube and avoid transferring any pellet that formed during step 3. Carefully mix the activated HSS by pipetting up and down 8–10 times
5. Add 9 μL HSS to the Binding Mix to initiate licensing (Licensing Mix). Mix thoroughly by pipetting up and down 30–40 times

6. At the desired timepoints withdraw 2.5 μL of Licensing Mix and add to 25 μL of Extraction Stop. Pipet up and down rapidly to ensure the full 2.5 μL of Licensing Mix is added to Extraction Stop. Flick the tube to ensure thorough mixing
7. Once the experiment is complete, combine 1 μL of p[*lacO-tet*-nick] (225 ng/μL) with 131 μL of Extraction Stop. Mix thoroughly and then remove a 27.5 μL aliquot to serve as a time 0 loading control
8. Add 1.25 μL of RNAse to all samples from steps 6–7 and incubate at 37 °C for at least 30 min
9. Add 1.25 μL of Proteinase K to all samples from steps 6–7 and incubate at 37 °C for 1 h, or overnight at room temperature
10. **Optional:** purify samples by phenol: chloroform extraction and ethanol precipitation as in (Kavlashvili & Dewar, 2022).
11. Add 0.2 volumes of DNA loading dye (6 ×) to each sample. Mix thoroughly
12. Prepare a 1% agarose gel in TBE buffer (1 ×) and submerge the gel in an electrophoresis tank containing 1 × TBE buffer
13. Load 6 μL of sample per lane
14. Perform electrophoresis at up to 2.5 V/cm for approximately 5 h or until the dye front has migrated approximately 10 cm
15. Submerge the gel in 1 × SYBR Gold stain solution in a light-protected plastic container. Stain for at least 30 min with agitation
16. Visualize the gel under UV light using a Gel Imager

5.3 Notes

- To reduce background from SYBR gold staining, thoroughly clean any glassware and plasticware that is used to cast the agarose gel and make sure to use filtered TBE buffer and water
- To help resolve nicked plasmids from covalently closed, relaxed plasmids, include 5 μg/mL chloroquine in the agarose gel and running buffer
- To maximize the activity of *Xenopus* egg extracts, we recommend thawing HSS rapidly in room temperature water
- This protocol is for only a single reaction but can be scaled up or down as necessary
- To perform a control reaction lacking TetR, substitute an equal volume of dialysis buffer

6. Replication fork collapse in *xenopus* egg extracts

When plasmid DNA is replicated in *Xenopus* egg extracts, characteristic DNA structures are formed at different stages of replication and in response to different DNA repair events (Dewar et al., 2015; Duxin et al., 2014; Räschle et al., 2008; Semlow et al., 2016). This allows a great deal of information to be gleaned simply by analyzing the pattern of DNA structures formed. In this section we outline how to detect the different DNA structures formed before, during, and after replication fork collapse (as in Fig. 3).

6.1 Materials and equipment

1. Items 1–15 from Section 5.1
2. LacR (18 μM) from Section 2
3. Dithiothreitol (DTT) stock: 1 M in water. Store at −20 °C
4. DTT: 50 mM in water. Prepare fresh from DTT stock and store at room temperature for up to 2 h
5. 10 × ELB Salts: 500 mM KCl; 25 mM $MgCl_2$; 100 mM HEPES-KOH, pH 7.7. Store at 4 °C and protect from light
6. 5 × ELB: 1.25 M sucrose in 1 × ELB Salts. Store aliquots at 4 °C for up to 1 month
7. [α-^{32}P]-dATP (3000 Ci/mmol)
8. NucleoPlasmic Extract (NPE): prepared as described (Lebofsky et al., 2009)
9. Whatman Paper
10. Hybond N+ membrane
11. Phosphorimaging Screen
12. Phosphorimager

6.2 Procedure

1. Combine 1 μL of p[*lacO-tet*-nick] (225 ng/μL), 2 μL of TetR (100 μM) His-TetR, and 2 μL LacR (18 μM) (Binding Mix). Mix gently by pipetting up and down. Incubate at room temperature for at least 1 h to allow TetR and LacR to bind DNA. Perform all subsequent steps at room temperature
2. Thaw a 33 μL aliquot of HSS then add 1 μL ARS and 0.2 μL Nocodozole. Mix by pipetting up and down 20–30 times
3. Centrifuge HSS at 13,000 × *g* for 5 min to remove any insoluble debris and provide time for the HSS to activate

4. Transfer 27 μL of HSS into a new tube and avoid transferring any pellet that formed during step 3. Carefully mix the activated HSS by pipetting up and down 8–10 times
5. Add 20 μL HSS to the Binding Mix to initiate licensing (Licensing Mix). Mix thoroughly by pipetting up and down 30–40 times. Incubate for 30 min at room temperature
6. Approximately 10 min after the start of Step 5 thaw a 20 μL aliquot of NPE then add 0.6 μL ARS, 0.825 μL DTT (50 mM), 2 μL [α-^{32}P] dATP, 4 μL ELB (5 ×), 4.22 μL of TetR (100 μM), and 10.6 μL H_2O (NPE Mix). Mix by pipetting up and down 20–30 times. Incubate for at least 15 min at room temperature
7. 30 min after the start of licensing (step 5) Initiate replication by adding 2 volumes NPE mix (step 6) to 1 volume licensing mix (step 5). Mix thoroughly by pipetting up and down 10–20 times
8. At the desired timepoints withdraw 2 μL of reaction and add to 20 μL of Extraction Stop. Pipet up and down rapidly to ensure the full 2 μL of reaction is added to Extraction Stop. Flick the tube to ensure thorough mixing
9. Add 1 μL of RNAse to all samples and incubate at 37 °C for at least 30 min
10. Add 1 μL of Proteinase K to all samples and incubate at 37 °C for 1 h, or overnight at room temperature
11. **Optional:** purify samples by phenol: chloroform extraction and ethanol precipitation as in (Kavlashvili & Dewar, 2022).
12. Add 0.2 volumes of DNA loading dye (6 ×) to each sample. Mix thoroughly
13. Prepare a 1% agarose gel in TBE buffer (1 ×) and submerge the gel in an electrophoresis tank containing 1 × TBE buffer
14. Load 6 μL of sample per lane
15. Perform electrophoresis at 4.5 V/cm for approximately 3 h or until the dye front has migrated approximately 6 cm
16. Dry the gel between Whatman paper and Hybond N+ membrane then perform phosphorimaging to detect radiolabeled DNA structures

6.3 Notes

1. To reduce background from unincorporated [α-^{32}P]dATP, cut the agarose gel immediately above the dye front before drying. Unincorporated [α-^{32}P]dATP runs below the dye front and can be discarded

2. To maximize the activity of *Xenopus* egg extracts, we recommend thawing NPE rapidly in room temperature water
3. This protocol is for only a single reaction but can be scaled up or down as necessary. We recommend a minimum reaction volume of 6 μL and a minimum sample volume of 1 μL to ensure accuracy of pipetting
4. To perform a control reaction lacking LacR, substitute an equal volume of dialysis buffer

7. Summary and conclusions

We have described key reagents and assays needed to induce and monitor replication fork collapse *in vitro*. This approach can be augmented with previously published assays to study DNA replication and repair in *Xenopus* egg extracts. For example, inactivation of specific proteins can be studied by immunodepletion of the target protein (Kavlashvili & Dewar, 2022). Additionally, DNA structures can be purified and then restriction digested to unambiguously determine the abundance of replication fork structures and assess ligation of the nascent DNA strands (Kavlashvili & Dewar, 2022). Nascent strands can also be analyzed using denaturing sequencing gels to monitor repair intermediates with nucleotide resolution (Knipscheer, Räschle, Schärer, & Walter, 2012). Finally, DNA-bound proteins can be readily isolated and analyzed by Western blotting (Kavlashvili & Dewar, 2022). The resultant data from these assays can be quantified by densitometry to allow kinetic analysis of the events that occur during and after fork collapse (Kavlashvili & Dewar, 2022). Collectively, these approaches should support detailed mechanistic analyses of replication fork collapse.

Acknowledgments

We thank Joseph Yeeles and Ken Marians for initially providing the LacR expression constructs and purification protocol. J.M.D. is supported by NIH grant R35GM128696. J.M.D. and S.C. are also supported by Vanderbilt-Ingram Cancer Center SPORE in Breast Cancer (P50CA098131).

References

Bhowmick, R., Minocherhomji, S., & Hickson, I. D. (2016). RAD52 facilitates mitotic DNA synthesis following replication stress. *Molecular Cell*, *64*(6), 1117–1126. https://doi.org/10.1016/j.molcel.2016.10.037.

Cortez, D. (2019). Replication-coupled DNA repair. *Molecular Cell*, *74*(5), 866–876. https://doi.org/10.1016/j.molcel.2019.04.027.

Deegan, T. D., Mukherjee, P. P., Fujisawa, R., Polo Rivera, C., & Labib, K. (2020). CMG helicase disassembly is controlled by replication fork DNA, replisome components and a ubiquitin threshold. *eLife, 9*. https://doi.org/10.7554/eLife.60371.

Dewar, J. M., Budzowska, M., & Walter, J. C. (2015). The mechanism of DNA replication termination in vertebrates. *Nature, 525*(7569), 345–350. https://doi.org/10.1038/nature14887.

Dewar, J. M., Low, E., Mann, M., Räschle, M., & Walter, J. C. (2017). CRL2(Lrr1) promotes unloading of the vertebrate replisome from chromatin during replication termination. *Genes & Development, 31*(3), 275–290. https://doi.org/10.1101/gad.291799.116.

Dewar, J. M., & Walter, J. C. (2017). Mechanisms of DNA replication termination. *Nature Reviews. Molecular Cell Biology, 18*(8), 507–516. https://doi.org/10.1038/nrm.2017.42.

Dias, M. P., Moser, S. C., Ganesan, S., & Jonkers, J. (2021). Understanding and overcoming resistance to PARP inhibitors in cancer therapy. *Nature Reviews. Clinical Oncology, 18*(12), 773–791. https://doi.org/10.1038/s41571-021-00532-x.

Duxin, J. P., Dewar, J. M., Yardimci, H., & Walter, J. C. (2014). Repair of a DNA-protein crosslink by replication-coupled proteolysis. *Cell, 159*(2), 346–357. https://doi.org/10.1016/j.cell.2014.09.024.

Goodhead, D. T. (1994). Initial events in the cellular effects of ionizing radiations: Clustered damage in DNA. *International Journal of Radiation Biology, 65*(1), 7–17. https://doi.org/10.1080/09553009414550021.

Hashimoto, Y., Puddu, F., & Costanzo, V. (2011). RAD51- and MRE11-dependent reassembly of uncoupled CMG helicase complex at collapsed replication forks. *Nature Structural & Molecular Biology, 19*(1), 17–24. https://doi.org/10.1038/nsmb.2177.

Jenkyn-Bedford, M., Jones, M. L., Baris, Y., Labib, K. P. M., Cannone, G., Yeeles, J. T. P., et al. (2021). A conserved mechanism for regulating replisome disassembly in eukaryotes. *Nature, 600*(7890), 743–747. https://doi.org/10.1038/s41586-021-04145-3.

Kang, Y. H., Galal, W. C., Farina, A., Tappin, I., & Hurwitz, J. (2012). Properties of the human Cdc45/Mcm2-7/GINS helicase complex and its action with DNA polymerase epsilon in rolling circle DNA synthesis. *Proceedings of the National Academy of Sciences of the United States of America, 109*(16), 6042–6047. https://doi.org/10.1073/pnas.1203734109.

Kavlashvili, T., & Dewar, J. M. (2022). Approaches to monitor termination of DNA replication using xenopus egg extracts. *Methods in Molecular Biology, 2444*, 105–124. https://doi.org/10.1007/978-1-0716-2063-2_7.35290634.

Knipscheer, P., Räschle, M., Schärer, O. D., & Walter, J. C. (2012). Replication-coupled DNA interstrand cross-link repair in xenopus egg extracts. *Methods in Molecular Biology, 920*, 221–243. https://doi.org/10.1007/978-1-61779-998-3_16.

Laskey, R. A., Mills, A. D., & Morris, N. R. (1977). Assembly of SV40 chromatin in a cell-free system from xenopus eggs. *Cell, 10*(2), 237–243. https://doi.org/10.1016/0092-8674(77)90217-3.

Lebofsky, R., Takahashi, T., & Walter, J. C. (2009). DNA replication in nucleus-free xenopus egg extracts. *Methods in Molecular Biology, 521*, 229–252. https://doi.org/10.1007/978-1-60327-815-7_13.

Liu, B., Xue, Q., Tang, Y., Cao, J., Guengerich, F. P., & Zhang, H. (2016). Mechanisms of mutagenesis: DNA replication in the presence of DNA damage. *Mutation Research, Reviews in Mutation Research, 768*, 53–67. https://doi.org/10.1016/j.mrrev.2016.03.006.

Low, E., Chistol, G., Zaher, M. S., Kochenova, O. V., & Walter, J. C. (2020). The DNA replication fork suppresses CMG unloading from chromatin before termination. *Genes & Development, 34*(21 – 22), 1534–1545. https://doi.org/10.1101/gad.339739.120.

Maric, M., Maculins, T., De Piccoli, G., & Labib, K. (2014). Cdc48 and a ubiquitin ligase drive disassembly of the CMG helicase at the end of DNA replication. *Science, 346*(6208), 1253596. https://doi.org/10.1126/science.1253596.

Moreno, S. P., Bailey, R., Campion, N., Herron, S., & Gambus, A. (2014). Polyubiquitylation drives replisome disassembly at the termination of DNA replication. *Science, 346*(6208), 477–481. https://doi.org/10.1126/science.1253585.

Petermann, E., Orta, M. L., Issaeva, N., Schultz, N., & Helleday, T. (2010). Hydroxyurea-stalled replication forks become progressively inactivated and require two different RAD51-mediated pathways for restart and repair. *Molecular Cell, 37*(4), 492–502. https://doi.org/10.1016/j.molcel.2010.01.021.

Räschle, M., Knipscheer, P., Enoiu, M., Angelov, T., Sun, J., Griffith, J. D., et al. (2008). Mechanism of replication-coupled DNA interstrand crosslink repair. *Cell, 134*(6), 969–980. https://doi.org/10.1016/j.cell.2008.08.030.

Semlow, D. R., Zhang, J., Budzowska, M., Drohat, A. C., & Walter, J. C. (2016). Replication-dependent unhooking of DNA Interstrand cross-links by the NEIL3 glycosylase. *Cell, 167*(2), 498–511.e414. https://doi.org/10.1016/j.cell.2016.09.008.

Sonneville, R., Bhowmick, R., Hoffmann, S., Mailand, N., Hickson, I. D., & Labib, K. (2019). TRAIP drives replisome disassembly and mitotic DNA repair synthesis at sites of incomplete DNA replication. *eLife, 8*. https://doi.org/10.7554/eLife.48686.

Sonneville, R., Moreno, S. P., Knebel, A., Johnson, C., Hastie, C. J., Gartner, A., et al. (2017). CUL-2(LRR-1) and UBXN-3 drive replisome disassembly during DNA replication termination and mitosis. *Nature Cell Biology, 19*(5), 468–479. https://doi.org/10.1038/ncb3500.

Sparks, J., & Walter, J. C. (2019). Extracts for analysis of DNA replication in a nucleus-free system. *Cold Spring Harbor Protocols, 2019*(3). https://doi.org/10.1101/pdb.prot097154.

Vrtis, K. B., Dewar, J. M., Chistol, G., Wu, R. A., Graham, T. G. W., & Walter, J. C. (2021). Single-strand DNA breaks cause replisome disassembly. *Molecular Cell, 81*(6), 1309–1318.e1306. https://doi.org/10.1016/j.molcel.2020.12.039.

Walter, J., & Newport, J. (2000). Initiation of eukaryotic DNA replication: Origin unwinding and sequential chromatin association of Cdc45, RPA, and DNA polymerase alpha. *Molecular Cell, 5*(4), 617–627. https://doi.org/10.1016/s1097-2765(00)80241-5.

Wu, X., & Malkova, A. (2021). Break-induced replication mechanisms in yeast and mammals. *Current Opinion in Genetics & Development, 71*, 163–170. https://doi.org/10.1016/j.gde.2021.08.002.

Wühr, M., Freeman, R. M., Jr., Presler, M., Horb, M. E., Peshkin, L., Gygi, S., et al. (2014). Deep proteomics of the Xenopus laevis egg using an mRNA-derived reference database. *Current Biology, 24*(13), 1467–1475. https://doi.org/10.1016/j.cub.2014.05.044.

Wühr, M., Güttler, T., Peshkin, L., McAlister, G. C., Sonnett, M., Ishihara, K., et al. (2015). The nuclear proteome of a vertebrate. *Current Biology, 25*(20), 2663–2671. https://doi.org/10.1016/j.cub.2015.08.047.

Zagnoli-Vieira, G., & Caldecott, K. W. (2020). Untangling trapped topoisomerases with tyrosyl-DNA phosphodiesterases. *DNA Repair (Amst), 94*. https://doi.org/10.1016/j.dnarep.2020.102900, 102900.

CHAPTER SIXTEEN

Measuring the contributions of helicases to break-induced replication

Zhenxin Yan[a,†], Liping Liu[b,†], Nhung Pham[a,‡], Pilendra K. Thakre[a], Anna Malkova[b,*], and Grzegorz Ira[a,*]

[a]Department of Molecular and Human Genetics, Baylor College of Medicine, Houston, TX, United States
[b]Department of Biology, University of Iowa, Iowa City, IA, United States
*Corresponding authors: e-mail address: anna-malkova@uiowa.edu; gira@bcm.edu

Contents

† Equal contribution.
‡ Present address: Department of Genomic Medicine, The University of Texas MD Anderson Cancer Center, 1881 East Road, Houston, TX 77054, USA.

Methods in Enzymology, Volume 672
ISSN 0076-6879
https://doi.org/10.1016/bs.mie.2022.02.025

Abstract

Break-Induced Replication (BIR) is a homologous recombination (HR) pathway that differentiates itself from all other HR pathways by involving extensive DNA synthesis of up to hundreds of kilobases. This DNA synthesis occurs in G2/M arrested cells by a mechanism distinct from regular DNA replication. BIR initiates by strand invasion of a single end of a DNA double-strand break (DSB) followed by extensive D-loop migration. The main replicative helicase Mcm2-7 is dispensable for BIR, however, Pif1 helicase and its PCNA interaction domain are required. Pif1 helicase was shown to be important for extensive repair-specific DNA synthesis at DSB in budding and fission yeasts, flies, and human cells, implicating conservation of the mechanism. Additionally, Mph1 helicase negatively regulates BIR by unwinding migrating D-loops, and Srs2 promotes BIR by eliminating the toxic joint molecules. Here, we describe the methods that address the following questions in studying BIR: (i) how to distinguish enzymes needed specifically for BIR from enzymes needed for other HR mechanisms that require short patch DNA synthesis, (ii) what are the phenotypes expected for mutants deficient in extensive synthesis during BIR, (iii) how to follow extensive DNA synthesis during BIR? Methods are described using yeast model organism and wild-type cells are compared side-by-side with Pif1 deficient cells.

1. Introduction

BIR repairs single-ended DSBs or two-ended DSBs, where homology is present or exposed by resection only at one side of the DSB (reviewed in Wu & Malkova, 2021). BIR repairs broken replication forks, particularly in situations where converging forks are not available (Mayle et al., 2015). BIR is also needed for extension and maintenance of telomeres in telomerase negative cells. Many types of genome instabilities and copy number variations result from BIR (Wu & Malkova, 2021).

Most of the mechanistic studies on BIR were done in budding yeast at HO or I-*Sce*I endonuclease induced two-ended DSBs. The advantage of these assays is the synchrony of break induction, allowing kinetic studies of each BIR step and understanding the function of proteins involved. In these assays, homologous sequences to the template molecule are present only on one side of the DSB. This DSB end invades the template and primes extensive DNA synthesis that can reach the end of the chromosome. DNA synthesis proceeds by D-loop migration whereby the newly synthesized

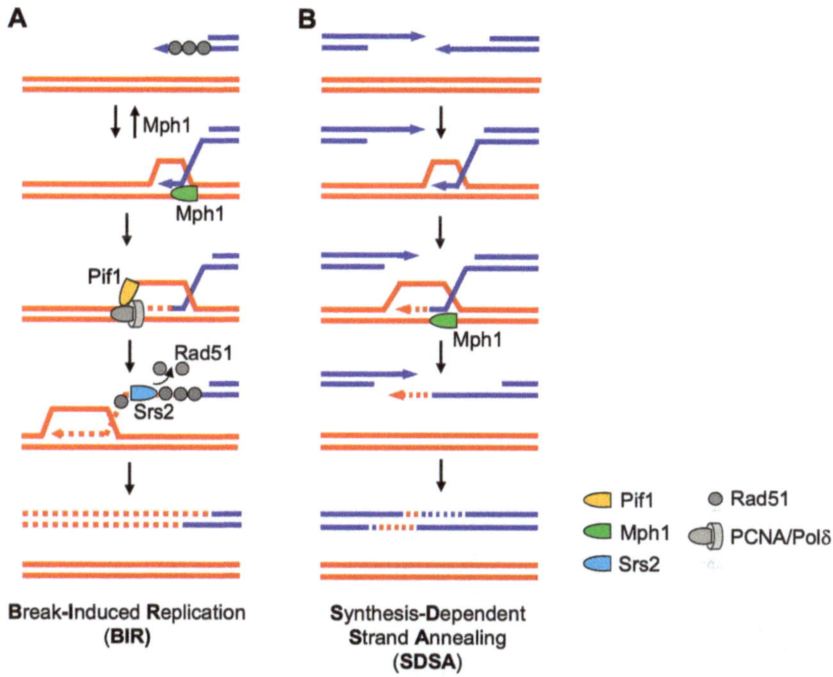

Fig. 1 The role of DNA helicases in BIR and SDSA. Mph1 disrupts the D-loop and therefore inhibits BIR (A) but promotes SDSA (B). Pif1 helicase promotes D-loop migration during BIR. Srs2 prevents formation of joint molecules between long ssDNA generated during BIR and the template.

leading strand is unwound from the template and the lagging strand is subsequently synthesized using the nascent leading strand as a template (Saini et al., 2013). Therefore, DNA synthesis in BIR results in conservative inheritance of newly synthesized strands (Fig. 1A). The migrating D-loop structure was demonstrated by 2D gel electrophoresis and by electron microscopy (Donnianni & Symington, 2013; Saini et al., 2013; Wilson et al., 2013). The consequences of the BIR mechanism are low fidelity DNA synthesis and a proclivity for template switches. The former results from decreased ability of mismatch corrections as the nascent strand is unwound from its template (Saini et al., 2013), and the latter results from overall low stability of the D-loop structure (Piazza et al., 2019; Smith, Llorente, & Symington, 2007). In contrast to BIR, extensive DNA synthesis is not required for repair of two-ended breaks. In synthesis-dependent strand annealing (SDSA), the predominant mechanism of DSB-induced gene conversion in mitotic cells, one end invades the template and primes new

DNA synthesis (Fig. 1B). This is followed by the unwinding of the D-loop and annealing of the newly-synthesized single-strand DNA to the second end of the break. Thus, unwinding of the D-loop prevents extensive D-loop migration, eliminating possible involvement of BIR in repair of two-ended DSBs (Mehta, Beach, & Haber, 2017; Pham et al., 2021).

Multiple DNA helicases are involved in BIR (Fig. 1A). Pif1 helicase translocates on ssDNA in the 5′ to 3′ direction and is very important for extensive DNA synthesis, specifically during BIR but is dispensable for SDSA (Wilson et al., 2013). Pif1 deficient cells can promote only limited leading strand DNA synthesis (Liu et al., 2021). In agreement with the phenotypes in mutant cells, purified yeast Pif1 is capable of stimulating extensive displacement DNA synthesis primed at Rad51-made D-loop (Wilson et al., 2013). Yeast Pif1 carries a PCNA interaction domain at its C-terminal region, which is important for extensive DNA synthesis (Buzovetsky et al., 2017). Pif1 is negatively controlled in the context of meiotic recombination, likely by other proteins interacting with PCNA (Vernekar et al., 2021). Pif1 is phosphorylated by Mec1, and this phosphorylation is likely essential for its function during BIR (Vasianovich, Harrington, & Makovets, 2014). A similar function of Pif1 homologs in promoting extensive DNA synthesis during HR was proposed in fission yeast, flies, and humans (Jalan, Oehler, Morrow, Osman, & Whitby, 2019; Kocak et al., 2019; Li et al., 2021). However, at least in the human experimental system studied so far, DNA synthesis during BIR rarely exceeds several kb. Besides Pif1 helicase, a nonessential component of the Polδ complex, called Pol32 in yeast, is essential for BIR (Lydeard, Jain, Yamaguchi, & Haber, 2007).

Additional DNA helicases also regulate the BIR pathway. Mph1 is capable of unwinding D-loops (Prakash et al., 2009) and therefore promotes SDSA but has a negative impact on BIR and stimulates template switches during BIR (Stafa, Donnianni, Timashev, Lam, & Symington, 2014) (Fig. 1A and B). During BIR, long ssDNA accumulates resulting from extensive 5′ strand resection and asynchronous synthesis of leading and lagging strands. This ssDNA may invade the template at many locations, thus impairing BIR completion (Saini et al., 2013; Wilson et al., 2013). Formation of extensive joint molecules between resected or nascent long ssDNA and the template during BIR is prevented by Srs2 helicase (Elango et al., 2017) (Fig. 1). Methods listed below are focused mostly on testing the role of Pif1 in DNA synthesis during BIR.

2. General method to grow yeast cells and induce DNA double-strand break

All methods presented below require the growth of yeast cells and induction of HO endonuclease that is under the control of galactose inducible promoter, *GAL10*. Yeast are grown in media containing raffinose sugar as the sole source of carbon. Addition of galactose causes induction of a DSB in all cells within 60 min. Cells are collected and frozen before or at different timepoints after DSB induction and repair. Individual steps of BIR are tested using different methodologies. In some assays, cells are plated on YEP-Gal plates and the final products of DSB repair are analyzed as described below. Here we describe the yeast media, growth, and HO induction.

2.1 Equipment

- Platform shaker (*e.g.*, Innova 2300, New Brunswick Scientific).
- Centrifuge (*e.g.*, LEGEND-RT, SORVALL).
- Benchtop centrifuge (*e.g.*, Beckman Coulter and Thermo Scientific).
- Hematocytometer

2.2 Buffers, reagents and media

- YEPD medium: 1% yeast extract, 2% Bacto Peptone, 2% dextrose, and 60 μg/mL adenine sulfate [pH 5.5], autoclave
- YEP-raffinose medium: 1% yeast extract, 2% Bacto Peptone, 2% raffinose, and 60 μg/mL adenine sulfate [pH 5.5], autoclave
- 20% galactose (filter sterilized)

2.3 Procedure

1. Grow yeast cells in 5 mL liquid YEPD in culture tube to saturation overnight on a roller drum or a shaker at 30 °C.
2. Collect 500 μL saturated cell culture, centrifuge at 6000 *g* for 30 s, discard the YEPD medium and wash the cells with YEP-raffinose twice and resuspend the cells in 500 μL YEP-raffinose.
3. Inoculate the washed cell culture to fresh YEP-raffinose medium in 1:10,000 ratio (*v*/v) for wild-type strains. Culture the cells with aeration in 30 °C for around 16–18 h (see note 1).

4. When cells reach the concentration of $0.5–1.5 \times 10^7\,mL^{-1}$, collect 50 mL of cells to a 50 mL tube, as "time 0" control, and induce HO break by adding 20% galactose solution to a final concentration of 2% (see note 2).
5. Spin down the cells at ~3450 g for 2 min, discard the liquid medium, suspend the cells in 500 μL ddH_2O and transfer the cells to a 1.5-mL EP tube. Spin down the cells at 6000 g for 30 s, discard the liquid and freeze the pellets.
6. Collect cells every 1 or 2 h until 6 or 8 h. Note: Most DSB repair pathways tested here are completed by 6 h after HO induction. Subsequent steps are described below depending on the assay used.

2.4 Notes

1. For mutant strains that have longer doubling time, inoculate more cells.
2. Cell concentration should be measured using a hemocytometer (it provides more accurate measurement than by spectrophotometer).

3. Analysis of protein contribution to BIR and its regulation in yeast cells

The yeast strain used here to study the kinetics of BIR repair is called yNP477 and carries H-0 intrachromosomal recombination assay, where the DSB is induced at *MAT***a** locus and is repaired by recombination with *HMLα*-inc template. *HMR* donor was deleted and *MAT***a** locus was modified to eliminate homology to *HMLα*-inc on the proximal end of a DSB (Pham et al., 2021). In this intrachromosomal BIR assay, the distal end of a DSB at *MAT***a** (Z-region) invades the *HMLα*-inc template and primes DNA synthesis that proceeds to the end of a chromosome, about 12 kb (Fig. 2A). The BIR product is inviable, as BIR forms an acentric chromosome fragment and leaves the unrepaired chromosome fragment carrying the centromere. Kinetics of repair or DNA synthesis is followed by PCR or Southern blot (Fig. 2B and C). This BIR assay shows fast repair kinetics and is suitable to test the role of proteins in BIR.

Another strain carrying H-1400 recombination assay (yNP574) is used to exclude proteins that are needed for simple gene conversion with short patch DNA synthesis (Fig. 2D). In this assay, both ends of the break at *MATa* share relatively long homology with the *HMLα*-inc donor. This assay is equivalent to regular yeast mating type switching, where ~700 bp *Ya* sequence is replaced with *Yα-inc* (cannot be cleaved by HO). BIR specific

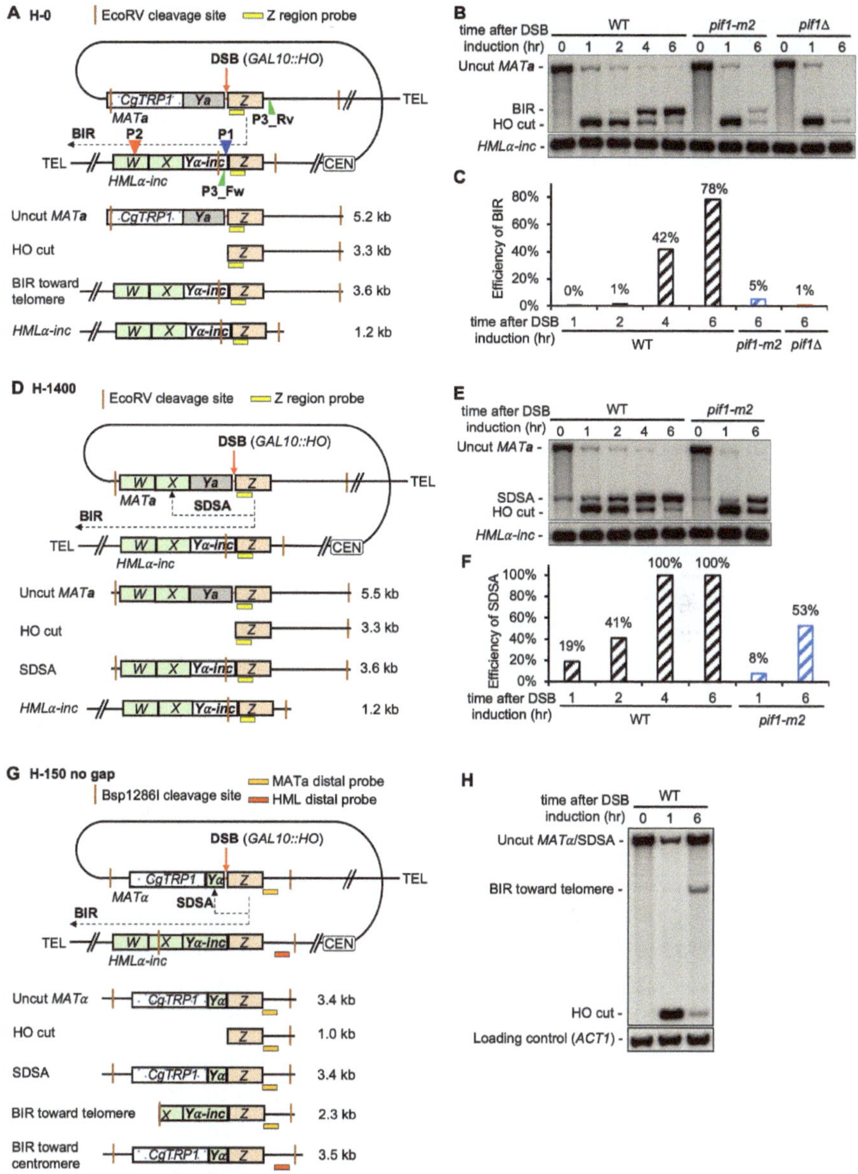

Fig. 2 Assays to analyze BIR mechanism and regulation. (A) Schematic of the H-0 BIR assay. DSB is induced at a modified *MAT***a** locus (Chr. III). Strand invasion occurs within the "Z" and DNA synthesis continues to the end of the chromosome. Restriction fragment size of all expected bands in parental and product DNA is shown. (B) Representative Southern blot showing DSB repair products in H-0 system in wild-type and *pif1* mutant cells. (C) Quantification of BIR product in wild-type and *pif1* mutants. (D) Schematic of the H-1400 assay. A DSB is induced at *MAT***a** and repaired by recombination with *HML*α-inc. There is ~1400 bp homology between *MAT***a** and *HML*α-inc sequences on the "second" end. (E and F) Representative Southern blot and quantification of DSB repair product (SDSA) in wild-type and *pif1* mutant cells. BIR does not contribute to repair in this system (Pham et al., 2021). (G) Schematic of H-150 no-gap assay, where homology is located next to the break site. (H) Representative Southern blot showing DSB repair products in wild-type cells carrying H-150 no-gap assay. BIR and SDSA products are indicated.

proteins should not be essential for this type of repair, as only ~700 bp long DNA must be synthesized during DSB repair. However, a slight decrease or delay in *pif1-m2* mutant in the repair can be attributed to ~700 bp gap that must be synthesized (Fig. 2E and F). In this assay, the DSB is repaired by gene conversion, where the "Z" end invades the template and primes DNA synthesis. Newly synthesized DNA is unwound and anneals to the second end of the break as proposed in SDSA mechanism (Fig. 1B) (Ira, Satory, & Haber, 2006). We note that BIR is not used in this situation because the second end carries ~1400 bp homology with the template making second end capture and SDSA very efficient.

The third strain carrying the H-150 no gap assay (yNP764) was designed in a way that the DSB could be repaired by either BIR or SDSA (Fig. 2G). In this intrachromosomal assay, a DSB induced at *MAT*α is repaired by either gene conversion or BIR. The *MAT*α locus proximal to HO cut site is partially deleted so that only 150 bp of Yα immediately next to the DSB is homologous to *HML*α-inc (Pham et al., 2021). As in the BIR strain described above, the distal end of a DSB at *MAT*α invades the *HML*α-inc template and primes DNA synthesis. We note that normally *MAT*α recombines with *HMR*a, but in this strain *HMR*a is deleted, leaving only template in *HML*α-inc locus. Two outcomes of the repair are possible. The successful capture of the short 150 bp second end leads to gene conversion of *MAT*α to *MAT*α-inc *via* SDSA mechanism and produces viable colonies. If the second end is not captured, DNA synthesis *via* migrating D-loop (BIR) continues to the end of the chromosome (12 kb distance). SDSA and BIR products can be distinguished by different restriction fragment sizes on a Southern blot (Fig. 2H). A detailed map of restriction cut sites and expected sizes for different repair products (BIR and SDSA) are shown (Fig. 2G). We note that crossovers were not observed in this system. This assay is suitable to test the impact of the enzymes processing D-loops, including many DNA helicases (Mehta et al., 2017; Pham et al., 2021). For example, decreased stability of the D-loop leads to more gene conversion, while decreased ability to capture the second end of a DSB increases BIR. In wild-type cells, repair by gene conversion dominates with >95% of total DSB repair, and the remaining 5% of cells repair DSB by BIR (Pham et al., 2021).

3.1 Equipment

- Cell disruptor (SI-D236, Scientific Industries)
- Glass disruptor beads, 0.5 mm (SI-BG05, Scientific Industries)
- Horizontal electrophoresis system (IBI Scientific)

- Slow-speed orbital shaker (Roto Mix, Barnstead International)
- Filter paper (46 × 57 cm, 3030–917, Whatman)
- Blotting pad (20 × 20 cm, 28,298–032, VWR)
- Transfer membrane: Positively charged nylon hybridization membrane (NEF988001PK PerkinElmer)
- Dri-Bath block (Type 17,600, Barnstead International)
- Hybridization tube (SciGene)
- Hybridization incubator (Model 400, SciGene)
- White light transilluminator (TW-43, UVP)
- Phosphor screen and exposure cassette (Cytiva)
- Typhoon Trio scanner (Cytiva)

3.2 Buffer, reagents, and yeast strains

- STES buffer: 0.2 M Tris-HCl [pH 7.6], 0.5 M NaCl, 0.1% SDS, 0.01 M EDTA [pH 8.0]
- Phenol/chloroform/isoamyl alcohol (25:24:1, *v*/v, Fisher BioReagents)
- Chloroform (Fisher BioReagents)
- TE buffer: 10 mM Tris-HCl [pH 8.0], 1 mM EDTA [pH 8.0]
- RNase A, 10 mg/mL
- NaAc, 3 M [pH 5.2]
- 20% SDS
- *Eco*RV and *Bsp*1286I (NEB)
- 5 × TBE buffer: 60.55 g Tris base, 30.9 g boric acid, and 3.35 g Na_2-EDTA in 1 L ddH_2O. Dilute with ddH_2O to 1 × TBE before use
- 0.8% agarose gel in 1 × TBE buffer. Add 20 μL 1% ethidium bromide to 1 L gel before use
- Denaturing buffer/transfer buffer: 0.4 M NaOH
- 20 × SSC: 172.3 g NaCl and 88.3 g Na-Citrate in 1 L ddH_2O
- Wash buffer: 2 × SSC, 0.2% SDS
- Hybridization buffer: Mix 500 mL 0.5 M Na_2HPO_4 [pH 7.2], 350 mL 20% SDS, 2 mL 0.5 M EDTA [pH 8.0], and 148 mL ddH_2O as 1 L buffer
- RadPrime DNA labeling system (18428–011, Invitrogen). Mix the 0.5 mM dCTP, dGTP, and dTTP in 1:1:1 ratio as the "CGT mix"
- [α-^{32}P] dATP, 3000 Ci/mmol 10 mCi/mL (BLU512H250UC, PerkinElmer)
- Isotope purification column: Micro Bio-Spin Chromatography Column (P-30, 732-6223, Bio-Rad)
- Membrane strip buffer: 1% SDS, 0.1% SSC
- NucleoSpin Gel and PCR Clean-up kit (740,609.250, MACHEREY-NAGEL)

Table 1 Yeast strains used in this study.

Strain name	Parental strain	Genotype	Source	Comment
yAM033		*ho ade3::GAL10::HO HMLα -inc MAT**a** hmr::ADE1 bar1::ADE3 nej1::KANMX ade1 leu2,3–112 trp1::hisG ura3–52 thr4 lys5*	Mehta et al. (2017)	
yNP477	yAM033	*MAT (WΔ XΔ::Cg-TRP1)*	Pham et al. (2021)	H-0
yNP489	yNP477	*pif1-m2*	This study	H-0
yNP901	yNP477	*pif1::HPHMX*	This study	H-0
yNP574	yAM033	*ho ade3::GAL10::HO HMLα -inc MAT**a** hmr::ADE1 bar1::ADE3 ade1 leu2,3–112 trp1::hisG ura3–52 thr4 lys5*	This study	H-1400
yNP888	yNP574	*pif1-m2*	This study	H-1400
yNP578	yAM033	*ho ade3::GAL10::HO HMLα -inc MATα hmr::ADE1 bar1::ADE3 nej1::KANMX ade1 leu2,3–112 trp1::hisG ura3–52 thr4 lys5*	Pham et al. (2021)	
yNP764	yNP578	*MAT (WΔ XΔ YαΔ150bp::Cg-TRP1)*	Pham et al. (2021)	H-150 no gap
AM1003		*MAT**a**-LEU2-tel/MATα-inc ade1 met13 ura3 leu2–3112/leu2 thr4 lys5 hml::ADE1/hml::ADE3 hmr::HYG ade3::GAL-HO FS2::NAT/FS2.*	Deem et al. (2008)	Disomic system

- Yeast strains used in this study are listed in Table 1. The *Cg-TRP1* (*Candida glabrata TRP1*) cassette was amplified from pYO2242 to remove some elements in *MATa* region (Mehta et al., 2017). Primers used for these modifications are listed in Table 2
- All media are prepared as in Section 2.2

3.3 Procedure A. Analysis of protein contribution to BIR

1. Collect cells and freeze the cell pellets as described in Section 2.3.
2. Isolate DNA by standard phenol/chloroform procedure. Add glass beads on top of the cell pellets to a total volume of 500 μL. Add 120 μL TE buffer, 300 μL STES buffer, and 350 μL chloroform/phenol/isoamyl

Table 2 Primers for constructing strains and making probes.

Name	5′ to 3′ sequence	Comment
H-0H-150 Fw	gatatcttttaatgttgtctcactatcttgccaataagactctacccagat TCGAG GTCGACGGTATC	To construct H-0 and H-150 no gap systems
H-0 Rv	acttttcctttttggaagttttatacagtttccccgaaagaacaaaaatcaag CGCTCTAGAACTAGTGGATC	To construct H-0 system
H-150 no gap Rv	gtaatatgctctaaaagccataaatgagtttaaatatttctttgaagatt CGCTCTAGAACTAGTGGATC	To construct H-150 no gap system
Z probe Fw	CGCAACAGTATAATTTTATAAA	To make Z region probe
Z probe Rv	ATGAGTGTATAAACAAACATTG	To make Z region probe
MATa distal Fw	ATTGGCTATACGGGACGG	To make MAT distal probe
MATa distal Rv	CTTAGCTTGTACCAGAGG	To make MAT distal probe
HML distal Fw	GGAAAATGCGTGTTCCTTGT	To make HML distal probe
HML distal Rv	TCATTGCAAACGTCGGGCTTT	To make HML distal probe
ACT1 probe Fw	TCTTCCCATCTATCGTCGGTAGAC	To make ACT1 probe (loading control)
ACT1 probe Rv	GGTCAATACCGGCAGATTCC	To make ACT1 probe (loading control)
ADE1 Fw	GGTTTGAAACAACCTCAAGGACTT	To check chromosome rearrangements
ADE1 Rv	AAGTCCTTGAGGTTGTTTCAAACC	To check chromosome rearrangements
ADE3 Fw	GCAGGGTTCGATTTCACTATGGGT	To check chromosome rearrangements
ADE3 Rv	ACCGTCATCATCGACTTCCACGGC	To check chromosome rearrangements
MAT probe Fw	GTTCTTTCGGGGAAACTGTA	To check chromosome rearrangements
MAT probe Fw	CCACATTTCTTTGCAACTTC	To check chromosome rearrangements

alcohol. Vortex the mixture by the cell disruptor for 6 min at 4 °C. Centrifuge the mixture at >16,000 *g* for 5 min, then transfer the aqueous top layer (~420 μL) to a 1.5 mL EP tube. Add another 350 μL chloroform/phenol/isoamyl alcohol, mix and centrifuge at 16,000 *g* for 5 min, then transfer the top aqueous layer (~400 μL) to a new 1.5 mL EP tube. In a hood, add 350 μL chloroform, mix and centrifuge at >16,000 *g* for 5 min, then transfer the top aqueous layer (~380 μL) to a new 1.5 mL EP tube. Add 40 μL 3 M NaAc and 1 mL pure ethanol to precipitate DNA. Keep at −20 °C freezer for at least 1 h.

3. Centrifuge the mixture at >16,000 *g* for 20 min. Remove the supernatant by aspiration. Then wash the pellet with 600 μL 70% ethanol, centrifuge for 5 min at >16,000 *g* and remove the supernatant by aspiration. Dry the pellet at room temperature for 30–60 min. Add 300 μL of TE to dissolve the pellet, add 2 μL RNase A (10 mg/mL) and incubate at 37 °C overnight. Next, precipitate the DNA by adding 30 μL 3 M NaAc and 900 μL pure ethanol. Mix, centrifuge at >16,000 *g* for 20 min and wash the pellet with 600 μL 70% ethanol. Dry the pellet and re-suspend it in 150 μL ddH_2O.
4. Digest all DNA by restriction enzyme(s) in a 300 μL reaction volume overnight at the temperature indicated by the vendor. Typically, 2–5 μL of restriction enzyme per sample is used. Restriction enzymes specific for monitoring BIR is *Eco*RV (Fig. 2A). After digestion, precipitate the DNA by adding 30 μL 3 M NaAc and 900 μL pure ethanol. Mix, centrifuge at >16,000 *g* for 20 min and wash the pellet with 600 μL 70% ethanol. Dry the pellet and re-suspend the pellet in 30–40 μL of TE buffer.
5. Check the quality of DNA digestion by loading and separating 2 μL (1–3 μg) of each sample in a small 0.8% agarose gel for 30 min at 80–120 V. Take a picture of the gel using the UV gel imaging system to make sure that the DNA is fully digested. Measure the amount of DNA in each sample using a fluorometer (*e.g.*, Qubit) and equalize the amount of DNA in samples with TE buffer. Add 4 μL DNA gel loading dye (6 ×) to 20 μL DNA (~10 μg), mix and centrifuge for a few seconds. Load 18–24 μL of each sample into a large (20 × 25 cm) 0.8% agarose gel. Run the gel at 40–60 V for 14–20 h until the DNA fragments between 2 and 10 kb are well separated. Take a picture of the gel using a UV gel-imaging system. Samples should be well equalized and separated.
6. Cut the DNA gel to the size of the blotting paper (20 × 20 cm) and place in a glass tray on an orbital shaker at ~30 rpm. Denature the DNA with

denaturing buffer for 20 min, then replace with new denaturing buffer and continue to shake for another 20 min. Prepare the filter papers and the transfer membrane. Cut four filter papers (Whatman) to 20 × 20 cm, four filter paper strips to 46 × 4 cm, and one filter paper to 16 × 18 cm. Cut the transfer membrane to 21 × 21 cm (1 cm longer and 1 cm wider than the gel). Label the corner of the membrane with the date and experiment number.

7. Transfer the DNA from the gel to the transfer membrane by assembling a "transfer sandwich" as shown in Fig. 3 (also known as a capillary blotting apparatus). In detail, place two packs of paper towels on the benchtop (7–10 cm high for each pack), then place 5–6 blotting pads on top of the paper towels. Wet two layers of filter paper (20 × 20 cm) and place on top of the blotting pad (avoid any bubbles). Wet the transfer membrane and put it beneath the gel. Hold the bottom of the membrane and put the membrane and gel together on top of the filter/blotting paper stack. Wet another two filter papers (20 × 20 cm) and put them

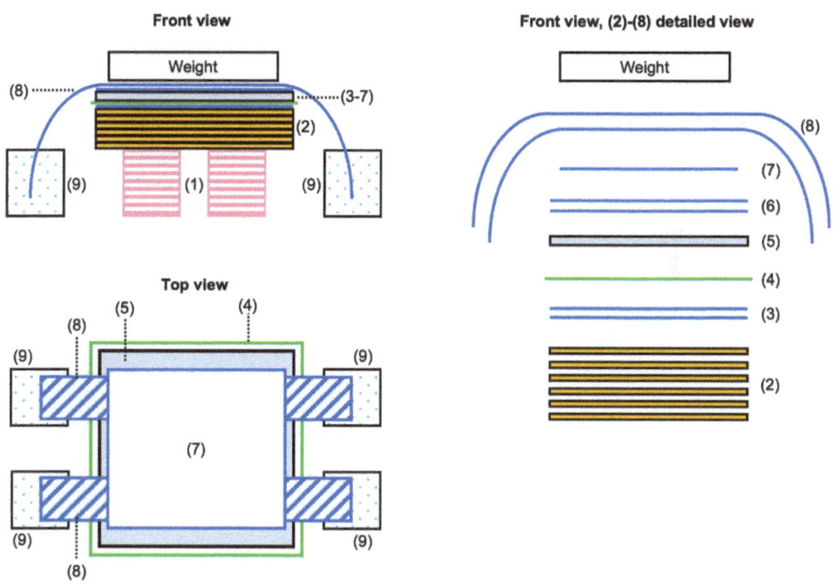

Fig. 3 Scheme of the Southern blot "transfer sandwich." (1) Two packs of paper towels; (2) Six layers of blotting pads, 20 × 20 cm; (3) Two layers of filter papers, 20 × 20 cm. (4) Nylon membrane; (5) Agarose gel with DNA; (6) Two layers of filter papers with the same size of the agarose gel; (7) Four beakers with denature/transfer buffer, each is ~250 mL; (8) Two layers of bridge filter paper strips, 46 × 4 cm, four strips in total; (9) One layer of filter paper, 16 × 18 cm.

on top of the gel. Roll out all the bubbles between each layer of the "sandwich" using a long plastic pipette. Put 2 two-layer "buffer bridge" wide strips in parallel on top of the stack, put the two ends of each bridge into the transfer buffer (the bridge acts as a wick for buffer transfer). Put another filter paper (16 × 18 cm) on top of the bridges to connect them together. Cover the stack with a piece of plastic wrap and place a weight (about 0.5 kg) on top of the stack. Cover the transfer sandwich with plastic wrap to prevent buffer evaporation. Leave the "transfer sandwich" for >20 h. Use tweezers to place the membrane in a glass tray and wash the membrane with wash buffer for 20 min on an orbital shaker at ~30 rpm. Dry the membrane on a paper towel at room temperature for at least 20 min. Put the membrane in the hybridization tube with 50 mL hybridization buffer. Put the hybridization tube in the hybridization incubator. Set the temperature to 65 °C. Incubate for at least 30 min for prehybridization.

8. DNA probes (length of 300–800 bp) for Southern blotting are prepared by PCR, purified by regular gel-extraction (NucleoSpin Gel and PCR Clean-up, MACHEREY-NAGEL) and diluted to a concentration of ~50 ng/μL. DNA probes specific for monitoring BIR product formation are specified in Table 2.
9. This and the next steps involving radioactive materials should be performed in area designated for work with radioactive materials. For one Southern blot, add 25–50 ng of the desired probe in a 1.5 mL EP tube, and ddH_2O to a total volume of 20 μL. Denature the DNA probe in a 100 °C Dri-Bath block for 5 min (hold the cap by a CapLock tube clip to keep the cap closed), then transfer immediately to ice for 5 min. Add 20 μL 2.5 × RadPrime buffer (Invitrogen), 3 μL CGT mix, 1 μL Klenow Fragment and 3 μL [α-^{32}P] dATP to the tube, mix gently, and spin down briefly. Incubate at 37 °C for 0.5–2 h. Purify the probe using a Micro Bio-Spin Chromatography Column (Bio-Rad) according to the manufacturer's instructions. Heat denature the purified sample at 100 °C for 5 min and transfer immediately to ice for 5 min. Pipette ~40 μL of the purified probe mix into the hybridization tube and incubate overnight.
10. Discard the hybridization buffer, add 50 mL wash buffer, and wash in the hybridization incubator for 30 min. Repeat the washing two additional times. Remove the membrane from the hybridization tube, place on a paper towel briefly to remove the remaining buffer and quickly wrap with plastic wrap (do not let it dry). Place the membrane face up inside the exposure cassette and put the phosphor screen on top

of the membrane. Expose the phosphor screen for several hours up to overnight depending on the strength of the signal. (Note: Erase the phosphor screen on a white light transilluminator for 30 min every time before use.)

11. Use Typhoon Trio scanner to scan the exposed phosphor screen. Align the screen to the top left. Open the Typhoon Trio Imager software, select the region to be scanned and follow the instructions to scan the screen at the desired resolution. Use ImageQuant TL (GE Healthcare) or ImageJ (NIH) software to quantify the intensity of the bands on a Southern blot. Export the band pixel intensity values to Excel for analysis. For quantification of repair efficiency in each time point, first normalize the pixel intensity of the BIR product bands, the SDSA product bands, and the parental bands by the intensity of the loading control (Fig. 2B, E, and H and Table 2). Then divide the normalized BIR or SDSA band intensity in each time point by the normalized band intensity of parental band at time 0 and multiply by 100 to obtain the percentage of BIR or SDSA efficiency over time (Fig. 2C and F).
12. After scanning, wrapped blots can be stored at 4 °C for a few months. To re-probe the same membrane, remove the plastic wrap, put the blot in a plastic tray, and immediately pour 400 mL of boiling strip buffer (use microwave to heat the buffer) on top of the blot. Gently shake the tray for 3 min then discard the strip buffer. Repeat the wash twice more to remove hybridized probe. Dry the blot by leaving it on paper towels at room temperature. *Re*-hybridize the blot with additional probes as needed
13. Example of the analysis is shown for *pif1* mutant cells in Fig. 2A–F. In BIR assay H-0, Pif1 is nearly essential to observe any repair. In gene conversion assay H-1400, Pif1 repair by 6 h is decreased by about half, likely due to the ~700 bp long gap that must be synthesized. However, *pif1*-m2 cells show viability comparable to wild-type (not shown, see note 5) in this system, meaning repair that appears low by 6 h is eventually completed. We note that in the hypomorphic *pif1*-m2 mutant, Pif1-m2 protein retains normal mitochondrial localization, while nuclear Pif1 is dramatically decreased

3.4 Procedure B. Analysis of contribution of gene conversion and BIR to DSB repair

As discussed in the introduction to Section 3, multiple assays can be used to compare the contribution of BIR and simple gene conversion by SDSA to the repair of a single DSB. Many mechanisms were described that control the respective contributions of BIR and SDSA, including several DNA helicases. Here, we present one of the most efficient and simple assays to compare BIR and SDSA contribution to repair of a DSB in strain yNP764.

1. Break induction, DNA isolation, and Southern blot analysis should be performed exactly as in Section 3.3 except that restriction enzymes (Bsp1286I, NEB) and probes are different. The break can be repaired by regular SDSA or by BIR and the products can be distinguished based on different restriction sizes as shown in Fig. 2G.
2. Quantification of the products is performed based on the intensity of the bands corresponding to SDSA and BIR: **(i)** Percentage of BIR among the products measured at 6 h after DSB induction is calculated as the proportion of the pixel intensity of the band corresponding to BIR to the sum of pixel intensity of the bands corresponding to both BIR and SDSA products. **(ii)** Repair efficiency among all cells for each mutant is calculated as the proportion of normalized pixel intensity of bands corresponding to both BIR and SDSA products at 6 h to the normalized pixel intensity of the parental band at 0 h. **(iii)** Repair efficiency compared to wild-type is measured as the proportion of normalized pixel intensity of bands corresponding to either BIR or SDSA at 6 h of a mutant to the normalized pixel intensity of bands corresponding to BIR and SDSA at 6 h of wild-type cells. Statistical comparison of the percentage of BIR and repair efficiency between different mutants can be performed using Welch's unpaired *t*-test (Prism 8 software). No quantification is shown in Fig. 2G–H as only a single experiment in wild-type was done here. Analyses of many DNA helicases in this and similar systems are presented in published manuscripts (Mehta et al., 2017; Pham et al., 2021).

3.5 Notes

1. The yNP764 strain was generated to identify enzymes regulating the contribution of BIR and gene conversion *via* SDSA to DSB repair. This assay was designed to make second end capture more challenging by reducing homology size to 150 bp. Additional assays can be used that have much longer homology on both sides of the DSB to make sure that phenotypes are not specific for recombination between short repeats as published (Pham et al., 2021).

2. Experiments using isotope should follow radiation safety regulations of the institute.
3. Alternative restriction enzymes can be used to detect BIR products (Pham et al., 2021).
4. In case any mutant shows decreased repair, it is important to make sure that this deficiency is not an indirect effect of spontaneous events (for example related to incomplete replication (Wang et al., 2004). To eliminate such possibility, BIR analysis should also be performed in G2/M arrested cells. G2/M arrest can be accomplished with nocodazole (0.015 mg/mL). This concern is particularly important when testing proteins essential for replication.
5. Viability is tested by plating cells from YEP-raffinose preculture on YEPD and YEP-Gal plates. Viability is calculated as the proportion of cells grown on YEP-Gal compared to control cell number grown on YEPD plates.

4. Analysis of genome instability in mutants deficient in extensive BIR synthesis

The BIR assay presented above (H-0) is ideal to follow kinetics of BIR repair; however, it cannot be used to follow any genome instabilities, such as mutagenesis associated with BIR, since no survivors can be obtained for analysis irrespective of whether the DSB is repaired or not. The alternative strain to follow BIR and different types of genome instability during BIR is a yeast disomic strain carrying two copies of chromosome III (Saini et al., 2013). Here the DSB is induced at *MAT***a** locus on a truncated chromosome III wherein the 100 kb centromere distal from *MAT***a** is replaced with *LEU2* followed by telomeric repeats. Only the proximal DSB end has extensive homology to the donor sequence, so that most cells employ BIR for repair. Following strand invasion, over 100 kb of the full-length template chromosome III donor is copied. In this assay, the chromosomes participating in repair are marked by either *LEU2*, *ADE1* or *ADE3* which allows determination of the repair pathway by growth on selective media (Fig. 4A). To avoid contribution of Ty retrotransposon repeats to repair, the nearest inverted Ty1 repeats were replaced with a *NAT* cassette. To confirm genome instability, all chromosomes were separated on agarose gels by pulsed-field gel electrophoresis (PFGE).

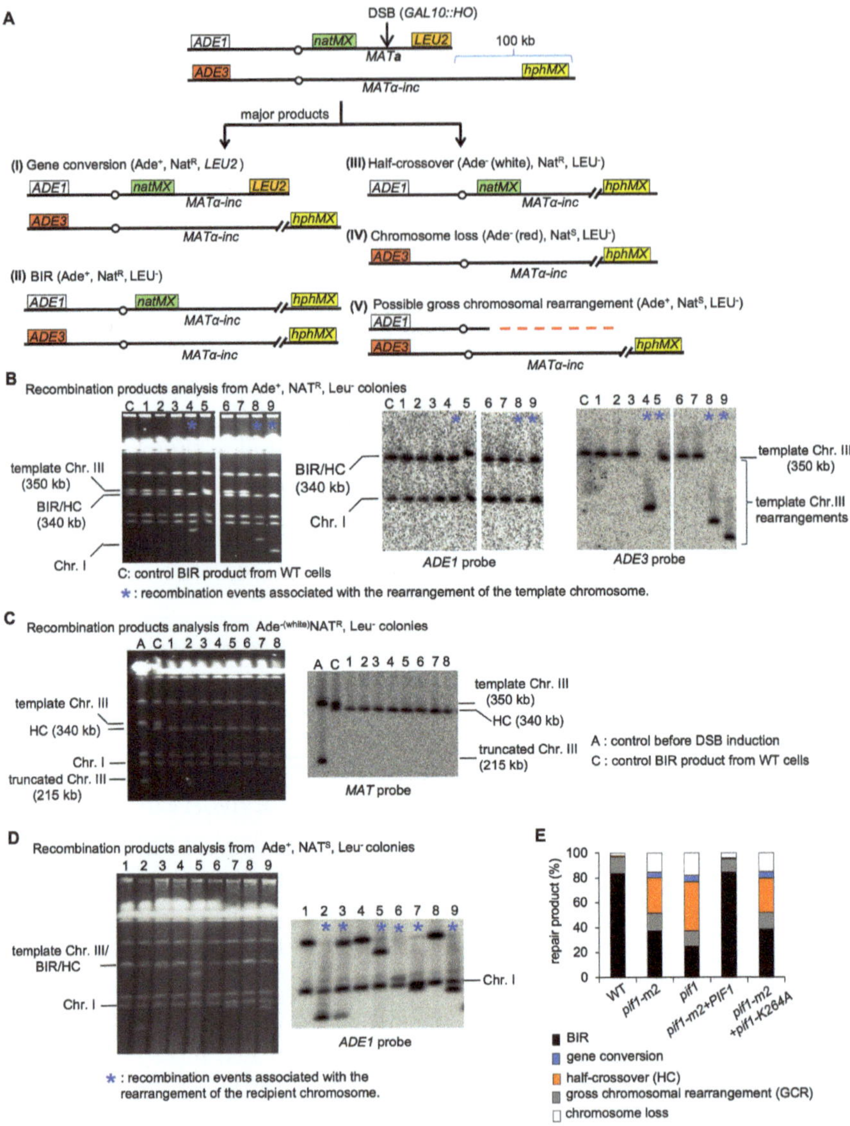

Fig. 4 Analysis of recombination products in allelic BIR assay in Pif1 deficient cells. (A) Illustration presents BIR assay and possible outcomes of the repair. Markers located on both recombining chromosomes provide means to score different DSB repair outcomes. The half-crossover outcome (iii) is typical for mutants deficient in extensive DNA synthesis. In half-crossovers, the part of the template chromosome centromere distal to the invasion site is fused to the broken chromosome, with the remainder of the template chromosome either becoming stabilized (examples shown in D) or lost (as shown in C). *natMX* and *hphMX* provide resistance for nourseothricin or hygromycin B, respectively. (B) Analysis of recombination products from Ade^{+} NAT^{R} Leu^{-} colonies. Examples

4.1 Equipment

- The CHEF-DR II system or CHEF Mapper XA System (Bio-Rad)
- Gel plug mold (Bio-Rad)
- Centrifuge (*e.g.*, LEGEND-RT, SORVALL)
- Benchtop centrifuge (*e.g.*, Beckman Coulter)
- Vortex (*e.g.*, Vortex-Genie 2)
- 5 mL conical tubes
- Water baths
- Gel imaging system

4.2 Reagents, media and strains

- Yeast strain used in this study is AM1003 (see Table 1).
- Standard yeast selective media plates lacking adenine, leucine, or YEPD plates with nourseothricin (NAT) (100 μg/mL). Dropout mixes used to prepare plates can be purchased from Bufferad, Inc. (ade^- #S0052, leu^- #S0064).
- Yeast growth media as in Section 2.
- Solution I: 100 mM EDTA, 50 mM Tris [pH 7.5].
- Zymolyase 100T 5 mg/mL as stock solution in 10 mM KPO_4, [pH 7.4] (keep 100 μL aliquots in −20 °C).
- 0.5% agarose (LE or other standard gelling temperature agarose) in Solution I (prepare fresh, melt and hold in 45 °C bath).
- PK/Sarkosyl solution: 5% Sarkosyl, 5 mg/mL proteinase K in 0.1 M EDTA [pH 7.5] (Store 1 mL aliquots at −20 °C).
- Pulsed-field certified agarose (162–0137, Bio-Rad).
- TBE 10 × buffer (890 mM Tris base, 890 mM boric acid, 20 mM EDTA [pH 8.0].
- Ethidium Bromide (EtBr) 10 mg/mL stock solution.
- 0.25 N HCl.
- Southern blot reagents as in Section 3, DNA probes are listed in Table 2.

Fig. 4—Cont'd are shown where rearrangements of the template chromosome are indicated by an asterisk. (C) Analysis of half-crossover products from Ade^{-white} NAT^R Leu^- colonies. (D) Analysis of rare Ade^+ NAT^S Leu^- colonies. Rearranged chromosomes are indicated by an asterisk. (E) Overall analysis of DSB repair products in wild-type and *pif1* mutants. At least 500 colonies were scored for each strain. *Panels (B–E) are reproduced from Wilson, M. A., Kwon, Y., Xu, Y., Chung, W. H., Chi, P., Niu, H., et al. (2013). Pif1 helicase and Poldelta promote recombination-coupled DNA synthesis via bubble migration.* Nature, 502, *393–396.*

4.3 Procedure

1. Inoculate AM1003 strain (Table 1) or its mutant derivative in 5 mL liquid YEPD as in Section 2.3. Transfer preculture to YEP-raffinose and grow overnight to the concentration of $\sim 1 \times 10^7$ cells/mL and plate 50–75 cells on YEPGal and YEPD plates.
2. Once the colonies are fully formed (3–4 days for wild-type cells), replica plate them on the following selective media plates: Ade-, Leu-, YEPD + NAT and incubate plates overnight at 30 °C.
3. Analyze data based on markers pattern. All or nearly all colonies replica plated from YEPD should maintain the same pattern of markers as parental strain, as no DSB was induced. BIR leads to the loss of the *LEU2* marker. When repair fails, *ADE1*, *NAT* and *LEU2* markers are lost, and the colonies appear red due to an Ade1 deficiency. When cells repair the DSB by gene conversion using the short homology on the right side of the break, all the markers are retained and mating type changes **a**/α to α. In rare cases in wild-type cells, part of the homologous template chromosome is lost due to a half-crossover that eliminates the *ADE3* marker, and colonies are Ade^- and appear white. Because the repair by BIR occurs in G2/M cells, two copies of each chromosome are present. Owing to random segregation of chromosomes, half of the half-crossovers (the major product in *pif1*Δ cells) segregate with an intact copy of the full-length chromosome III and are genetically and structurally indistinguishable from BIR. Therefore, the number of Ade^- white colonies scored on selective media as half-crossovers represent only half of these events. To correct for this, the number of half-crossovers (Ade^-, white colonies) is multiplied by two, and consequently the number of BIR events is adjusted by subtracting the number of observed Ade^-, white colonies. Finally, Ade^+ colonies that have lost the distal *NAT* marker can correspond to genomic rearrangement, or BIR, where strand invasion occurred proximal to the *NAT* marker. To distinguish between the two, chromosome size is analyzed in random 20–30 colonies that lost NAT marker using CHEF. An example of BIR analysis is shown in Fig. 4 for WT and *pif1* mutants. While the marker analysis provides an overall view of major categories, additional verification of the products by CHEF is recommended for mutants that show poor repair and/or increased sectored colonies. Some of the products that are scored as BIR products may represent genome rearrangements as shown in *pif1-m2* mutants below.

4. To perform analysis of chromosome sizes for the products of DSB repair, inoculate yeast colonies from YEPGal plates in 5 mL of YEPD culture and grow overnight to saturation. In case the colony carries several sectors on minimal media, streak the colony on selective media to make sure the analysis is done for a pure single product. All sectors of a single colony can be analyzed separately as previously published (Vasan, Deem, Ramakrishnan, Argueso, & Malkova, 2014).
5. Spin down yeast cultures at 4000 RPM for 2 mins, resuspend cell pellets in 1 mL of solution I. Transfer cells to EP tubes, spin down for 1 min at 6000 RPM and discard supernatant. The pellets can be stored at −80 °C or kept on ice until the next step. Use the equivalent of $\sim 1 \times 10^8$ stationary phase cells per plug sample (~100 μL). Fewer cells are fine, but more cells might overload the lane and perturb chromosome separation.
6. Make 0.5% agarose in solution I in EP tubes and hold in 45 °C bath. Add 6.3 μL of Zymolyase (to digest the yeast cell wall) per 100 μL of 0.5% agarose solution to make Zymolyase/agarose mix. Note: 1 mL of Zymolyase/agarose mix is sufficient for about 10 plugs. Calculate the appropriate volume of Zymolyase/agarose mix needed for all samples. Vortex briefly and prewarm the tube containing the loosened pellet for ~7–10 s in the 45 °C water bath (no longer), add 100 μL of Zymolyase/agarose mix to the pellet and mix slowly by pipetting (avoid bubbles).
7. Pipet the mixture quickly to the agarose plug mold taped on the bottom (avoid any bubbles). Allow agarose plugs to harden in the mold for ~5 min at 4 °C. Take off the tape and push agarose plugs from single sample (1–2 plugs) to 800 μL of Solution I in a 1.5 mL EP tube or 15 mL conical tube and incubate at 37 °C for 30 min (Note: Rinse the plug mold well with water after pushing plugs out, dry and reuse if needed).
8. Add 200 ul of PK/Sarcosyl solution or 5% Sarcosyl alone to each sample and incubate overnight at 37 °C without agitation. PK/Sarcosyl solution is used when chromosomes need to be digested with restriction enzymes. Next day, wash the plugs once in Solution I and keep in Solution I at 4 °C for up to 1 month. If restriction enzyme digestion is needed, wash the plugs 4 times for 1 h each with gentle agitation.
9. Prepare 1% pulsed-field certified agarose gel in 0.5 × TBE in CHEF casting stand. After the gel solidifies, using a micro-spatula, load agarose plugs (half or less of the plug is recommended for better separation)

containing chromosome size DNA fragments on the bottom of the wells and seal the wells using melted 1% agarose. After 30 min, disassemble the gel-casting stand while leaving the gel adhered to the black plate and install the gel on a plate in an electrophoresis tank filled with 0.5× TBE buffer (~2.2 L) and run at 6 V/cm for 30 h (initial time = 10 s, final time = 35 s). The buffer should be circulated in the CHEF apparatus electrophoresis tank and cooled down to 14 °C before electrophoresis is started.

10. After electrophoresis, stain the gel by shaking it gently for 30 min in electrophoresis buffer with EtBr (1 μg/mL). Remove the buffer, destain the gel for 60 min in 0.5× TBE, and document it using gel imaging system.
11. Perform Southern blotting analysis as described above with following modification: incubate the gel in a tray with 0.25 N HCl for 30 min gently shaking. Rinse the gel with water and treat with DNA denaturing solution as described above. Proceed as described above with setting transfer sandwich and probing. The following probes should be used: *ADE1* gene-specific, *ADE3* gene-specific, *MAT* locus specific. Primers to amplify the probes are listed in Table 3.
12. Analysis of the data: The size of the BIR product is slightly smaller compared to the template chromosome III; so, it can be distinguished on EtBr-stained gel and/or after probing with *ADE1* gene. *ADE3* gene probe is used to observe template chromosome rearrangements.

Table 3 Primers for detecting BIR synthesis by ddPCR in the H-0 system.

Name	5′ to 3′ sequence	Dye
P1	Forward: CAGTTTGGCTCCGGTGTAA Reverse: GCGCGAAGTAGTCCCATATT Probe: TTCTCTGCTCGCTGAAGAATGGCA	FAM
P2	Forward: TGTGCGGCAGCATCTTAT Reverse: AGAGGGTCACAGCACTACTA Probe: AAATCGTCCTGTCCCATTACGAGCTG	FAM
P3	Forward: GCACGGAATATGGACTACTT Reverse: ATTTGTCATCCGTCCCGTATAG Probe: AGATAAACAACCTCCGCCACGACC	FAM
ACT1	Forward: GCCTTCTACGTTTCCATCCA Reverse: GGTAGAGAGAAACCAGCGTAAA Probe: TTCCGGTGATGGTGTTACTCACGT	HEX

Initial inspection of the EtBr-stained gel shows whether template chromosome III has the proper size (by comparison to the sample without DSB). Probing with *ADE3* either confirms the proper chromosome size or identifies a rearranged chromosome. If the recipient (broken) chromosome was rearranged during repair, *ADE1*-specific probe detects most products as shown in Fig. 4. *MAT*-specific probe can be used as well to detect half-crossovers. Analysis in Fig. 4 shows dramatic BIR deficiency and half-crossover increase in Pif1 deficient cells including helicase dead *pif1*-K264A mutant.

4.4 Notes

1. CHEF gel separation conditions used are good to separate BIR product (345 kb), template chromosome III (355 kb), truncated chromosome III (~200 kb) and any chromosome up to ~400 kb. Rare longer chromosome rearrangements can still be detected by Southern blot, but the exact' size of these cannot be estimated. To separate all chromosomes, set initial pulse time to 10 s and final to 90 s and the run for 30 h.

5. Analysis of DNA synthesis during BIR by ddPCR

Previous methods detecting BIR synthesis mainly focused on either early product that only involves up to several kilobases of synthesis or final products (Donnianni & Symington, 2013; Lydeard et al., 2007, 2010; Malkova, Naylor, Yamaguchi, Ira, & Haber, 2005). It was not possible to follow BIR synthesis along the whole template with one single method. A recently developed Assay for Monitoring BIR Elongation Rate (AMBER) (Liu et al., 2021) can be applied to any BIR assay to follow initial and extensive DNA synthesis. The basic principle is based on analysis of copy number increase during DNA synthesis using droplet digital PCR (ddPCR), which is a technique that allows detection of very small changes of DNA copy number. Not only does AMBER allow for monitoring of the progression of efficient BIR synthesis along the entire template in wild-type cells, but also for the detection of residual levels of synthesis in various BIR-deficient mutants. Furthermore, unlike other methods, AMBER can be used to distinguish real BIR products from half-crossover products, which are often formed when BIR is interrupted (Deem et al., 2008; Smith, Lam, & Symington, 2009; Vasan et al., 2014). In addition, the DNA preparation is optimized in

AMBER to increase ssDNA stability, allowing for detection of BIR leading strand synthesis in mutants where lagging strand synthesis is compromised (Liu et al., 2021).

5.1 Equipment

General equipment

- 37 °C and 60 °C water baths
- Microcentrifuge
- 1.5 mL Eppendorf tubes (EP tubes)
- Pipettes
- Vortex
- RAININ tips

Equipment for ddPCR

- QX200 Droplet Digital PCR (ddPCR) System and related supplies including DG8 Cartridges (1864008, Bio-Rad), ddPCR 96-well plates (12001925, Bio-Rad), PCR plate heat seal foil (1814040, Bio-Rad), ddPCR Reader oil (1863004, Bio-Rad), DG8 Gaskets (1863009, Bio-Rad).
- 50 and 200 μL multiple channel pipettes.
- RAININ micro pipette tips (1000, 200, and 10 μL) (NOTE: using RAININ tips for ddPCR is important to get good quantity and quality of droplets).
- 8-strip PCR tubes.

5.2 Reagents

- Zymolyase buffer: 0.1 g/mL 20 T zymolyase (can use 100 T as well) (08320921, MP Biomedicals) dissolved in 2% glucose, 50 mM Tris–HCl [pH 7.5], sterilize by filtration and store at 4 °C. (Best if used within a month).
- Spheroplasting buffer: 0.4 M sorbitol, 0.4 M KCl, 40 mM sodium phosphate buffer [pH 7.2], 0.5 mM $MgCl_2$. Sterilize by filtration and keep at 4 °C. (Best if used within a month).
- 10 × CutSmart buffer: 0.5 M potassium acetate, 0.2 M Tris-acetate, 0.1 M magnesium acetate, 1 mg/mL bovine serum albumin (BSA) [pH 7.9] at 25 °C. Store at −20 °C.
- 10% sodium dodecyl sulfate (SDS) buffer: Store at room temperature.
- 10% Triton X-100 buffer: Sterilize by filtration and store at room temperature.

- 10 mg/mL RNase A: boil for 10 min, then cool down at room temperature for 1 h. Store at −20 °C.
- 10 mg/mL proteinase K: dissolve 10 mg proteinase K powder in 1 mL ddH_2O (always make fresh before use).
- 3 M sodium acetate buffer [pH 5.2].
- Phenol:chloroform:isoamyl alcohol (25:24:1) (Fisher, #15593-049).
- Isopropanol.
- 80% Ethanol.
- RNase and DNase-free water.

Reagents for ddPCR

- Droplet generation oil (1863005, Bio-Rad)
- 2 × Supermix for probe (186-3026, Bio-Rad)
- RNase and DNase-free water
- Primer sets with probe (primers to probe at 4:1 ratio) ordered from IDT (Table 3)

5.3 Procedure

1. Yeast growth and DSB break induction should be performed as in Section 2 with the following modifications: HO induction should be done when cell concentration reaches 0.5×10^7 mL^{-1}. 10–30 mL of culture should be collected for each time point and stocked at −80 °C.
2. Remove cells from −80 °C and thaw by incubating in a 37 °C water bath for ~2 min.
3. Make digestion buffer by mixing 1 mL spheroplasting buffer with 5 μL zymolyase buffer per sample.
4. Quickly pellet cells by centrifugation at 3000 rpm for 1 min, remove all liquid, and resuspend cells with 1 mL of digestion buffer per sample. Digest cells by incubating at 37 °C for 10–15 min, with gentle mixing every 5 min.
5. After digestion, briefly shake by hand to suspend cells and then keep them on ice. Count the cells by hemocytometer, take about 5×10^5 cells from each sample (~10–50 μL), centrifuge at 3000 rpm for 2 min, and remove liquid.
6. Resuspend cells in 360 μL 1 × CutSmart buffer, add 4 μL 10% SDS and mix by vortexing.
7. Incubate at 60 °C for 15 min. Vortex every 5 min.
8. Cool solution on ice for 5 min.

9. Add 46 μL 10% Triton X-100 to neutralize SDS; mix gently by vortexing.
10. Add 20 μL of 10 mg/mL RNase, mix by vortexing, and then incubate at 37 °C for 30 min.
11. Add 20 μL of 10 mg/mL Proteinase K. Mix by vortex and then incubate at 60 °C for 30 min. Cool on ice for 5 min.
12. Add 500 μL phenol:chloroform:isoamyl alcohol (25:24:1) and then vortex for 30 s. Centrifuge at 13,000 rpm for 15 min. Carefully transfer the top layer of liquid to a new tube.
13. Repeat step 12 once more.
14. Add 1/10 volume (~35 μL) of 3 M sodium acetate [pH 5.2] and 1 volume of isopropanol (~350 μL). Mix thoroughly by inverting, then incubate at room temperature for 30 min. Precipitate DNA by centrifuging at 13,000 rpm for 20 min and drain the liquid as much as possible without touching the pellet.
15. Add 500 μL 80% ethanol and centrifuge for another 5 min at 13,000 rpm. Then remove all liquid. Dry the pellet, then add 100 μL water to dissolve DNA. DNA is stored at −20 °C for later use.
16. DNA samples are quantified by Qubit then diluted to ~0.2 ng/μL for ddPCR experiment in step 17–20.
17. PCR samples are mixed using the following recipe: 10 μL supermix for probe, 6 μL water, 1 μL primer set for target locus, 1 μL primer set for the *ACT1* locus, 2 μL diluted DNA sample (~0.2 ng/μL). Mix thoroughly by gentle vortexing for 20 s, then briefly centrifuge to collect all liquid.
18. The PCR mix is then used to generate ddPCR droplets using the QX200 Droplet Digital PCR (ddPCR) System. The droplets are transferred to 96 well plates and sealed with foil.
19. PCR is conducted using the following program: step 1: 95 °C for 10 min; step 2: 94 °C for 30 s (2 °C change per *sec*), 60 °C for 2.5 min (2 °C change per sec); step 3, repeat step 2 for 39 cycles; step 4: 98 °C for 10 min; step 5: 12 °C for 30 min.
20. After carrying out PCR, the samples are analyzed using the QX200 Droplet Digital PCR (ddPCR) System.

5.4 Analysis of results

In the experiment shown in Fig. 5, three sets of primers were used to detect DNA synthesis during BIR using the H0 system described in Section 3.

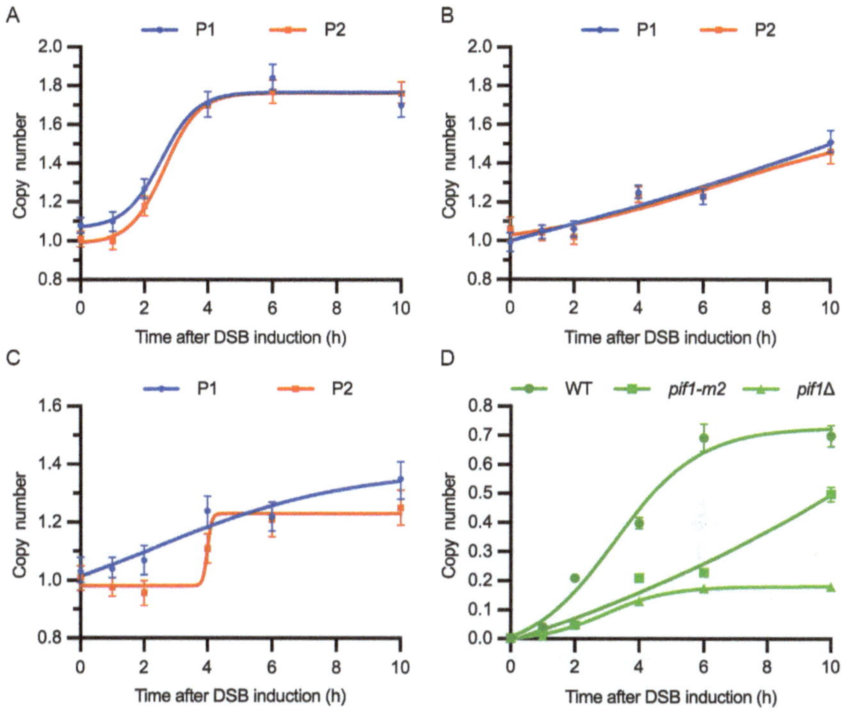

Fig. 5 BIR synthesis detected by AMBER. DNA copy number increase detected by set P1 and set P2 in wild-type (A), *pif1-m2* (B), and *pif1Δ* (C). (D) DNA copy number increase detected by set P3 (junction DNA) in different strains. Mean values of ratios between target to reference (*ACT1*) loci are calculated by Poisson distribution based on 10,000–20,000 droplets. The error bars represent Poisson 95% confidence intervals (CI).

Primer sequences are listed in Table 3 and their position is shown in Fig. 2A. Two pairs of primers, set P1 and set P2, are located on the template. The copy number of the PCR products produced by each primer set was expected to increase from 1× to approximately 2× but is often ≤2× during repair by BIR (Fig. 5A–C). Another set of primers, P3_Fw and P3_Rv, is designed to recognize the newly formed junction resulting from BIR synthesis. Unlike other primer sets, P3 has one primer specifically recognizing a sequence on the recipient DNA and the other primer specifically recognizing the template sequence. Therefore, copy number of P3 was expected to be 0 before BIR, and increase to ≤1× during BIR (Fig. 5D).

In wild-type strain, BIR synthesis indicated by a DNA copy number increase from 1× to ~1.1× can be readily detected at ~1 h after DSB induction using P1 primer set (Fig. 5A). DNA copy number at P1 position constantly increases and reaches ~1.8× by 6h. The kinetics of DNA synthesis during BIR at P2 position is similar to that at P1 position (Fig. 5A). Copy number of the BIR junction detected by P3 primer set in this strain equals ~0× before BIR and increases to 0.04× by 1h and to 0.2× at 2h (Fig. 5D). The copy number for this BIR junction increases to 0.7× at 6 and 10h after DSB induction, consistent with the DNA copy number increases detected by P1 and P2 primer sets (Fig. 5A).

As the major helicase during BIR (Wilson et al., 2013), Pif1 plays an important role in promoting BIR synthesis. Two alleles of *pif1* mutation, *pif1-m2* and *pif1Δ* were demonstrated to cause BIR defect (Saini et al., 2013; Wilson et al., 2013). Here, we observed that in *pif1-m2,* DNA copy number at P1 and P2 positions begins to increase to 1.1× between 2 and 4h, indicative of a slower initiation and elongation of BIR synthesis in *pif1-m2* compared to wild-type strain. The DNA copy number at both P1 and P2 position increases to ~1.5× at late time points (10h) (Fig. 5B). In *pif1Δ*, the defect of BIR synthesis is more severe than in *pif1-m2*. DNA synthesis detected in *pif1Δ* using P1 and P2 is delayed compared to both *PIF1* and *pif1-m2* strains (Fig. 5C). In particular, copy number does not increase in *pif1Δ* to ≥1.1× until 4h post DSB at P1 and P2, and only increases to ~1.3 at 10h post DSB, which is much lower than in *pif1-m2* and *PIF1* strains (Fig. 5A and B). The kinetics of copy number increase in both *pif1-m2* and *pif1Δ* detected by P3 (measuring formation of BIR junction) are consistent to the kinetics detected by P1 and P2. In particular, in *pif1-m2,* the junction DNA is first detected at 1h after DSB induction (0.02× increase) and reaches 0.2× at 4h and 0.5× by 10h, while in *pif1Δ* the product detected by P3 appears as 0.05× at 2h, and reaches 0.18× by 10h (Fig. 5D).

5.5 Notes

1. Buffers including Spheroplasting buffer, Zymolyase buffer, and 10% Triton X-100 can be contaminated with bacteria, fungi, or molds, and therefore must be well stored and checked before use.
2. After digestion in step5, check the cells under the microscope to make sure that cells are efficiently digested. The digested cells look like "ghosts."
3. It is very important to make sure that samples are mixed thoroughly in this experiment.
4. DNA pellets in steps 14 and 15 may not be visible by eye. Therefore, labeling the outer side of the EP tube before centrifuging is useful.
5. In step 17, do not vortex harshly.

6. After carrying out the PCR reaction, if not scanning right away, keep the plate in 4 °C fridge for 1–2 days but covered to avoid light.
7. Additional treatment of DNA with S1 nuclease during purification can eliminate ssDNA and therefore indicate mutants that fail or delay synthesis of lagging strand (Liu et al., 2021).

6. Summary and conclusions

Yeast is so far the best model organism to study BIR mechanism. However, similar methods as described here should be developed in other model organisms or human cells. We suggest developing such assays at subtelomeric loci, as we noted that BIR is most efficient close to telomeres. Of particular interest are future studies in human cells carrying many more DNA helicases compared to yeast systems. Many of these helicases are expected to contribute to regulation of the of BIR.

Acknowledgments

We would like to thank Dr. James Haber for sharing strains and for discussions. We thank Jordan Lee for critical reading of this manuscript. This work was supported by NIH grants to G.I. (GM080600 and GM125650) and to A.M. (R35GM127006). N.P. is currently supported by CPRIT Training Award (RP210028).

References

Buzovetsky, O., Kwon, Y., Pham, N. T., Kim, C., Ira, G., Sung, P., et al. (2017). Role of the Pif1-PCNA complex in pol delta-dependent strand displacement DNA synthesis and break-induced replication. *Cell Reports*, *21*, 1707–1714.

Deem, A., Barker, K., Vanhulle, K., Downing, B., Vayl, A., & Malkova, A. (2008). Defective break-induced replication leads to half-crossovers in Saccharomyces cerevisiae. *Genetics*, *179*, 1845–1860.

Donnianni, R. A., & Symington, L. S. (2013). Break-induced replication occurs by conservative DNA synthesis. *Proceedings of the National Academy of Sciences of the United States of America*, *110*, 13475–13480.

Elango, R., Sheng, Z., Jackson, J., DeCata, J., Ibrahim, Y., Pham, N. T., et al. (2017). Break-induced replication promotes formation of lethal joint molecules dissolved by Srs2. *Nature Communications*, *8*, 1790.

Ira, G., Satory, D., & Haber, J. E. (2006). Conservative inheritance of newly synthesized DNA in double-strand break-induced gene conversion. *Molecular and Cellular Biology*, *26*, 9424–9429.

Jalan, M., Oehler, J., Morrow, C. A., Osman, F., & Whitby, M. C. (2019). Factors affecting template switch recombination associated with restarted DNA replication. *eLife*, *8*.

Kocak, E., Dykstra, S., Nemeth, A., Coughlin, C. G., Rodgers, K., & McVey, M. (2019). The Drosophila melanogaster PIF1 helicase promotes survival during replication stress and processive DNA synthesis during double-strand gap repair. *Genetics*, *213*, 835–847.

Li, S., Wang, H., Jehi, S., Li, J., Liu, S., Wang, Z., et al. (2021). PIF1 helicase promotes break-induced replication in mammalian cells. *The EMBO Journal*, *40*, e104509.

Liu, L., Yan, Z., Osia, B. A., Twarowski, J., Sun, L., Kramara, J., et al. (2021). Tracking break-induced replication shows that it stalls at roadblocks. *Nature, 590*, 655–659.

Lydeard, J. R., Jain, S., Yamaguchi, M., & Haber, J. E. (2007). Break-induced replication and telomerase-independent telomere maintenance require Pol32. *Nature, 448*, 820–823.

Lydeard, J. R., Lipkin-Moore, Z., Sheu, Y. J., Stillman, B., Burgers, P. M., & Haber, J. E. (2010). Break-induced replication requires all essential DNA replication factors except those specific for pre-RC assembly. *Genes & Development, 24*, 1133–1144.

Malkova, A., Naylor, M. L., Yamaguchi, M., Ira, G., & Haber, J. E. (2005). RAD51-dependent break-induced replication differs in kinetics and checkpoint responses from RAD51-mediated gene conversion. *Molecular and Cellular Biology, 25*, 933–944.

Mayle, R., Campbell, I. M., Beck, C. R., Yu, Y., Wilson, M., Shaw, C. A., et al. (2015). DNA REPAIR. Mus81 and converging forks limit the mutagenicity of replication fork breakage. *Science, 349*, 742–747.

Mehta, A., Beach, A., & Haber, J. E. (2017). Homology requirements and competition between gene conversion and break-induced replication during double-strand break repair. *Molecular Cell, 65*, 515–526. e513.

Pham, N., Yan, Z., Yu, Y., Faria Afreen, M., Malkova, A., Haber, J. E., et al. (2021). Mechanisms restraining break-induced replication at two-ended DNA double-strand breaks. *The EMBO Journal, 40*. e104847.

Piazza, A., Shah, S. S., Wright, W. D., Gore, S. K., Koszul, R., & Heyer, W. D. (2019). Dynamic processing of displacement loops during recombinational DNA repair. *Molecular Cell, 73*, 1255–1266. e1254.

Prakash, R., Satory, D., Dray, E., Papusha, A., Scheller, J., Kramer, W., et al. (2009). Yeast Mph1 helicase dissociates Rad51-made D-loops: Implications for crossover control in mitotic recombination. *Genes & Development, 23*, 67–79.

Saini, N., Ramakrishnan, S., Elango, R., Ayyar, S., Zhang, Y., Deem, A., et al. (2013). Migrating bubble during break-induced replication drives conservative DNA synthesis. *Nature, 502*, 389–392.

Smith, C. E., Lam, A. F., & Symington, L. S. (2009). Aberrant double-strand break repair resulting in half crossovers in mutants defective for Rad51 or the DNA polymerase delta complex. *Molecular and Cellular Biology, 29*, 1432–1441.

Smith, C. E., Llorente, B., & Symington, L. S. (2007). Template switching during break-induced replication. *Nature, 447*, 102–105.

Stafa, A., Donnianni, R. A., Timashev, L. A., Lam, A. F., & Symington, L. S. (2014). Template switching during break-induced replication is promoted by the Mph1 helicase in Saccharomyces cerevisiae. *Genetics, 196*, 1017–1028.

Vasan, S., Deem, A., Ramakrishnan, S., Argueso, J. L., & Malkova, A. (2014). Cascades of genetic instability resulting from compromised break-induced replication. *PLoS Genetics, 10*, e1004119.

Vasianovich, Y., Harrington, L. A., & Makovets, S. (2014). Break-induced replication requires DNA damage-induced phosphorylation of Pif1 and leads to telomere lengthening. *PLoS Genetics, 10*, e1004679.

Vernekar, D. V., Reginato, G., Adam, C., Ranjha, L., Dingli, F., Marsolier, M. C., et al. (2021). The Pif1 helicase is actively inhibited during meiotic recombination which restrains gene conversion tract length. *Nucleic Acids Research, 49*, 4522–4533.

Wang, X., Ira, G., Tercero, J. A., Holmes, A. M., Diffley, J. F., & Haber, J. E. (2004). Role of DNA replication proteins in double-strand break-induced recombination in Saccharomyces cerevisiae. *Molecular and Cellular Biology, 24*, 6891–6899.

Wilson, M. A., Kwon, Y., Xu, Y., Chung, W. H., Chi, P., Niu, H., et al. (2013). Pif1 helicase and Poldelta promote recombination-coupled DNA synthesis via bubble migration. *Nature, 502*, 393–396.

Wu, X., & Malkova, A. (2021). Break-induced replication mechanisms in yeast and mammals. *Current Opinion in Genetics & Development, 71*, 163–170.

CHAPTER SEVENTEEN

A RADAR method to measure DNA topoisomerase covalent complexes

Alice Meroni and Alessandro Vindigni*
Division of Oncology, Department of Medicine, Washington University in St. Louis, St. Louis, MO, United States
*Corresponding author: e-mail address: avindigni@wustl.edu

Contents

Abstract

DNA topoisomerases resolve topological stress by introducing transient single- or double-strand breaks into the DNA duplex. This reaction requires the covalent binding of topoisomerases to DNA while the topological stress is being released. This transient intermediate is known as topoisomerase-covalent complex and represents the target of many anti-cancer drugs. Here, we describe a protocol to quantitatively detect topoisomerase-covalent complexes in vivo, called RADAR (rapid approach to DNA adduct recovery). DNA and protein-DNA covalent complexes are rapidly isolated from cells through chaotropic extraction. After normalization, samples are loaded on a slot blot, and the covalent complexes are detected using specific topoisomerase antibodies.

Methods in Enzymology, Volume 672
ISSN 0076-6879
https://doi.org/10.1016/bs.mie.2022.03.036

In addition to being fast and robust, this assay produces quantitative results. Consequently, the RADAR assay can be applied to investigate the topoisomerase-covalent complex biology, including the effect of specific topoisomerase inhibitors. Finally, the same assay can be more generally applied to study covalent complexes of other enzymes with DNA.

1. Introduction

DNA topoisomerases are a family of enzymes essential to maintain the topology of the DNA double helix (Champoux, 2001; Wang, 2002). Several DNA transactions, including DNA replication, transcription, and recombination, transiently unwind the DNA double helix, altering its supercoiling state (Liu & Wang, 1987; Masse & Drolet, 1999; Wu, Shyy, Wang, & Liu, 1988). If not resolved, positive supercoiling will stall DNA replication and transcription, whereas negative supercoiling will favor the stabilization of aberrant DNA secondary structures. Physical constraints in eukaryotic chromosomes or circular genomes prevent the resolution of these supercoiling states (Schoeffler & Berger, 2008). However, DNA topoisomerases act promptly on the supercoiled DNA to re-establish the appropriate DNA topology and ensure faithful completion of these DNA transactions (Pommier, Sun, Huang, & Nitiss, 2016).

Topoisomerase enzymes quickly resolve topological stresses by introducing transient breaks into the DNA helix, which allow the dissolution of the supercoiled structures. Those breaks are then re-ligated by the topoisomerase before dissociation from the DNA (Champoux, 2001). Topoisomerases are classified as type I or II according to the number of DNA strands that they cleave (Forterre, Gribaldo, Gadelle, & Serre, 2007). Type I Topoisomerases cleave only one DNA strand in a reaction that does not require ATP, and they are further classified as type IA, IB, and IC, according to their structure and reaction mechanism. Conversely, type II Topoisomerases cleave both DNA strands in an ATP-dependent reaction, and they are further classified as type IIA and IIB, according to their structures.

During the cleavage reaction, topoisomerases form a transient covalent bond between a conserved tyrosine residue in their catalytic site and one end of the DNA break (Champoux, 2001). These complexes are commonly known as topoisomerase-covalent complexes. Topoisomerase inhibitors are a widely used class of anticancer agents that act by stabilizing the topoisomerase-covalent complexes on DNA, preventing the re-ligation of the broken DNA ends (Nitiss, 2009; Pommier, 2006; Venditto & Simanek, 2010). As a consequence, cells accumulate DNA-protein crosslinks and DNA breaks, ultimately leading to cell death (Nitiss, 2009; Pommier, 2006; Strumberg et al., 2000). Studying the kinetics of formation and resolution of topoisomerase-covalent complexes is crucial to understanding the topoisomerase mechanism of action and testing the efficacy of new potential inhibitors.

While in vitro studies with purified topoisomerases are relatively established and straightforward (Nitiss, Soans, Rogojina, Seth, & Mishina, 2012), the detection of cellular topoisomerase-covalent complexes is technically challenging because of their transient nature. Moreover, the number of topoisomerase-covalent complexes is extremely low compared to the free soluble topoisomerase enzymes, further complicating their detection. The first assay for the topoisomerase-covalent complex detection was developed by Muller and colleagues and it is called ICE assay (In vivo Complex of Enzyme) (Subramanian, Kraut, Staubus, Young, & Muller, 1995). In this assay, the DNA, together with the DNA-bound proteins, is purified by CsCl gradient and topoisomerase-covalent complexes are then visualized by slot or dot blot using specific anti-topoisomerase antibodies. Although this method yields the highest DNA purity, it is time-consuming and necessitates the use of ultracentrifuges. Nevertheless, the ICE assay is still used by several laboratories (Anand, Sun, Zhao, Nitiss, & Nitiss, 2018). More recently, the Meizels' group developed a faster approach to detect DNA-protein covalent complexes, called RADAR assay (rapid approach to DNA adduct recovery) (Kiianitsa & Maizels, 2013). The DNA and the DNA-bound complexes are extracted from cells by a chaotropic extraction and then ethanol-precipitated. Similar to the ICE method, the RADAR assay also uses specific antibodies to visualize topoisomerase-covalent complexes by slot or dot blot. This method has gained popularity because it is relatively fast and does not necessitate any special equipment or a large amount of starting material.

In this chapter, we provide a detailed protocol for the RADAR assay, which we specifically optimized to detect Topoisomerase 1 adducts. The assay described here can be more generally applied to detect other topoisomerases complexes, as well as other types of DNA-protein crosslinks. Briefly, cells are lysed using a solution containing a chaotropic salt, which allows for rapid extraction of nucleic acids (DNA and RNA). DNA is then ethanol precipitated and resuspended in NaOH. After accurate DNA quantification and normalization, samples are deposited on a nitrocellulose membrane by slot blotting. Topoisomerase-covalent complexes are detected by incubation with topoisomerase-specific primary antibodies, and subsequently by quantitative binding to IRDye® conjugated secondary antibodies, which utilize an infrared (IR) developing technology. Following this protocol, topoisomerase-covalent complexes can be visualized and quantified in a highly reproducible manner. The quantification is particularly helpful to study changes in the levels of topoisomerase-covalent complexes, their kinetics of formation or degradation, and to compare efficacies of different topoisomerase inhibitors.

2. Before you begin

1. Label 1.5 mL tubes.
2. Prepare all the buffers.
3. Refrigerate a bench-top centrifuge.

3. Key resources table

Reagent or resource	Source	Identifier
Antibodies		
Rabbit polyclonal Anti-Topoisomerase 1	Abcam	Cat# ab28432; RRID:AB_778545
IRDye® 800CW Goat anti-Rabbit IgG Secondary Antibody	LI-COR	Cat# 926-32211; RRID:AB_621843
Chemicals, peptides, and recombinant proteins		
Camptothecin	Selleck Chemicals	Cat# S1288
DNAzol™ Reagent, for isolation of genomic DNA from solid and liquid samples	Thermo Fisher Scientific	Cat# 10503027
Experimental models: Cell lines		
Human: hTERT RPE-1		
Software and Algorithms		
ImageStudio™	LI-COR	https://www.licor.com/bio/image-studio/ RRID:SCR_013715

4. Materials and equipment

4.1 Materials

- DNAzol reagent (Thermo Fisher Scientific)
- 100% ethanol
- 75% ethanol
- 8 mM NaOH
- Cell scrapers
- 6 cm cell culture plates
- 1.5 mL tubes
- Glass Pasteur pipette
- Tris-Buffered Saline (TBS)
- Tris-Buffered Saline (TBS) + 0.1% Tween-20
- 3% blotting milk in TBS 0.1% Tween-20
- Nitrocellulose membrane (0.45 μm)
- Blotting paper (Pure Cellulose Chromatography Paper, Thick)
- IRDye® conjugated secondary antibodies (LI-COR)

4.2 Equipment

- Nanodrop
- Slot blot apparatus (Bio-Rad)
- LI-COR Odyssey imaging system
- Microcentrifuge (refrigerated)
- Vortex

4.3 Software

- ImageStudio™ (LI-COR)

5. Step-by-step method details

5.1 Topoisomerase covalent complexes extraction (timing: 1–1.5 h)

1. Plate 6–7 × 10^5 mammalian cells in a 6 cm plate (1 plate/experimental condition, including the untreated sample) (see *Note 1*).
2. Treat cells with the topoisomerase inhibitor of choice (see *Note 2*).
3. Quickly aspirate the media and lyse cells with 1 mL of DNAzol reagent.

4. Collect by scraping and transfer the cell lysate into pre-labeled 1.5 mL tubes.
5. Add ½ volume of 100% cold ethanol (0.5 mL) to precipitate DNA and DNA-protein crosslinks.
6. Vortex briefly.
7. Incubate the samples at −20 °C for 10 min.
8. Pellet at 10,000 rpm for 10 min at 4 °C.
9. Carefully aspirate the supernatant with a glass Pasteur pipette and add 1 mL of 70% cold ethanol to wash the pellets.
10. Vortex and pellet at 14,000 rpm for 5 min at 4 °C.
11. Repeat steps 8 and 9.
12. Air dry pellets for 3 min at RT (see *Note 3*).
13. Add 0.3 mL of freshly prepared 8 mM NaOH at RT and resuspend either by vortexing or pipetting up and down in the case of a harder pellet. Keep samples always on ice from here.

Note 1: For our experiment, we used human hTERT-RPE1 cells. However, this protocol can be applied to any mammalian cell line. Different cell lines may respond differently to topoisomerase inhibitors and an initial optimization is required to choose the appropriate topoisomerase inhibitor concentration. Further modifications would be necessary to apply the same protocol to bacteria and yeast, according to previously published methods (Aldred, Payne, & Voegerl, 2019; Anand et al., 2018).

Note 2: Topoisomerase 1-covalent complexes are specifically targeted by camptothecin (CPT) and its clinical derivatives: topotecan and irinotecan. Non-camptothecin inhibitors have been also developed and are currently under investigation, including indenoisoquinoline, phenanthridines, and indolocarbazoles. Topoisomerase 2-covalent complexes are inhibited by etoposide and teniposide, doxorubicin, and its derivatives. For a detailed review of Topoisomerase inhibitors see Delgado, Hsieh, Chan, and Hiasa (2018).

Note 3: Air drying the pellet is important to ensure that the remaining ethanol evaporates from the samples. On the other hand, over-drying the samples will make pellets very difficult to resuspend in the next step. For this reason, we recommend aspirating as much ethanol as possible with the Pasteur pipette during step 9.

5.2 Samples normalization (timing: 20–40 min)

1. Quantify DNA using a Nanodrop instrument.

2. Normalize DNA to 30 ng/μL in 0.3 mL TBS.
3. Quantify DNA again with the Nanodrop and proceed with additional normalization, if needed (see *Note 3*).
4. Add 0.6 mL TBS to obtain a final DNA concentration of 10 ng/μL. We suggest not normalizing the samples directly to 10 ng/μL because it is a low concentration and could lead to a quantification error.

Critical: The normalization step is critical for further data quantification and interpretation. Repeat this step until the samples are exactly at the same concentration (tolerate maximum 1–2 ng/μL of variation, which corresponds to the instrument error).

Note 3: The samples are normalized based on their DNA content immediately before their use. DNA can be quantified using a Nanodrop instrument. Typically, one 6 cm plate gives a DNA yield of 70–200 ng/μL of DNA. Alternatively, samples can be loaded on a second slot blot for DNA quantification. This method requires UV-crosslinking of the DNA that is subsequently visualized by incubation with anti-dsDNA antibodies or SYBR green staining (Fielden et al., 2020; Nierzwicki-Bauer, Gebhardt, Linkkila, & Walsh, 1990; Sun et al., 2020; Xiao, Lin, & Lin, 2021). Because this procedure is frequently used, we verified that the Nanodrop quantification was as accurate as of the quantification with the anti-dsDNA antibodies, and the results are shown in Fig. 1. The advantages of using the Nanodrop quantification relative to the other methods are (1) there is no need to perform a slot blot and process a second membrane; (2) the DNA quantification corresponds exactly to the same samples that will be blotted and probed for the quantification of the topoisomerase-covalent complexes; (3) it requires a low amount of DNA (1–2 μL), allowing multiple measurements, if necessary; (4) it measures the concentration of DNA directly (ng/μL) rather than using arbitrary units.

5.3 Slot blotting (timing: 1.5 h)

1. Cut nitrocellulose membrane (0.45 μm) according to the size of your slot blot apparatus.
2. Equilibrate the nitrocellulose membrane for 15 min in TBS.
3. Assemble the slot blot apparatus according to the manufacturers' instructions and connect it to a vacuum system.
4. Wash the wells 2 × with ~0.3–0.5 mL TBS by applying the vacuum and make sure there is no leaking of the wells.
5. Once the wells are completely empty from TBS, stop the vacuum.

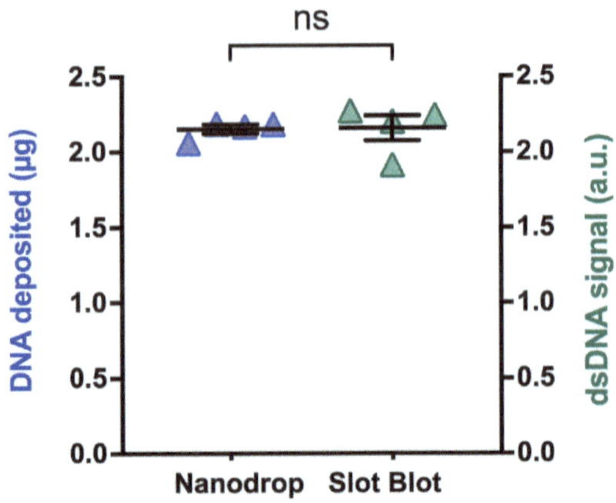

Fig. 1 Nanodrop versus slot blot DNA quantification. For the Nanodrop quantification, the DNA is extracted from 4 independent plates of RPE-1 cells, quantified with the Nanodrop, and normalized according to the protocol described above. For the slot blot quantification, 2 µg of each sample are slot blotted on nitrocellulose, UV-crosslinked, and probed with anti-dsDNA antibodies. The signal from the Nanodrop quantification is expressed in µg of DNA deposited (left y-axis, blue triangles), the signal from the anti-dsDNA antibodies is expressed as arbitrary units (a.u., right y-axis, green triangles). The difference between the two groups is non-significant, as calculated by the paired *t*-test.

6. Load 3 different DNA quantities per sample. The sample DNA concentration is 10 ng/µL, thus load 50, 100, and 300 µL to have 0.5, 1, and 3 µg of DNA, respectively (see *Note 4*). Fill the empty wells with 0.3 mL of TBS.
7. Apply the vacuum at medium speed.
8. When all the wells are completely empty, stop the vacuum and open the slot blot apparatus.
9. Wash the membrane 1 × in TBS for 2 min.
10. Block with 3% milk in TBS + 0.1% Tween-20 for 1 h at RT.
11. Probe with specific anti-topoisomerase antibodies O/N at 4 °C (1:1500 dilution for rabbit polyclonal anti-Topoisomerase 1).
12. Wash 3 × with TBS + 0.1% Tween-20.
13. Incubate with 1:10,000 IRDye® 800CW Secondary Antibody (Goat anti-Rabbit IgG).
14. Wash 3 × with TBS + 0.1% Tween-20.
15. Put the membrane in TBS and scan it using a LI-COR Odyssey imaging system.

Note 4: Loading 3 different amounts of DNA will ensure more consistent and reproducible results. The best range of concentrations has to be determined according to the type and concentration of the inhibitor and

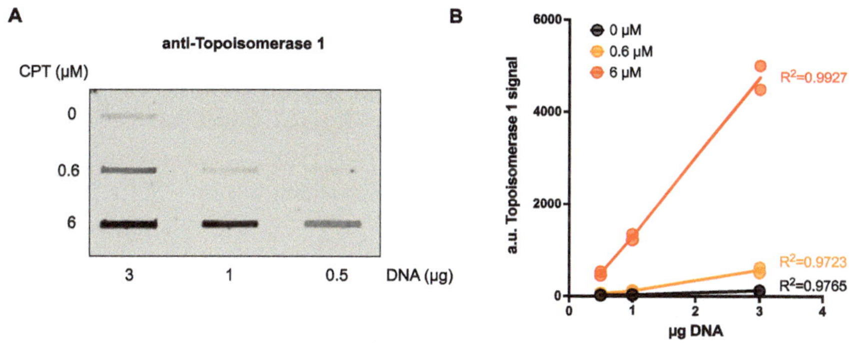

Fig. 2 RADAR assay results and signal analysis. (A) Representative result of the RADAR assay to detect Topoisomerase 1-covalent complexes that are stabilized by CPT. Cells are treated for 1 h with two different CPT concentrations, 0.6 and 6 µM. After proper normalization, three DNA amounts are deposited on the nitrocellulose membrane (0.5, 1, and 3 µg) and probed with anti-Topoisomerase 1 antibodies. (B) XY chart representing the signal quantification from A). The µg of DNA is plotted versus the a.u. Topoisomerase 1 signal for the three different conditions. The data are then interpolated with linear regression, and the R^2 values are shown.

the quality of the primary antibody. When performing this protocol for the first time, load 6–12 different amounts of DNA to identify a linear range for quantification. The topoisomerase-covalent complex signal originating from the 3 selected DNA amounts should fall into a linear range ($R^2 > 0.90$, see Section 7).

Fig. 2A represents a typical blot for Topoisomerase 1-covalent complexes. Cells are plated in 6 cm plates and treated with two different concentrations of CPT, 0.6 and 6 µM, for 1 h. After the treatment, cells are collected, and the DNA is extracted according to the protocol. Three different DNA quantities are blotted (0.5, 1, and 3 µg), which were previously assessed to fall into a linear range for such CPT treatments. The signal of each band is quantified with ImageStudio™ (see Section 7) and plotted as XY chart, with the µg of DNA versus the a.u. Top1 signal (Fig. 2B). The linearity of the range is estimated by calculating the R^2 value of the linear regression line of each sample (Fig. 2B).

6. Expected outcomes

The RADAR assay allows the detection of topoisomerase-covalent complexes extracted from living cells in a dose-dependent manner, as shown in Fig. 2A. Moreover, the covalent complexes can be quantified, making this protocol suitable for statistical analysis.

7. Quantification and statistical analysis

IRDye® conjugated secondary antibodies (LI-COR) are our method of choice to develop and quantify signals from blots. This is considered the best quantification method, as it is more linear than traditional ECL methods (Janes, 2015; Wang et al., 2007).

To quantify the slot blot results, proceed as follows:

1. Open the slot blot file in ImageStudio™ and draw a rectangle on the first slot band from the analysis pane. Then, copy and paste the same rectangle on each slot band.
2. Copy and paste the Intensity signal data into a new spreadsheet (i.e., Microsoft Excel).
3. Normalize each Intensity signal value on the corresponding untreated control sample, according to the DNA quantity.
4. Plot the data as a bar plot showing the mean ± SEM of two or three independent biological replicates. The topoisomerase-covalent complex fold change with respect to the untreated control is represented on the y-axis, while the experimental conditions are shown on the x-axis (Fig. 3).
5. Calculate the statistical significance using the *t*-test (in the case of two conditions) or the One-Way ANOVA test for multiple comparisons.

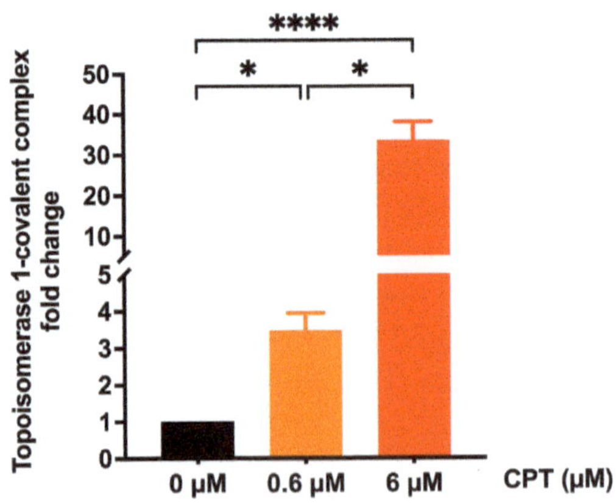

Fig. 3 RADAR assay quantification plot. Bar plot of the RADAR assay quantification for Topoisomerase 1-covalent complexes accumulation in cells treated with two different concentrations of CPT, 0.6 and 6 μM. The values represent the mean ± SEM and are normalized to the untreated control, which results as 1. Increasing the CPT concentration 10-fold also led to a 10-fold increase in the Topoisomerase 1-covalent complexes.

Using different amounts of DNA per sample is critical for determining the amount of covalent complexes with greater accuracy. The signal originating from poorly defined bands should be excluded from the analysis. Each experiment should be repeated in duplicate or triplicate for proper statistical analysis. We recommend changing the loading scheme of the samples in each replicate to avoid well-related artifacts.

8. Advantages

The RADAR assay is a quick and quantitative method to study topoisomerase-covalent complexes.

The assay is specific for the topoisomerase of choice and can be applied to other protein-DNA covalent complexes. Loading multiple DNA amounts and performing biological replicates of the experiments will ensure strong data reproducibility and the possibility to perform statistical analysis.

Compared to the ICE assay, the RADAR assay does not require any special equipment and does not include time-consuming steps. Many conditions can be tested at the same time because of the low amount of starting material needed for the experiment ($6–7 \times 10^5$ cells). The Nanodrop quantification also saves time and minimizes the amount of material required for a single blot.

9. Limitations

This assay requires initial optimization according to the type of inhibitor used to trap the topoisomerase, the dose and duration of the treatment, and the mammalian cell line used in the experiments. Although the RADAR assay can detect changes in topoisomerase-covalent complexes, the sensitivity of the assay might preclude the identification of slight changes in the amount of these complexes that are below the detection limit of the western blot. Moreover, the detection of the covalent complexes of interest is dependent on the availability of good primary antibodies.

10. Optimization and troubleshooting

Here are reported the most common problems that might require troubleshooting while using this protocol.

Problem	Potential solution or explanation
Bands are not sharp and well-defined.	The apparatus is not well sealed, and this leads to leaking of the samples during the blotting step.
The membrane is dirty and there is high background.	Increase the washing steps either after primary or secondary antibodies incubation.
Low DNA yield after extraction.	Few cells were plated, or the treatment conditions are too toxic and lead to cell death.

During the production of this volume, a complementary description of the RADAR assay was published. Nitiss et al. Curr Protoc. 2021. doi: 10.1002/cpz1.250.

Acknowledgments

This work was supported by a fellowship from the American Italian Cancer Foundation to A.M; the National Cancer Institute (NCI) grants R01CA237263 and R01CA248526 to A.V.; the U.S. Department of Defense (DOD) Breast Cancer Research Program (BRCP) Expansion Award BC191374 to A.V.; the Alvin J. Siteman Cancer Center Siteman Investment Program (supported by The Foundation for Barnes-Jewish Hospital, Cancer Frontier Fund) to A.V.; and the Barnard Foundation to A.V.

Reference

Aldred, K. J., Payne, A., & Voegerl, O. (2019). A RADAR-based assay to isolate covalent DNA complexes in bacteria. *Antibiotics (Basel)*, *8*(1). https://doi.org/10.3390/antibiotics8010017.

Anand, J., Sun, Y., Zhao, Y., Nitiss, K. C., & Nitiss, J. L. (2018). Detection of topoisomerase covalent complexes in eukaryotic cells. In M. Drolet (Ed.), *DNA topoisomerases: Methods and protocols* (pp. 283–299). New York, NY: Springer New York.

Champoux, J. J. (2001). DNA topoisomerases: Structure, function, and mechanism. *Annual Review of Biochemistry*, *70*, 369–413. https://doi.org/10.1146/annurev.biochem.70.1.369.

Delgado, J. L., Hsieh, C.-M., Chan, N.-L., & Hiasa, H. (2018). Topoisomerases as anticancer targets. *The Biochemical Journal*, *475*(2). https://doi.org/10.1042/BCJ20160583.

Fielden, J., Wiseman, K., Torrecilla, I., Li, S., Hume, S., Chiang, S.-C., et al. (2020). TEX264 coordinates p97- and SPRTN-mediated resolution of topoisomerase 1-DNA adducts. *Nature Communications*, *11*(1), 1–16. https://doi.org/10.1038/s41467-020-15000-w.

Forterre, P., Gribaldo, S., Gadelle, D., & Serre, M. C. (2007). Origin and evolution of DNA topoisomerases. *Biochimie*, *89*(4), 427–446. https://doi.org/10.1016/j.biochi.2006.12.009.

Janes, K. A. (2015). An analysis of critical factors for quantitative immunoblotting. *Science Signaling*, *8*(371), rs2. https://doi.org/10.1126/scisignal.2005966.

Kiianitsa, K., & Maizels, N. (2013). A rapid and sensitive assay for DNA-protein covalent complexes in living cells. *Nucleic Acids Research*, *41*(9), e104. https://doi.org/10.1093/nar/gkt171.

Liu, L. F., & Wang, J. C. (1987). Supercoiling of the DNA template during transcription. *Proceedings of the National Academy of Sciences of the United States of America*, *84*(20), 7024–7027. https://doi.org/10.1073/pnas.84.20.7024.

Masse, E., & Drolet, M. (1999). Relaxation of transcription-induced negative supercoiling is an essential function of Escherichia coli DNA topoisomerase I. *The Journal of Biological Chemistry*, *274*(23), 16654–16658. https://doi.org/10.1074/jbc.274.23.16654.

Nierzwicki-Bauer, S. A., Gebhardt, J. S., Linkkila, L., & Walsh, K. (1990). A comparison of UV cross-linking and vacuum baking for nucleic acid immobilization and retention. *BioTechniques*, *9*(4), 472–478. Retrieved from https://www.ncbi.nlm.nih.gov/pubmed/2257145.

Nitiss, J. L. (2009). Targeting DNA topoisomerase II in cancer chemotherapy. *Nature Reviews. Cancer*, *9*(5), 338–350. https://doi.org/10.1038/nrc2607.

Nitiss, J. L., Soans, E., Rogojina, A., Seth, A., & Mishina, M. (2012). Topoisomerase assays. *Current Protocols in Pharmacology*. https://doi.org/10.1002/0471141755.ph0303s57. Chapter 3, Unit 3.3.

Pommier, Y. (2006). Topoisomerase I inhibitors: Camptothecins and beyond. *Nature Reviews. Cancer*, *6*(10), 789–802. https://doi.org/10.1038/nrc1977.

Pommier, Y., Sun, Y., Huang, S. N., & Nitiss, J. L. (2016). Roles of eukaryotic topoisomerases in transcription, replication and genomic stability. *Nature Reviews. Molecular Cell Biology*, *17*(11), 703–721. https://doi.org/10.1038/nrm.2016.111.

Schoeffler, A. J., & Berger, J. M. (2008). DNA topoisomerases: Harnessing and constraining energy to govern chromosome topology. *Quarterly Reviews of Biophysics*, *41*(1), 41–101. https://doi.org/10.1017/S003358350800468X.

Strumberg, D., Pilon, A. A., Smith, M., Hickey, R., Malkas, L., & Pommier, Y. (2000). Conversion of topoisomerase I cleavage complexes on the leading strand of ribosomal DNA into 5′-phosphorylated DNA double-strand breaks by replication runoff. *Molecular and Cellular Biology*, *20*(11), 3977–3987. https://doi.org/10.1128/MCB.20.11.3977-3987.2000.

Subramanian, D., Kraut, E., Staubus, A., Young, D. C., & Muller, M. T. (1995). Analysis of topoisomerase I/DNA complexes in patients administered topotecan. *Cancer Research*, *55*(10), 2097–2103. Retrieved from https://cancerres.aacrjournals.org/content/canres/55/10/2097.full.pdf.

Sun, Y., Miller Jenkins, L. M., Su, Y. P., Nitiss, K. C., Nitiss, J. L., & Pommier, Y. (2020). A conserved SUMO pathway repairs topoisomerase DNA-protein cross-links by engaging ubiquitin-mediated proteasomal degradation. *Science Advances*, *6*(46). https://doi.org/10.1126/sciadv.aba6290.

Venditto, V. J., & Simanek, E. E. (2010). Cancer therapies utilizing the camptothecins: A review of the in vivo literature. *Molecular Pharmaceutics*, 7(2), 307–349. https://doi.org/10.1021/mp900243b.

Wang, J. C. (2002). Cellular roles of DNA topoisomerases: A molecular perspective. *Nature Reviews. Molecular Cell Biology*, *3*(6), 430–440. https://doi.org/10.1038/nrm831.

Wang, Y. V., Wade, M., Wong, E., Li, Y. C., Rodewald, L. W., & Wahl, G. M. (2007). Quantitative analyses reveal the importance of regulated Hdmx degradation for p53 activation. *Proceedings of the National Academy of Sciences of the United States of America*, *104*(30), 12365–12370. https://doi.org/10.1073/pnas.0701497104.

Wu, H. Y., Shyy, S. H., Wang, J. C., & Liu, L. F. (1988). Transcription generates positively and negatively supercoiled domains in the template. *Cell*, *53*(3), 433–440. https://doi.org/10.1016/0092-8674(88)90163-8.

Xiao, Y., Lin, F.-T., & Lin, W.-C. (2021). ACTL6A promotes repair of cisplatin-induced DNA damage, a new mechanism of platinum resistance in cancer. *Proceedings of the National Academy of Sciences of the United States of America*. https://doi.org/10.1073/pnas.2015808118.

FSC
www.fsc.org
MIX
Papier | Fördert
gute Waldnutzung
FSC® C083411

Kolibri 360 GmbH
Ferdinand-Jühlke-Straße 7
99095 Erfurt, Deutschland
produktsicherheit@kolibri360.de